A transect of the southern Canadian Cordillera from Calgary to Vancouver

May 2003

R.A. Price and J.W.H. Monger

Geological Association of Canada,
Cordilleran Section
Vancouver

CORDILLERAN SECTION
GAC
GEOLOGICAL ASSOCIATION OF CANADA

Canadian Cataloguing in Publication Data

Price, Raymond A. (Raymond Alexander), 1933-
A transect of the southern Canadian Cordillera from Calgary to Vancouver / R.A. Price and J.W.H. Monger. – Rev. ed.

Paper presented at the 96th Annual Meeting, Geological Society of America, Cordilleran Section, Vancouver, Apr. 2000.
Includes bibliographical references.
ISBN 0-9687005-3-5

1. Geology, Structural—British Columbia. 2. Geology, Structural—Alberta. 3. Geology—British Columbia. 4. Geology—Alberta. 5. Geology, Structural—Rocky Mountains, Canadian (B.C. and Alta.)* I. Monger, J. W. H. II. Geological Association of Canada. Cordilleran Section. III. Geological Society of America. Cordilleran Section. Meeting (96th : 2000 : Vancouver, B.C.) IV. Title.
QE628.5.B7P74 2003 557.11 C2003-910734-5

This is one of two volumes of field trip guidebooks produced for the Annual Meeting of the Geological Society of Canada, the Mineralogical Society of Canada, and the Society of Economic Geolgists, held in Vancouver in May 2003

The other volume contains nine trip guides under the title:
Geological field trips in southern British Columbia

Copies of both books are available from:

Geological Association of Canada Cordilleran Section
P.O. Box 398
Station A, Bentall Centre
Vancouver, B.C. V6C 2N2
Canada

and from:

Geological Survey of Canada Sales Office
101 - 605 Robson Street
Vancouver, BC V6B 5J3
Tel. 604-666-0529
Fax 604-666-1124

Typeset by Joy Woodsworth

Front and back covers: Route of field trip, superimposed upon the Tectonic Assemblage Map of the Canadian Cordillera by Wheeler and McFeely (1991).

CONTENTS

A transect of the southern Canadian Cordillera from Calgary to Vancouver

R.A. Price[1] and J.W.H. Monger[2]

INTRODUCTION

This introduction provides general background material for a three-day field trip across the southern Canadian Cordillera between Calgary and Vancouver. The field trip follows the Trans-Canada Highway (Highway 1) from Calgary westward to the village of Cache Creek. From Cache Creek southwestward to Vancouver the route is mainly along Highway 99 (Fig. 1). The guidebook was written in 2000 and minor corrections have been made for text errors in this reprinted version. Overnight stops are in Rogers Pass and Cache Creek. As the trip must traverse a distance of just over 1000 km in three days, emphasis is placed on localities that provide specific information bearing on the geological architecture and tectonic evolution of the southern Canadian Cordillera as a whole. Price is responsible for the eastern part of the road log, from Calgary to Sicamous, and Monger for the western part, from Sicamous to Vancouver.

The geology of the southern Canadian Cordillera is relatively well known, having been studied for nearly 130 years. Most of the earliest geological studies in the Canadian Cordillera were done in the south (e.g. Selwyn, 1872; Dawson, 1879, 1881; McConnell, 1887). Since then, mapping by the Geological Survey of Canada and the British Columbia Geological Survey Branch has continued intermittently in this region, and in the last five decades topical studies have been carried out by many university groups. In addition, the southern Cordillera has been the focus of integrated geophysical, geochemical and geological studies that were carried out under the aegis of Lithoprobe, Canada's National Geoscience project, and provide data for a three-dimensional interpretation of the architecture of the crust and upper mantle within the region (Cook, 1995).

Physiographic Divisions and Active Tectonics of the Southern Canadian Cordillera

The southern Canadian Cordillera is nearly 900 km wide, measured from the easternmost deformed strata overlying the stable continental platform (or *craton*) ~15 km west of Calgary, to the base of the continental slope ~150 km west of southern Vancouver Island. From east to west, the major physiographic divisions are: (1) the Rocky Mountains; (2) the Columbia Mountains (comprising the Purcell, Selkirk and Monashee mountains) that westward decrease in relief and grade through (3) the Shuswap and Okanagan Highlands into (4) the Interior Plateaus; (5) the Cascade and Coast Mountains; (6) the Coastal depressions (represented by the Strait of Georgia); (7) the Insular Mountains (of Vancouver Island); and (8) the continental shelf and slope, which is bounded on the west by the Cascadia subduction zone (Fig. 2; Holland, 1964; Mathews, 1986). The summit heights of most mountains in the region are between 2000–3000 m, the elevation of the Interior Plateaus is about 1000 m, and the toe of the continental slope is in water depths of ~2000 m. The area of land that is over ~1000 m in elevation decreases from east to west.

The Cordillera is the leading edge of the North American plate, and seismicity, recently active volcanism, and rapid vertical uplift attest to tectonic activity along the convergent boundary with oceanic lithosphere of the *Juan de Fuca Plate* (Fig. 3; Riddihough and Hyndman, 1991). The contemporary plate boundary is relatively sharp, but as is the case with most active orogens (Gordon, 1998), the Cordillera is a "soft plate margin." Deformation extends into the converging plates, and the limit of plate margin deformation is diffuse. The *Cascadia subduction zone* (or *Cascadia megathrust*) extends under the Cordillera between northern Vancouver Island (latitude 51°N) and northern California (40°N). The

[1] Department of Geological Sciences and Geological Engineering, Queen's University, Kingston, Ontario K7L 3N6, email price@geol.queensu.ca

[2] Geological Survey of Canada, 605 Robson Street, Vancouver, B.C. V6B 5J3, and Department of Earth Sciences, Simon Fraser University, Burnaby, B.C. V5A 1S5, email jimonger@nrcan.gc.ca

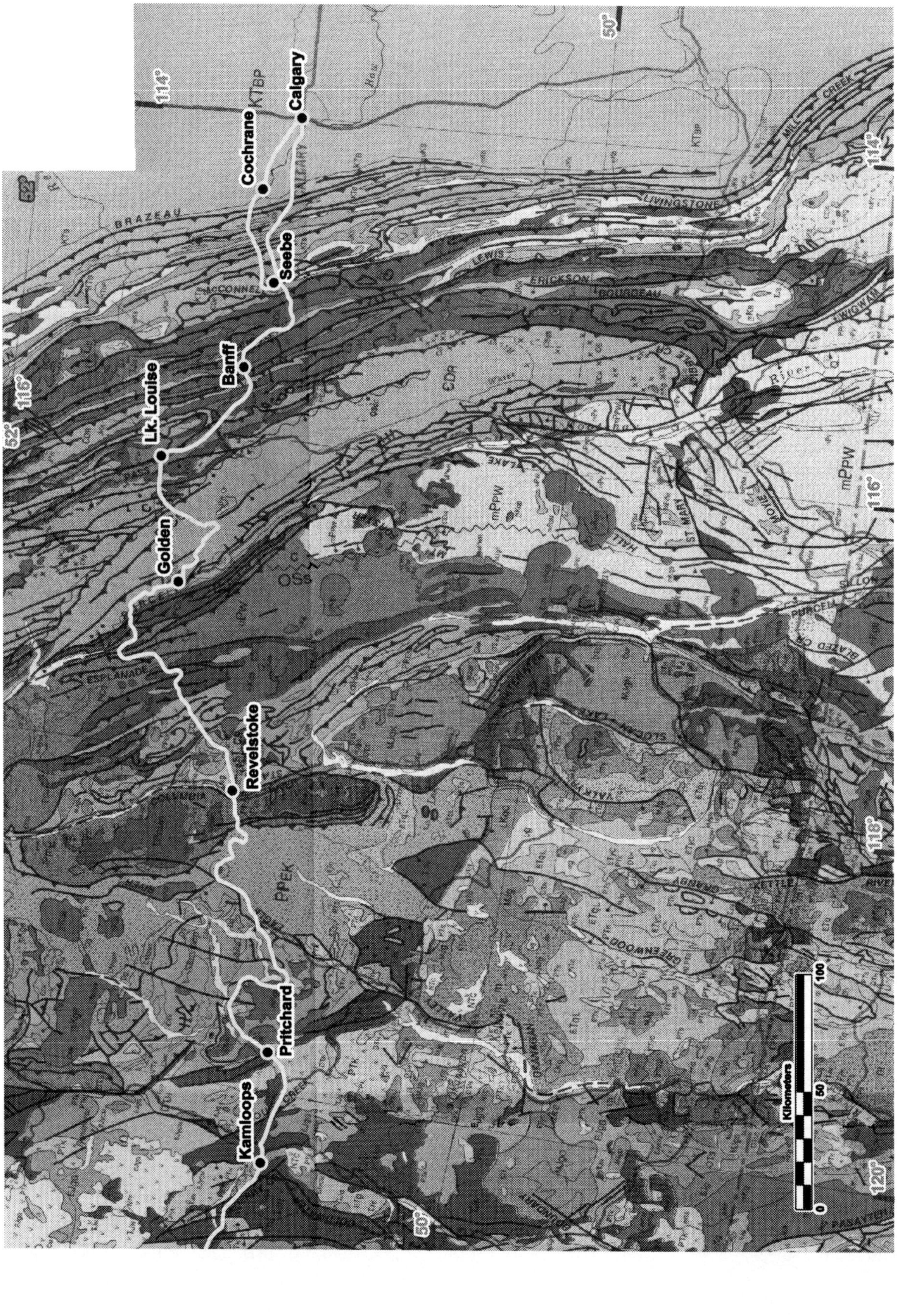

Figure 1. Route of field trip, superimposed upon the Tectonic Assemblage Map of the Canadian Cordillera by Wheeler and McFeely (1991). 1a: eastern part; 1b: western part. A colour version of these figures appears on the cover.

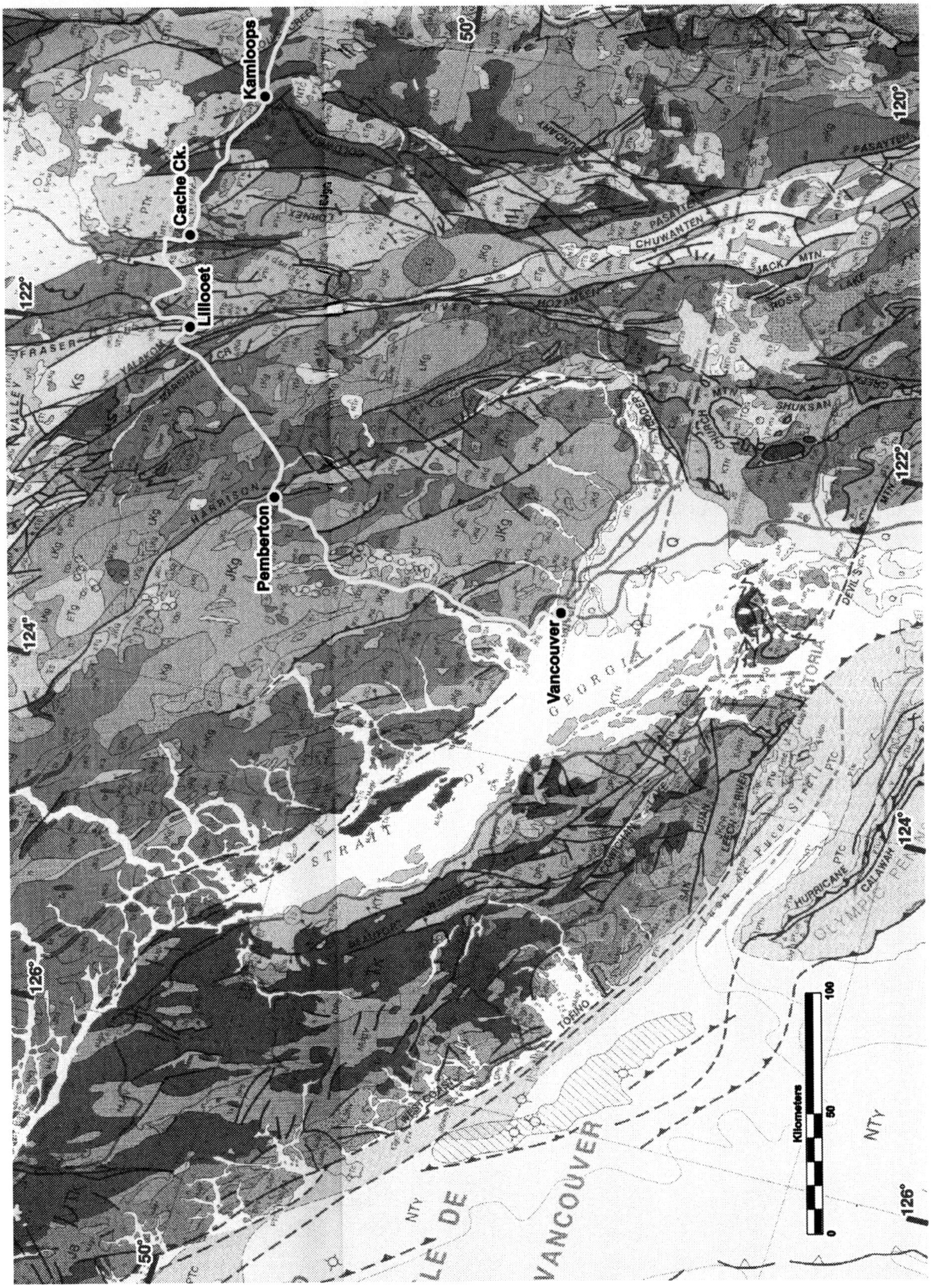
Kamloops
Cache Ck.
Lillooet
Pemberton
Vancouver
122°
124°
126°
120°
50°
Kilometers
0
50
100

small, young (≥6 Ma), oceanic Juan de Fuca Plate is being subducted (or underthrust) beneath the Cordilleran margin at rates of 4.0–4.6 cm/y. Above it, and located ~300 km inboard of the continental margin, the *Cascade magmatic arc* comprises the chain of large stratovolcanoes extending from northern California, through Oregon and Washington, and ending in the Coast Mountains of southwestern British Columbia, north of the city of Vancouver. North of latitude 51°N, the plate boundary is the offshore Queen Charlotte-Fairweather dextral transform fault, west of which is the enormous northward-moving Pacific Plate.

Figure 2. Physiographic map of the Canadian Cordillera (modified from Mathews, 1986). The smaller physiographic divisions in the southern Canadian Cordillera are shown by: Ca = Cascade Mountains; Cb = Cariboo Mountains; M = Monashee Mountains; O = Okanagan Highland; P = Purcell Mountains; Sh = Shuswap Highland; Se = Selkirk Mountains.

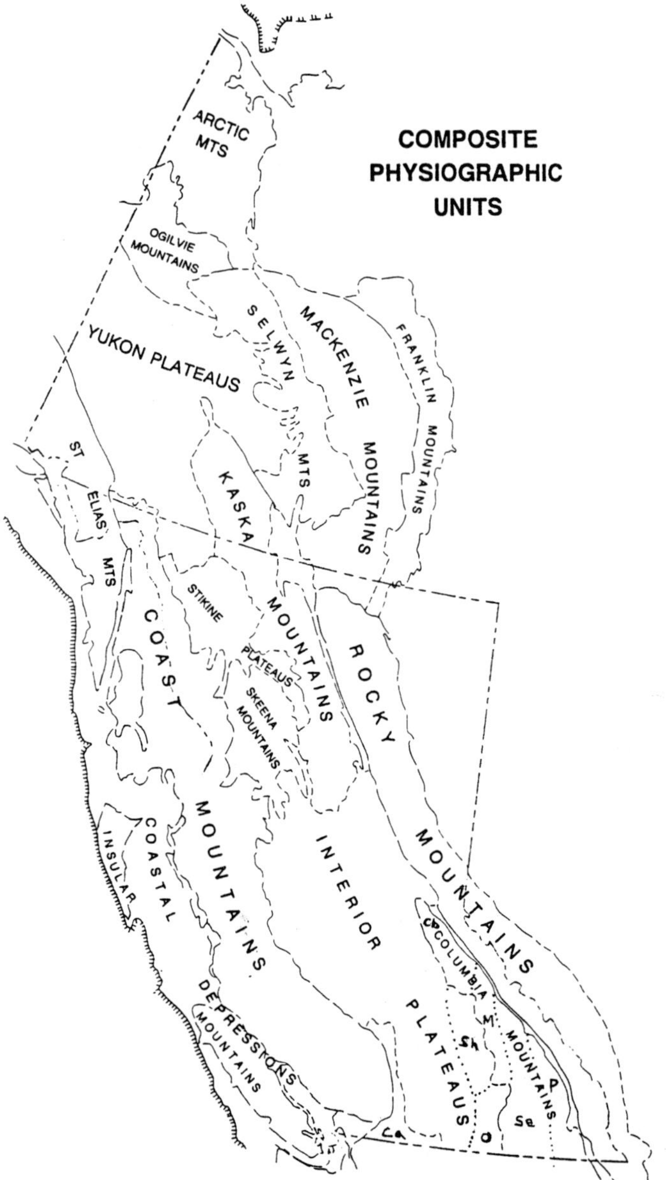

Geological evidence for active deformation is clearly expressed adjacent to the plate boundary where a southwestward-tapering accretionary wedge of thrust faulted and folded sediments is being scraped off the subducting Juan de Fuca plate. However, geological evidence of Quaternary deformation on land is more elusive. This is probably due in part to erosion by the continental glaciation that prior to 12,000 years B.P. extended just south of latitude 49°N. It may have obliterated the geomorphic expression of small-scale tectonic features. In addition, glacial erosion has created oversteepened topography in some mountainous areas which is unsupported after ice retreat, thus causing gravitational collapse and faults (*sackungen*) that may be difficult to distinguish from faults caused by plate margin deformation. Geodetic monitoring, carried out because of concern for a potential *megathrust earthquake* caused by locking and catastrophic release of the Cascadia subduction zone, shows that the coastal regions on Vancouver Island are undergoing short-term tectonic uplift at rates of 1–4 mm/y and east-west shortening at rates of 10 mm/y (Hyndman, 1995).

Morphogeological Belts

Traditionally, the Canadian Cordillera has been divided into five orogen-parallel, linear morphogeological belts, each of which features a distinctive combination of physiography, rock types, metamorphic grade and structure (Dawson, 1881; Gabrielse et al., 1991). Although the belts reflect the sum of all processes that from Proterozoic to Holocene time have shaped the Cordillera, they are dominated by features formed during Middle Jurassic to early Tertiary mountain building. From east to west, they have come to be called Foreland, Omineca, Intermontane, Coast and Insular belts. Although the match between physiography and bedrock geology is not everywhere perfect, largely because of late Tertiary and Quaternary differential vertical movements, the five belts provide a simple spatial and tectonic framework within which to discuss the geology and tectonic evolution of the southern Canadian Cordillera. Their distribution is shown in Figure 4A and their distinguishing attributes and relationship to physiographic units are summarized in Table 1. The road log in the guidebook is divided into sections corresponding to: (1) Foreland; (2) Omineca; (3) Intermontane; and

(4) Coast belts. The Insular Belt is not examined on this field trip.

One difference between the belts, which was recognized by the earliest geologists, is that the Foreland Belt is composed mainly of marine sedimentary rocks, whereas the four other belts contain abundant marine volcanic and sedimentary rocks, as well as plutonic rocks. Prior to the application of plate tectonic concepts to Canadian Cordilleran geology in ~1970, recognition of this difference led to the division of the Cordillera into an eastern *miogeosyncline* and a western *eugeosyncline*. These terms still linger on in the literature largely because of their convenience.

A second difference between the belts is that the high-grade metamorphic and granitic rocks that

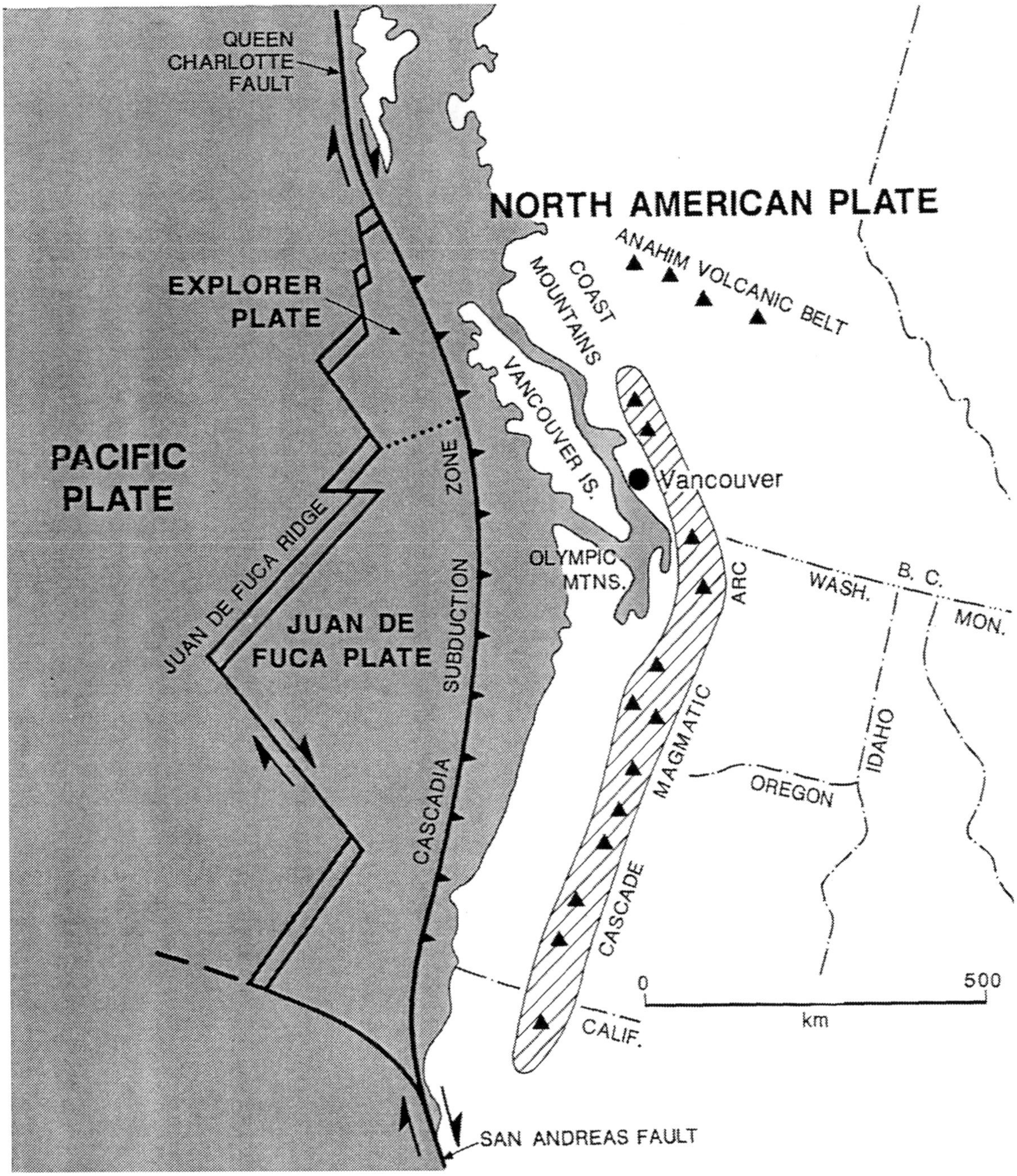

Figure 3. Distribution of active tectonic features in the southern Canadian Cordillera and adjoining parts of the conterminous United States (from Riddihough and Hyndman, 1991).

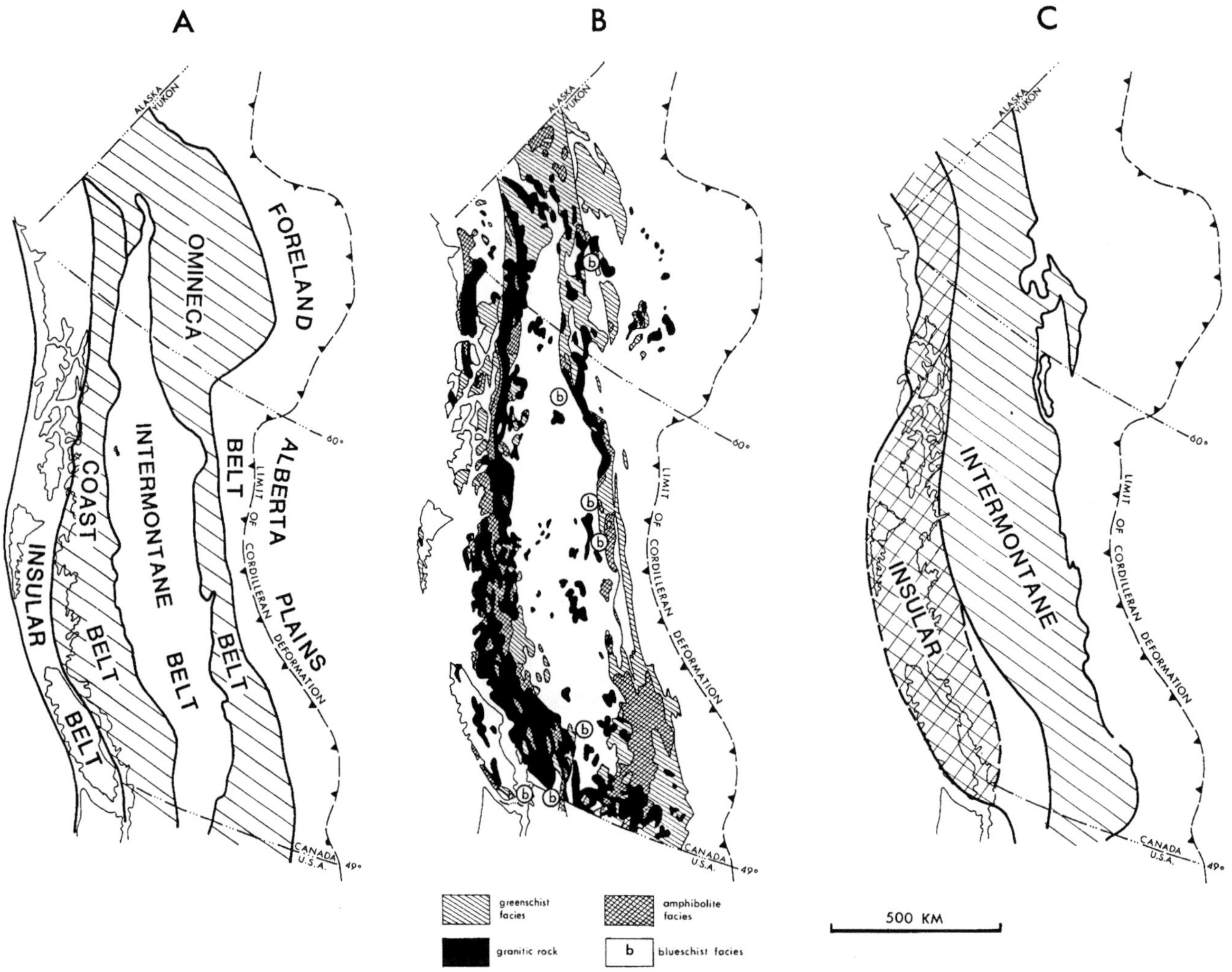

Figure 4. Distinctive features of the Canadian segment of the North American Cordillera.
A. Morphogeological belts.
B. Metamorphic map.
C. Spatial distribution of the stratified rocks belonging to terranes of the Intermontane and Insular belts; note that their boundaries lie mostly within the Omineca and Coast belts.

characterize Omineca and Coast belts separate little-metamorphosed rocks in the other three belts (Fig. 4B). Many rocks in Omineca and Coast belts once were buried to depths of 25 km or more, and their present exposure at surface reflects differential and far greater crustal thickening, uplift, and erosion in these two belts than in the flanking Foreland, Intermontane, and Insular belts. In general style and scale, the crustal structure of the Omineca and Coast belts is comparable with that of orogens such as the Alps and Pyrenees, which are recognized as having formed by continent-continent collision (c.f. Varsek et al., 1993, and Cook, 1995, with Pfiffner et al., 1991). However, Canadian Cordilleran orogenesis appears to be dominated by the subduction of oceanic plates and the accompanying arc magmatism, with no evidence for continent-continent collision. The boundary between the rocks deposited on the ancient North American continental margin and the mainly magmatic arc-related, isotopically juvenile rocks of the Intermontane Belt is in the Omineca Belt. The boundary between the Intermontane Belt and the isotopically juvenile rocks of the Insular Belt is in the Coast Belt (Fig. 4C). This relationship led to the suggestion that many of the structural and metamorphic characteristics of the Omineca and Coast belts reflect sequential Jurassic and Cretaceous accretions of the rocks in, respectively, Intermontane and Insular belts to the ancient North American continental margin (Monger et al., 1982).

Table 1.
Morphogeological belts of the Canadian Cordillera

Foreland Belt

Physiographic units. Includes Rocky, Mackenzie, and Franklin mountains.

Geology. Mainly sedimentary rock: (1) Mesoproterozoic (locally Paleoproterozoic) clastics, carbonates and minor magmatic rocks deposited in continental basins; (2) Neoproterozoic-earliest Cambrian clastics, minor carbonate and local magmatic rock formed during rifting and initial break up of "Rodinia"; (3) Cambrian to Jurassic shelf and slope, passive margin, carbonate and shale deposited near the craton margin; (4) Late Jurassic to early Cenozoic marine to non-marine clastics eroded from the emerging Omineca and Foreland belts; rocks folded and thrust eastwards over craton margin in Late Jurassic-early Tertiary time.

Omineca Belt

Physiographic units. Includes Purcell, Selkirk, Monashee, Cariboo, Omineca, Cassiar and Selwyn mountains, Okanagan and Shuswap highlands.

Geology. Sedimentary and volcanic rocks variously metamorphosed up to high grades; complexly deformed, mainly by compression between Middle Jurassic and early Tertiary time, and by extension in early Tertiary time: (1) local Paleoproterozoic continental crust; (2) Neoproterozoic rift-related clastics and volcanics; (3) Paleozoic pericratonic, mainly fine-grained clastic and volcanic rocks; (4) local accreted late Paleozoic to Early Jurassic volcanic and sedimentary rocks formed in island arcs and marginal basins; (5) early Cenozoic continental volcanic and sedimentary rocks; (6) Paleozoic to early Tertiary granitic rocks.

Intermontane Belt

Physiographic units. Includes Interior, Stikine, and Yukon plateaus and Skeena Mountains.

Geology. Little-metamorphosed volcanic and sedimentary strata and granitic plutons deformed mainly by compression in Mesozoic and extension-transtension in early Cenozoic: (1) Devonian to Early Jurassic sedimentary and volcanic rocks formed in island arcs and accretionary complexes; (2a) Middle Jurassic to early Cenozoic volcanic rocks formed mainly in continental arcs and (2b) marine and non-marine clastics eroded mainly from the emerging Omineca Belt; (3) granitic rocks ranging in age from Devonian to Cenozoic.

Coast Belt

Physiographic units. Coast and Cascade mountains.

Geology. (1) Mainly Jurassic through Cenozoic granitic rock; (2) remainder is latest Proterozoic(?), Paleozoic to Holocene volcanic and sedimentary rocks formed mainly in magmatic arcs but locally in accretionary complexes, metamorphosed up to high grades and complexly deformed mainly in mid-Cretaceous through early Cenozoic time.

Insular Belt

Physiographic units. Insular Mountains, Saint Elias Ranges, coastal depressions, continental shelf and slope.

Geology. Little-metamorphosed volcanic and sedimentary strata and granitic plutons deformed mainly in Late Cretaceous to Holocene time: (1) latest Proterozoic to mid-Cretaceous volcanic and sedimentary rock formed mainly in island arc or less commonly oceanic plateau settings; (2) mid-Cretaceous and younger clastics eroded from the Coast Belt; (3) Paleozoic to Tertiary granitic intrusions; (4) Late Jurassic to Holocene clastic-rich accretionary complexes.

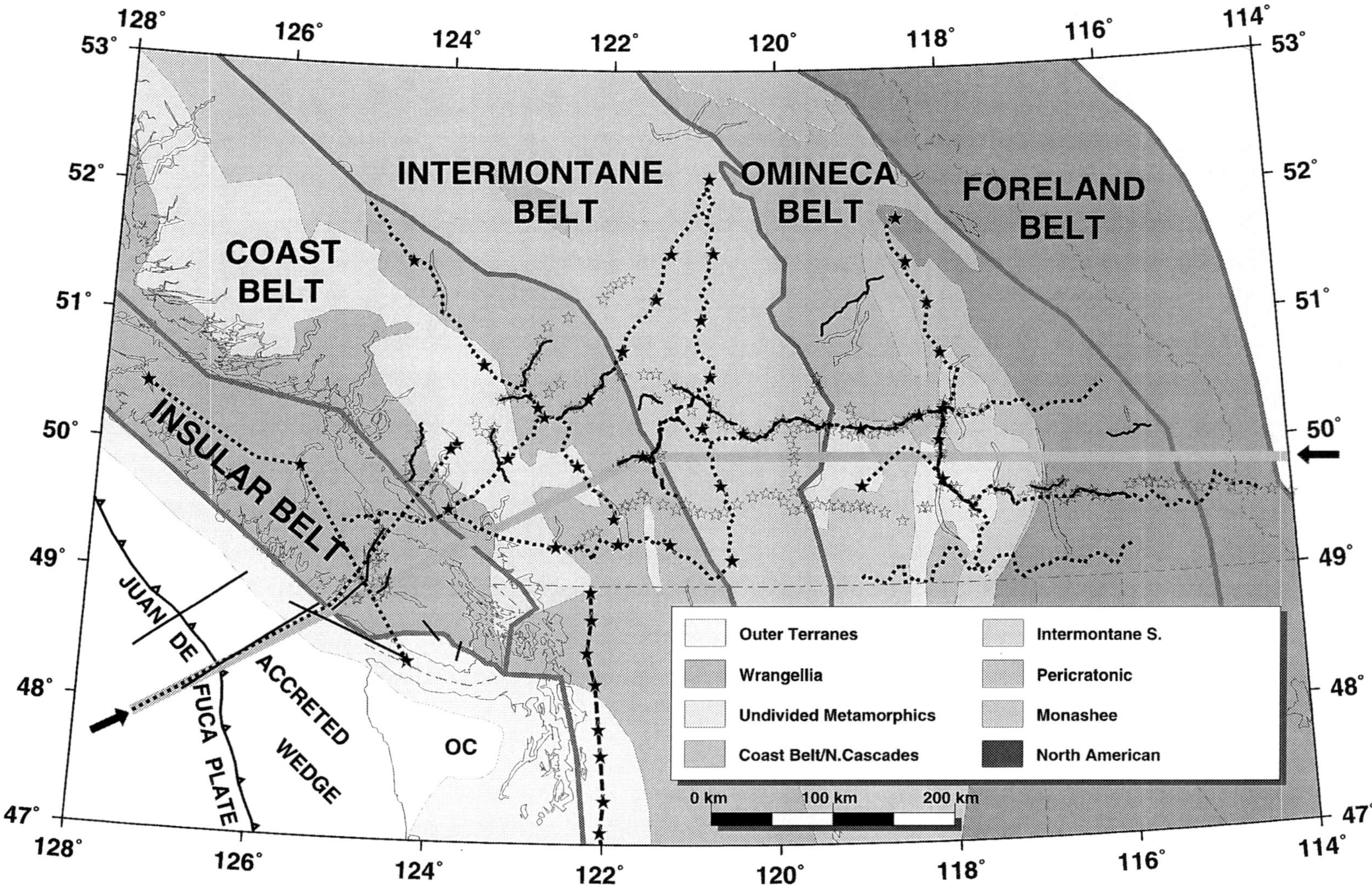

Figure 5a. Map of the southern Canadian Cordillera showing Lithoprobe seismic reflection and refraction profiles (Clowes et al., 1995). Dotted lines show locations of seismic refraction-wide angle reflection (R-WAR profiles) with shot points indicated by stars. The thick grey line (between heavy arrows) shows the general location of the cross-sections in Fig. 5b. On this simplified map "North American craton" includes the Belt-Purcell and Windermere rocks; "Pericratonic" includes Kootenay terrane (see Fig. 7); "Intermontane superterrane" includes Quesnellia, Stikinia and Cache Creek terranes; Coast Belt (N. Cascades) includes Methow, Cadwallader and Bridge River terranes; OC is the core of the Olympic Mountains.

Figure 5b. Lithospheric cross-sections from the Juan de Fuca Plate to the Foreland thrust and fold belt. The location of the section is given by the thick grey line between the arrows on Fig. 5a. (a) Continuous velocity-depth structure section along the cross-section; (b) interpreted lithospheric cross section based on seismic refraction-wide angle reflection data, seismic reflection data and other Lithoprobe studies; (c) cross-section is similar to (b) but is an interpretation made only from seismic reflection data.

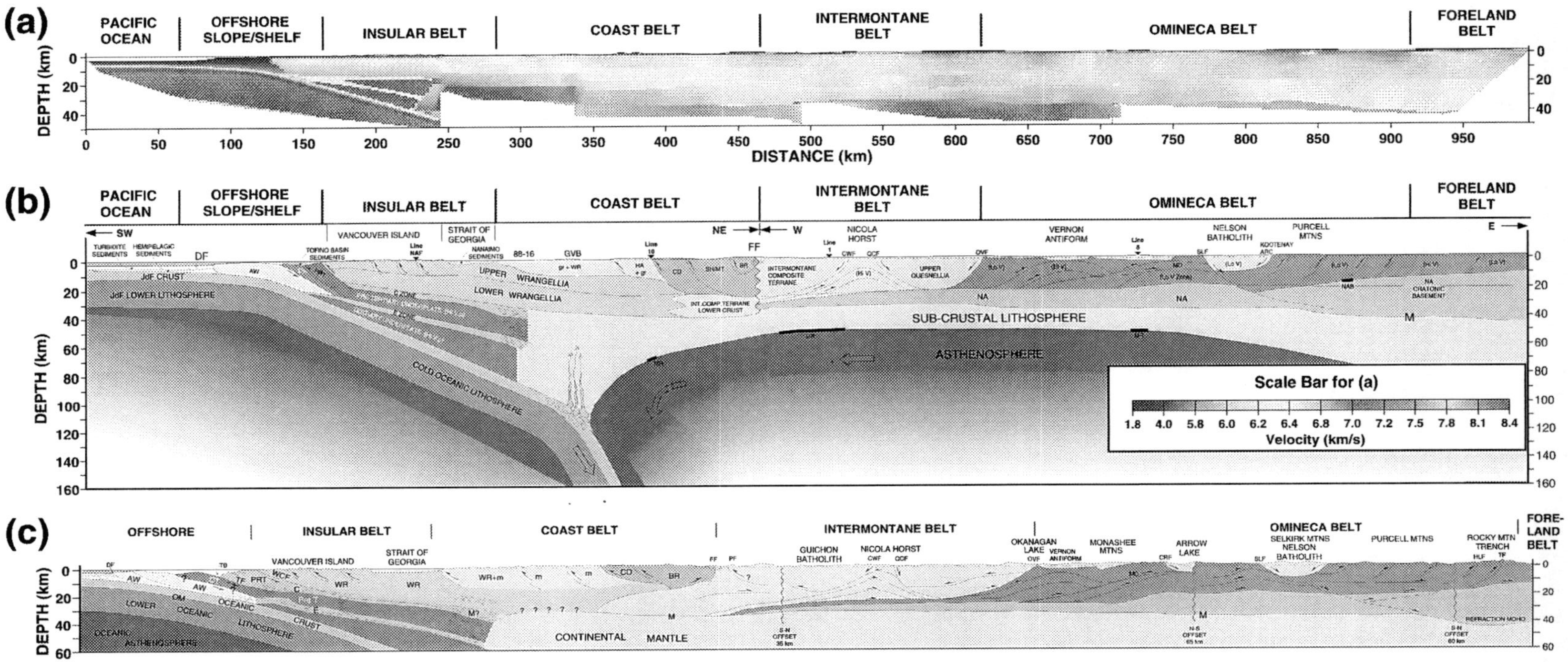

The Crust and Upper Mantle of the Southern Canadian Cordillera

Lithoprobe seismic refraction and reflection experiments provide important new insights on the structure of the crust and upper mantle of the southern Canadian Cordillera (Fig. 5; Clowes et al., 1995; Cook, 1995). The crust increases in thickness from ~40–45 km east of the Rocky Mountains (Foreland Belt) to a maximum of 50–55 km below the western Rocky Mountains and eastern Columbia Mountains (Omineca Belt) near Golden, about 200 km west of Calgary. (Note: the profile in Figure 5 is located ~140 km south of Golden, near Kimberley, where the Moho is shallower). The increase in crustal thickness is due to the formation of the Foreland thrust and fold belt, beneath which the underlying Paleoproterozoic crust, which is not involved in the later folding and thrusting, has been depressed by ~10 km. Farther west, beneath the Columbia Mountains (Omineca Belt), the crust thins rapidly to ~36 km. The thinning there is due to Eocene (55–45 Ma) crustal stretching, which has exhumed parautochthonous crustal basement from a depth of >20 km to where it is now exposed in the high Monashee Mountains west of Revelstoke. Westward from the Monashee Mountains, across the Interior Plateaus (Intermontane Belt) to the vicinity of the Fraser Fault on the eastern margin of the Coast Mountains (Coast Belt), the base of the crust is relatively planar, and the crust thins gradually westward from ~36 to ~32 km. West of the Fraser Fault, the crust thickens to about 40 km under the southern Coast Mountains. West of these, in the Insular Belt, it tapers to a thickness of ~15 km on the west side of Vancouver Island, below which are the Jura-Cretaceous, Eocene and Neogene accretionary wedges that underlie the continental slope and extend below Vancouver Island, above subducting oceanic lithosphere of the Juan de Fuca Plate (Fig. 5).

Variations in upper mantle seismic velocity and electrical conductivity, and in surface heat flow, indicate that the subcrustal lithosphere is abnormally thin and hot beneath the western Omineca, Intermontane and eastern Coast belts (Clowes et al., 1995; Hyndman and Lewis, 1995). The deep seated thermal anomalies, which underlie a north-south trending belt of widespread Eocene (55–45 Ma) transtensional crustal stretching, were attributed to a present day, north-trending belt of symmetrical mantle upwelling by Gough and Majorowicz (1992). The present day high elevation and rugged topography of the southern Foreland and Omineca belts may be dynamically supported by the upward mantle flow.

The crust of the North American craton in western Canada is an orogenic collage of Paleoproterozoic juvenile magmatic belts and accreted terranes that was assembled and consolidated by ~1.8 Ga (Hoffman, 1989; Ross, 1991; Villeneuve et al., 1993). Individual components of the collage, which have distinctive signatures on aeromagnetic and gravity anomaly maps, can be tracked under the Rocky Mountains as far west as the Rocky Mountain Trench (Fig. 6, modified after Gehrels and Ross, 1998, and Crowley, 1999).

The sea-floor spreading and continental separation that initiated formation of the proto-Pacific Ocean (or *Panthalassa*) and miogeoclinal deposition along the newly created continental margin were preceded by two episodes of late Neoproterozoic and earliest Cambrian crustal stretching and rifting: one in the mid-Neoproterozoic (~750 Ma) and the other in the latest Proterozoic and earliest Cambrian (~555–530 Ma; Sears and Price, 2000). The location of the eastern limit of crustal stretching, as recorded by thickness and facies changes within contemporaneous and overlapping supracrustal rocks, coincides with the boundary between the North American craton and the miogeocline. However, the supracrustal, miogeoclinal strata that contain this record were detached from the basement on which they were deposited and were displaced eastward during formation of the Cordilleran fold and thrust belt. A recent three-dimensional palinspastic reconstruction (Price and Sears, 2000; and Fig. 2-23) outlines the original location of the boundary between the Cordilleran miogeocline and the North American craton. It also suggests that near Highway 1, the northeastern limit of Neoproterozoic-earliest Cambrian stretching coincides with a crustal-scale ramp that is outlined by Lithoprobe seismic reflection imaging along the west side of the Monashee Complex. Along this ramp, the basal décollement of the Foreland fold and thrust belt, which in this region is called the Monashee décollement, descends rather rapidly westward from the top of a panel of Paleoproterozoic continental crust that is 35–40 km thick in the Monashee Complex west of Revelstoke, to a crustal level that is 20–30 km lower. The change in crustal level of the basal décollement west of the Monashee Complex coincides with the palinspastically restored western margin of the miogeocline; and therefore evidently was controlled by a crustal-scale ramp produced by Neoproterozoic and earliest Cambrian crustal stretching.

Lithoprobe seismic reflection and refraction crustal imaging (Fig. 5) and Sr_i isotopic signatures (Fig. 6) of Tertiary magmatic rocks in the Intermontane Belt, indicate that attenuated Paleoproterozoic continental crust forms a westward-tapering layer

~10–2 km thick in the lowermost crust. The layer extends westward to the Fraser River and the eastern margin of the Coast Mountains (Fig. 5; Clowes et al., 1995). This layer of old continental crust presumably was initially created by latest Neoproterozoic-earliest Cambrian crustal stretching, and has been modified by Jurassic through Paleocene (~180–60 Ma) contraction and by substantial Eocene (~55–45 Ma) crustal stretching. The overlying crust of the Intermontane Belt is mainly isotopically juvenile volcanic and plutonic rock, and sedimentary rock that gives mostly juvenile and but locally evolved isotopic signatures (Armstrong, 1988; Samson and Patchett, 1991; Ghosh, 1995; Patchett and Gehrels, 1998). West of the Fraser River, in the southern Coast and Insular belts, all igneous rocks are isotopically juvenile (Friedman et al., 1995). The addition of the juvenile material, which mostly originated in offshore magmatic arcs, to the old continental crust, amounts to about half the total volume of the crust of the southern Canadian Cordillera. Just how much of the juvenile material formed more-or-less in place, and just how much formed elsewhere and later was relocated to its present position, are questions explored on the field trip.

The juvenile material now mostly occurs as *tectonic flakes* ≤10 km thick that have been delaminated

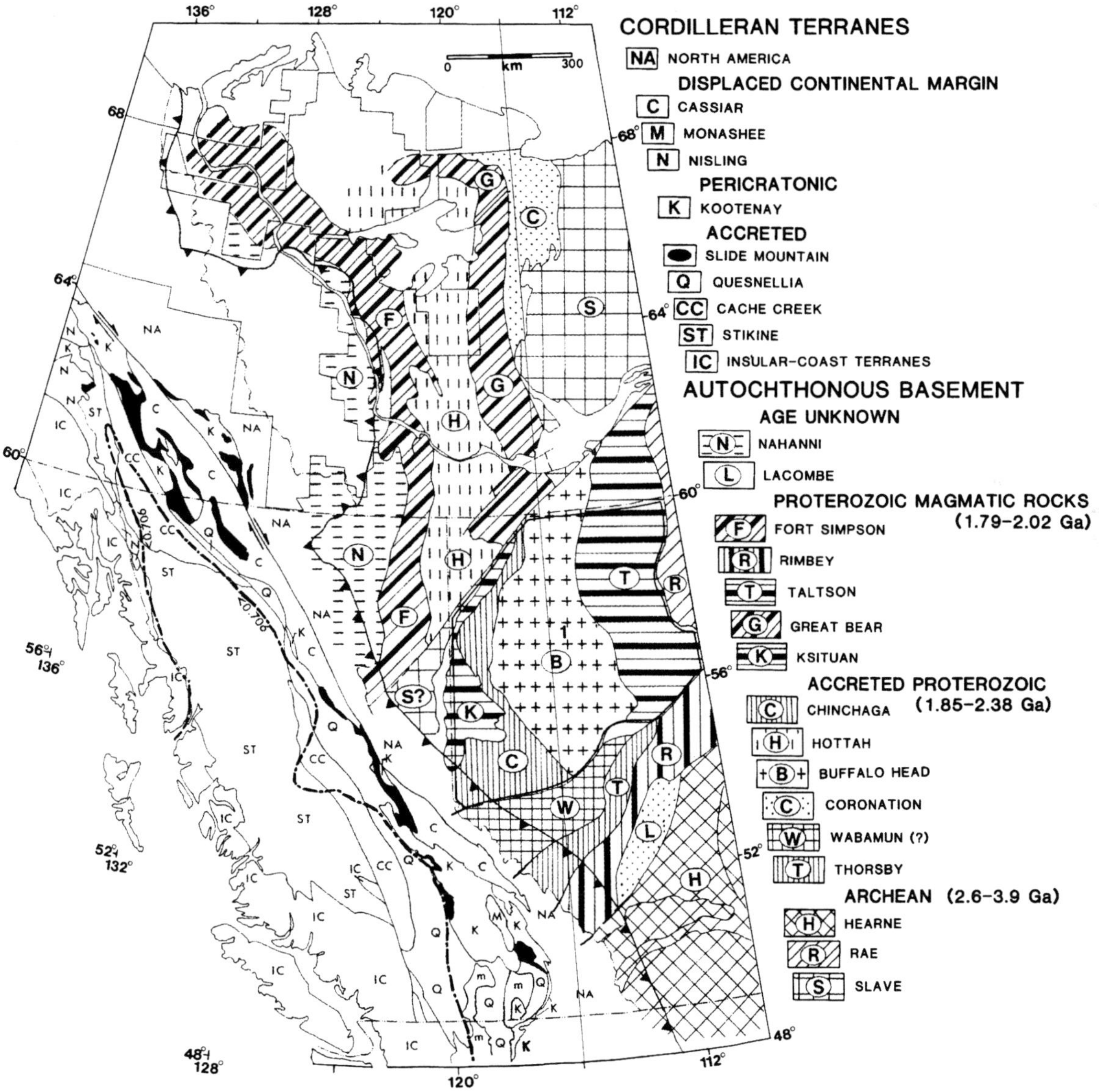

Figure 6. The Paleoproterozoic orogenic collage of Archean lithospheric fragments, Paleoproterozoic accreted terranes and magmatic arcs that forms the cratonic basement of western Canada. Line with barbs indictes eastern limit of Cordilleran deformation. The broken line labelled "<0.706" marks the boundary between Sr_i values of 0.705 and 0.707 in Cenozoic igneous rocks (from Armstrong, 1988). Modified from Ross (1991).

from their lower crustal and mantle roots. The tectonic flakes have been thrust over and in places structurally imbricated with the attenuated old continental crust and its overlying deposits to form new, specifically Cordilleran, crust (Monger et al., 1985; Clowes et al., 1995). The addition of this material resulted in oceanward migration of the continental margin for ~450 km between Early Jurassic (~185 Ma) and Late Cretaceous (90 Ma) time, from an initial location in the Paleozoic ~100 km west of Revelstoke, to a location just west of Vancouver Island.

The Cordilleran Orogenic Collage and Its Analysis: Tectonic Assemblages and Terranes

Helwig (1974) proposed the term *orogenic collage* for the geological disorder of a region created by plate tectonic processes acting for a long time. The Canadian Cordillera is a prime example of an orogenic collage. It is located on a continent-ocean boundary that has existed through virtually all of Phanerozoic time (≥530 Ma), and features arc magmatism recording plate convergence and subduction beneath the North American Plate that probably is at least as old as Middle Devonian (387 Ma). In the late Paleozoic, a back-arc basin separated the arc from the ancient continental margin. An early Mesozoic subduction complex is in the middle of the collage, between coeval arc terranes. Older arc-subduction complex configurations are overprinted by younger magmatic arcs, and further disrupted by large, late Mesozoic-Tertiary orogen-parallel strike-slip faults and related crustal extension, with exhumation of rocks from mid-crustal levels.

Two parallel methodologies evolved to aid analysis of the orogenic collage. The first has been used since about 1970, when plate tectonic concepts were first applied to Cordilleran geology. Different rock types associated in space-time are grouped on the basis of actualistic models of the tectonic settings in which they formed, such as passive continental margin deposits, magmatic arcs, and accretionary (or subduction) complexes. These groupings, which have been called *tectonic assemblages,* became the major units on geological maps (tectonic assemblage maps) of the entire Canadian Cordillera (Fig. 1; Wheeler and McFeely, 1991, scale 1:2 million).

The second methodology is analogous to that used long ago by Argand (1924) in the Alps and central Asia. It involves the identification of regions called *terranes* that are distinguished by having geological records distinct from those of other parts of the collage, including strata deposited on and near the craton margin (Fig. 7, Table 2). Some terranes contain paleontological and/or paleomagnetic records that are very different from those of others, and from rocks of similar age now at the same latitude on the craton margin. In these cases, it is inferred that the terranes originally were widely separated from one another and from their positions in the present continental margin. They were called *suspect terranes* by Coney et al. (1980) in order to emphasize the uncertain paleogeographic positions of their homelands. Other qualifiers that may be useful include *accreted* for terranes composed mainly of juvenile material added later to the craton margin, and *pericratonic* for terranes that formed near the craton but in uncertain paleogeographic relationship to it. Qualifiers that emphasize the dominant tectonic setting in which rocks of a terrane formed, such as arc or accretionary complex terranes, merge the two methodologies (Fig. 7, Table 2).

Terranes are the building blocks of the Cordilleran collage. Their amalgamation with one another and their accretion to the craton margin are recorded by stratigraphic units that overlap two or more terranes, by plutonic belts common to two or more terranes, or by clastic rocks identified as having been eroded from one terrane and deposited upon another (Fig. 8).

Global and Other Considerations Bearing on Canadian Cordilleran Evolution

Global Paleogeography

The Cordillera is located on a persistent continent-ocean boundary that was initiated at ≥530 Ma, when rifting of a hypothetical Neoproterozoic supercontinent called *Rodinia* evolved into the sea-floor spreading and continental drift that produced the ancestor of the present Pacific Ocean basin, called *Panthalassa.* Panthalassa eventually expanded to occupy more than a hemisphere in latest Paleozoic time. The formerly dispersed continental fragments were concentrated in the other hemisphere and amalgamated to form the supercontinent called *Pangea.* Subsequently, during Jurassic and later break-up of Pangea, Panthalassa contracted in size to become the Pacific Ocean. A relationship between Cordilleran orogenesis and the North Atlantic sea-floor spreading that accompanied the break up of Pangea has long been suspected (Daly, 1926; Coney, 1972).

Pacific Ocean Floor Record

Vast areas of oceanic lithosphere have been consumed by subduction along the Cordilleran margin of North America. The present distribution of the ages of Pacific Ocean lithosphere is markedly asymmetric. The main spreading centre (the East Pacific

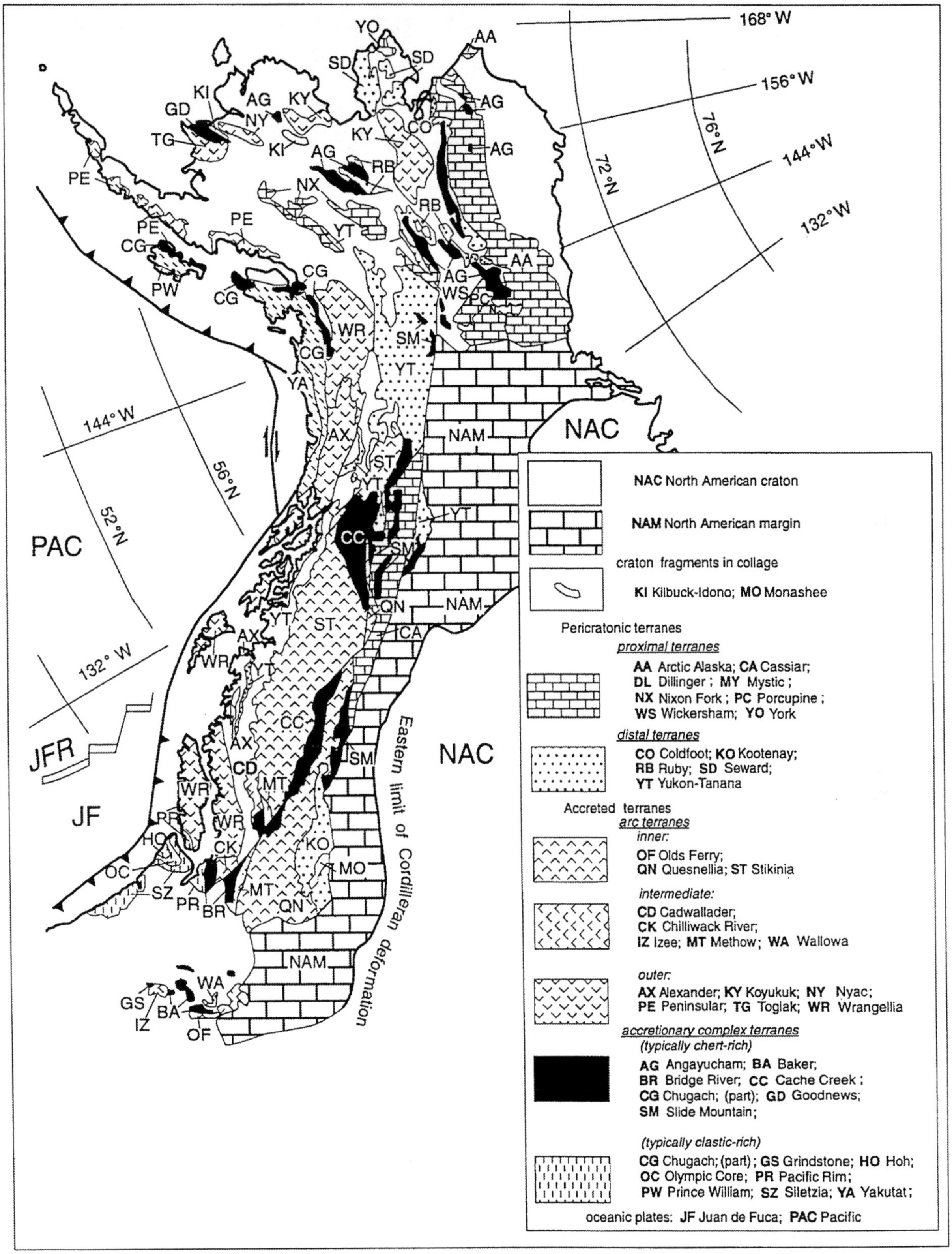

Figure 7. Terrane map of the northern Cordillera; patterns show the predominant tectonic setting within which each terrane formed. JFR = Juan de Fuca ridge. Note that much of Slide Mountain terrane is not an accretionary complex sensu stricto, but rather is a collapsed marginal basin. From Monger and Nokleberg (1996).

Table 2. Summary descriptions of terranes in the Canadian Cordillera

Abbreviations are those on Figure 7. Terranes crossed on field trip are in bold face. Accreted terranes are denoted by (A), pericratonic terranes by (P).

AX = ALEXANDER TERRANE (A). Latest Proterozoic to Triassic mafic to felsic volcanic rocks, terrigenous clastic and carbonate rocks, and early Paleozoic granitic rock; pre-Devonian rocks arc-related; Paleozoic paleomagnetic data discordant with those of North America.

BR = BRIDGE RIVER TERRANE (A). Disrupted, variably metamorphosed, Mississippian to Late Jurassic chert, argillite, sandstone, basalt, ultramafic rocks, and minor carbonate rocks; contains Triassic blueschist; early to late Mesozoic accretionary complex.

CA = CASSIAR TERRANE (P). Late Precambrian to Devonian platformal carbonate rocks, sandstone, and graptolitic shale; a sliced-off fragment of craton-margin deposits transported northwards 500–1000 km on Tintina, Northern Rocky Mountain Trench strike-slip faults.

CC = CACHE CREEK TERRANE (A). Disrupted Mississippian to Early Jurassic chert, argillite, basalt, carbonate, and ultramafic rocks; local Triassic blueschist. Late Permian shallow water faunas resemble those of Paleo-Tethys and western Panthalassa; early Mesozoic accretionary complex accompanying Quesnellian and Stikinian arcs.

CD = CADWALLADER TERRANE (A). Permian basalt, gabbro and ultramafic rock, overlain by Triassic arc-related basalt, carbonate and clastics, overlain by Upper Triassic to mid-Cretaceous clastic rocks; overlying Late Cretaceous arc volcanics carry paleomagnetic signatures indicating ~35° northward displacement with respect to craton.

CG = CHUGACH TERRANE (A). Landward: metachert, basalt and local carbonate with Early Jurassic blueschist; oceanward: Cretaceous greywacke, argillite, and melange having a matrix of Late Jurassic to Early Cretaceous cherty argillite and blocks of basic volcanic rocks, chert, limestone, and ultramafic rocks; interpreted as a long-lived Mesozoic accretionary complex.

CK = CHILLIWACK TERRANE (A). Devonian to Permian clastics, carbonate, and dacitic to basaltic volcanic rocks overlain by Late Triassic and Early Jurassic argillite, siltstone and volcanic sandstone, and (?) by Jurassic felsic to intermediate volcanics and Jurassic to mid-Cretaceous clastics (latter section on west side Harrison Lake); Paleozoic and Mesozoic rocks arc related; Permian fusulinid faunas resemble those of Quesnellia and Stikinia and faunas in southwestern United States.

CR = CRESCENT TERRANE (A). Eocene basalt flows, breccia, tuff, and volcanic sandstones; cut by gabbro and diabase intrusions.

HA = HARRISON TERRANE (A). Triassic basalt and chert, overlain by Early and Middle Jurassic arc volcanics and clastics; maybe related to WR

HO = HOH TERRANE (A). Middle Miocene to Late Oligocene melange and broken formation.

JF = JUAN DE FUCA OCEANIC PLATE

KO = KOOTENAY TERRANE (P). Variably metamorphosed, Neoproterozoic and Paleozoic strata comprising continent-derived clastic rocks, mafic to felsic volcanic rocks, and Devonian intrusions; these rocks probably deformed prior to deposition of late Paleozoic pelite, conglomerate, sandstone, limestone, and basic volcanic rocks; overlapped(?) by Slide Mountain and(?) Quesnellian strata.

MO = MONASHEE TERRANE (P). Paleoproterozoic, high-grade metamorphic rocks comprising paragneiss and orthogneiss dated 2.0 Ga, mantled by Paleoproterozoic to early Paleozoic quartzitic, semi-pelitic and calc-silicate gneiss and minor marble and metaquartzite; probably partly reworked craton and its platformal cover.

Table 2. Summary descriptions of terranes in the Canadian Cordillera (cont'd.)

MT = Methow terrane (A). Permian basalt, gabbro and ultramafics; Early Jurassic to Cretaceous marine clastic sedimentary strata containing local Middle Jurassic arc volcanic rocks; may be equivalent to Cadwallader terrane, but lacks the Triassic arc rocks.

OZ = Ozette terrane (A). Eocene melange and broken formation.

PA = Pacific oceanic plate

PR = Pacific Rim terrane (A). Disrupted, mainly Late Jurassic and Early Cretaceous greywacke, argillite, chert, and basalt.

QN = Quesnel terrane, or Quesnellia (A). Late Devonian to Permian clastics, mafic to felsic arc related volcanic rocks and carbonate, Ordovician(?) to Permian ultramafics, basalt, chert, pelite and carbonate marginal basin or ocean floor rocks (=? Slide Mountain terrane); dominant Late Triassic and Early Jurassic arc related volcanic and intrusive rocks, argillite, volcanic sandstone, and local carbonate; Permian, Triassic and Early Jurassic faunas differ from coeval faunas at same latitude on craton and closer to those in SW United States and NW Mexico; overlying mid-Cretaceous continental arc volcanics on southwestern edge displaced ~10°? northward with respect to craton.

SM = Slide Mountain terrane (A). Late Paleozoic mafic volcanics, ultramafic and local mafic intrusions, chert, clastics and local carbonate; Pennsylvanian and Permian paleomagnetic record suggests northward displacements of 20° with respect to the craton. Probably mostly formed in a back-arc basin setting.

ST = Stikine terrane, or Stikinia (A). Devonian to Jurassic mafic to felsic, arc-related volcanic rocks and clastic, and late Paleozoic carbonate; Permian, Triassic and Jurassic faunas differ from those at same present latitude on craton, but are similar to those farther south in the western United States; however paleomagnetic studies on Permian, Late Triassic and Early Jurassic rocks suggest little or no displacement with respect to craton.

WR = Wrangell terrane, or Wrangellia (A). Middle Paleozoic to Jurassic mafic to felsic volcanic rocks and comagmatic intrusions, limestone, pelite, with conspicuous Middle(?) and Late Triassic plume-related(?) tholeiitic basalt overlain by carbonate. Paleomagnetic data on Triassic strata on Vancouver Island indicate little or no displacement, whereas those from southern Alaska indicate about 20° of northward displacement relative to North America; this is interpreted as late separation between northern and southern regions.

YA = Yakutat terrane (A and P). Late Mesozoic pelite, greywacke, and melange, and Cenozoic marine and continental clastic rocks.

YT = Yukon-Tanana terrane (A and P). Heterogenous metamorphic terrane comprising sedimentary and magmatic rocks of Late Proterozoic, Paleozoic, and Mesozoic protolith ages. Similar to Kootenay terrane in many aspects but may include Slide Mountain terrane rocks.

Rise) is in the southeastern Pacific Ocean, and its northern extensions (Juan de Fuca and Gorda ridges) either lie close to the North American continent or are replaced by transform plate boundaries (the Queen Charlotte and San Andreas faults). At latitude 40°N, the Pacific Ocean is ~7000 km wide and floored entirely by the Pacific Plate, which increases in age westward from Oligocene off northern California to Jurassic (~180 Ma) southeast of Japan. During the last 150 million years a strip of oceanic lithosphere ~13,000 km wide (a length equal to *one-third of Earth's circumference*) has flowed down into the mantle in subduction zones that were adjacent to and under the western margin of the North American plate, according to reconstructions of global plate kinematics derived from magnetic anomaly patterns in oceanic lithosphere and hot spot tracks (Engebretson et al., 1992).

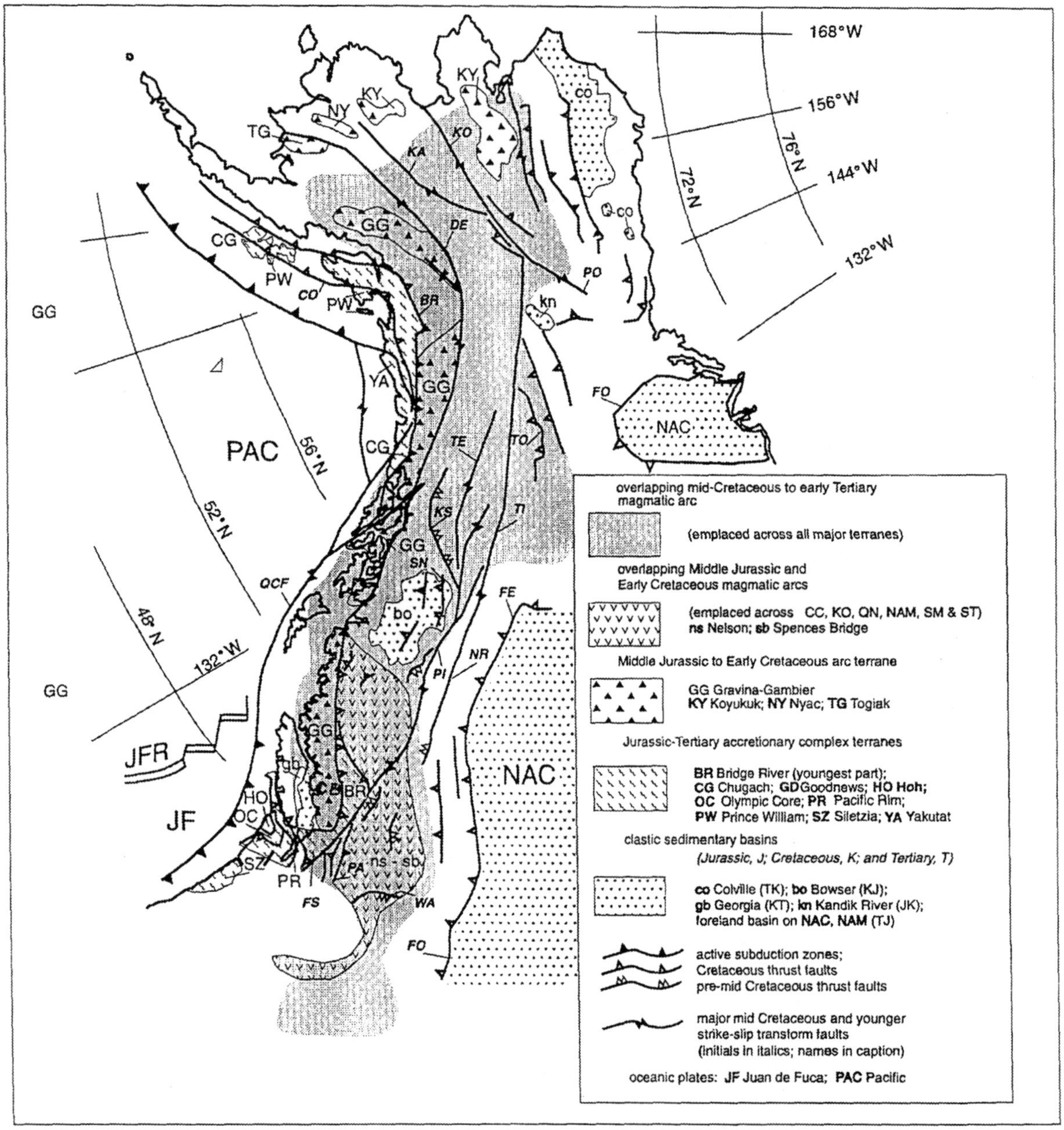

Figure 8. Middle Jurassic and younger features of the consolidating northern Cordilleran orogenic collage: (a) widespread mid-Cretaceous to early Tertiary continental margin arc; (b) Middle Jurassic continental-margin arc overlaps previously accreted terranes and continental-margin deposits; (c) Middle Jurassic—Early Cretaceous island arcs (notably Gravina-Gambier arc); (d) accretionary complex terranes near the present continental margin; (e) sedimentary basins with clastic deposits eroded from regions uplifted during compression; and (f) major compressional and strike-slip faults, most of which are of Late Cretaceous-early Tertiary age. Those encountered on trip are shown boldface below. Note that these faults form many terrane boundaries, so that the present distribution of components of the Cordilleran orogenic collage results largely from late displacements *within* the North American plate margin. BR = Border Ranges thrust fault system; CB = Coast Belt fold and thrust system; CO = contact thrust fault; De Denali Fault; FO = Foreland fold and thrust system; FS = Fraser-Straight Creek dextral strike-slip fault; KA = Kaltag dextral strike-slip fault; KO = Kobuk-South Fork strike-slip fault; KS = King Salmon Fault; NF = Nixon Fork dextral strike-slip fault; NR = Northern Rocky Mountain Trench dextral strike-slip fault; PA = Pasayten sinistral strike-slip and reverse fault; PI = Pinchi dextral strike-slip fault; PO = Porcupine Fault; QCF = Queen Charlotte-Fairweather dextral transform; SN = Skeena fold and thrust system; TE = Teslin dextral slip fault; TO = Tombstone thrust fault; WA = Waneta thrust fault; JFR = Juan de Fuca ridge. From Monger and Nokleberg (1996).

Exotic and Displaced Rocks

The ≤180 Ma record of sea-floor spreading preserved in Pacific Ocean lithosphere involves less than half of the probable duration (~390 Ma) of the period of subduction beneath the western edge of the North American plate. This raises questions about the total amount of oceanic lithosphere subducted beneath the North American plate margin in the last 390 Ma, and about the potential for very far-travelled (*exotic*) material having been brought into the plate margin on oceanic plates. Such material includes components of accretionary complexes (such as parts of Cache Creek terrane), and probably the pre-Jurassic arc rocks of Wrangellia and Alexander terranes (Fig. 7; Monger and Nokleberg, 1996; Butler et al., 1997). The remaining accreted terranes (Quesnellia, Stikinia and Slide Mountain) probably always were part of the North American plate; however, there are paleobiogeographic and paleomagnetic grounds for believing that they were *displaced* relative to the craton margin prior to their final accretion to the North American continent.

Arc Facing Directions

Subduction of oceanic lithosphere beneath the North American plate is recorded by arc magmatism that is at least as old as Middle Devonian and has persisted more-or-less continuously until the present day, although at some times being more widespread than at others (Fig. 9-1; Armstrong, 1988; Monger and Nokleberg, 1996). The facing directions of arcs potentially can be determined from lateral variations in the petrology and geochemistry of the magmatic rocks within the arc, and from the lateral relationships of the arc to coeval accretionary complexes on one side and to back arc basins and/or the craton margin on the other. Most mid-Mesozoic and younger arcs appear to have faced away from the North American craton; coeval accretionary complexes lie oceanward of the arcs. The facing directions of Triassic and Jurassic arc rocks of Quesnel terrane (Mortimer, 1987) and the Middle Jurassic arc of Stikinia (Anderson, 1993) were away from the continent, but facing directions of early Mesozoic and older arcs of Stikine, Wrangellia and Alexander terranes are unknown. The facing direction of the late Paleozoic arc of Quesnel terrane is away from the craton (Roback et al., 1994); contrarily that of a Permian arc in Yukon-Tanana terrane (north of latitude 60°N) is thought to be towards the craton (Mortensen, 1992). Late Devonian arc rocks occur from California to Alaska within the pericratonic and craton margin deposits.

Continental Arcs and Island Arcs

Even though a case can be made for more-or-less continuous subduction beneath the North American plate margin, the late Paleozoic through Triassic arc magmatism differed from the preceding and succeeding arc magmatism in its spatial relationship to the ancient continental margin.

The differences between the ancient arc types may be explained by observations on modern arcs (e.g. Hyndman, 1972; Uyeda and Kanamori, 1979; Jarrard, 1986). These suggest that an important factor contributing to the different types of arcs is the direction of relative motion of the upper plate, either toward or away from the oceanic trenches, which are the surface traces of the subduction zones "anchored" deep within the mantle. More recently, seismic tomography suggests that major circum-Pacific subduction zones probably descend to the core-mantle boundary (~-2900 km; Davis and Richards, 1992; Grand et al., 1997), so that the mantle "anchor" may be set even more firmly than the earlier workers realized! When the upper plate retreats from the trench, island arcs form, like those in the western Pacific Ocean. When the upper plate advances towards the trench, continental arcs form, like those in the mountainous regions bordering the present eastern Pacific Ocean. In addition, most island arcs tend to be in a state of extension, whereas most continental arcs tend to be under compression. Compression in the hot weak arc lithosphere causes crustal thickening by folding, thrusting and flow, and eventually builds mountains.

There appear to be three main stages in arc evolution in the Cordillera (Fig. 9-2):

1. **Middle and Late Devonian** arc rocks were emplaced in and on continental margin deposits and pericratonic terranes from California to Arctic Alaska, and coeval tuff is found locally in strata deposited on the craton edge (Richards et al., 1994). The arc presumably formed where the oldest, coldest, part of Panthalassa's oceanic lithosphere sank into the mantle, as it was located just outboard of the Cordilleran miogeocline where sea-floor spreading had begun by about 530 Ma (Sears and Price 2000).

2. **Carboniferous, Permian and Triassic** arc rocks in Quesnel and Stikine terranes in Canada and in the conterminous U.S. north of latitude ~40°N (but not in the southwestern United States) were separated from the continental margin by basins of unknown width. In the late Paleozoic the basins contained radiolarian chert and basalt with MORB chemistry (represented by Slide Mountain terrane), and in the Triassic, fine-grained clastics. The basins probably

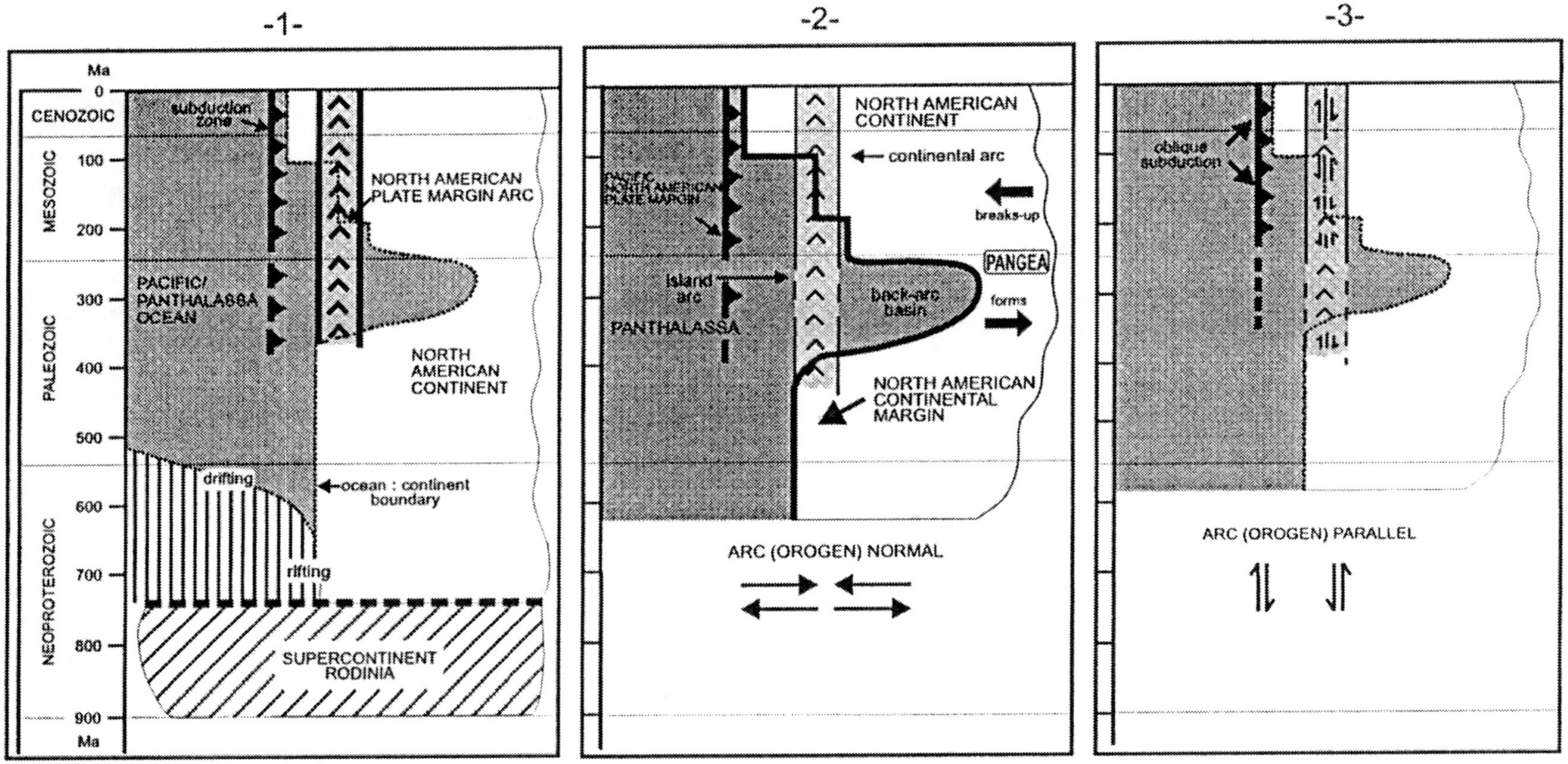

Figure 9. Time-space cartoons showing some controls on the evolution of the Canadian Cordillera.
1. Transition from (1) rift to (2) passive margin to (3) North American plate-margin arc.
2. Location of the North American plate-margin arc relative to the ancient continental margin, and the times (J, K) of change from an island arc regime to a dominantly continental arc regime.
3. Time range and sense of known and possible intra-arc strike-slip faults, related(?) to oblique subduction.

formed by back-arc rifting and spreading during subduction and arc extension, although there is local structural and stratigraphic evidence for compression in Devono-Mississippian and Permo-Triassic times. The Devonian arc evolved into a late Paleozoic-early Mesozoic island arc chain (represented by Quesnel and Stikine terranes) which was located on the margin of the North American plate, oceanward of an expanding back-arc basin (Slide Mountain terrane). The setting of the late Paleozoic-early Mesozoic arcs probably was similar to that of the present-day Japanese and Philippine islands. Globally, the change possibly occurred as the continental part of the North American plate retreated from the trench during amalgamation of Pangea.

3. **Middle Jurassic and younger** stratified arc rocks are mainly continental and formed across older island arc rocks previously accreted to the ancient continental margin, and in places on the margin itself. The exception is the Middle Jurassic-Early Cretaceous Gravina-Gambier island arc rocks of the northern Insular Belt and southwestern Coast Belt (Fig. 8). Tuff as old as Early Jurassic is interbedded with sedimentary strata deposited on the craton margin (Hall et al., 1998). Onset of Cordilleran-wide uplift and erosion is recorded first by the Middle Jurassic (~175 Ma) clastic strata of Bowser Basin, northern British Columbia (Fig. 8). By the mid-Cretaceous (~95 Ma), deposition across the entire Canadian Cordillera was non-marine except in fringing parts. The structural record shows that between the Early Jurassic and late Paleocene (~185–58 Ma) the arc region was under compression and transpression, although extension, transtension and dextral translation prevailed during the early Tertiary. As the North American plate rifted and eventually separated from Africa and Eurasia during Pangean break-up, presumably it advanced towards and overrode the trench on its western margin. It *collided* (Russo and Silver, 1996) with the oceanic plate, with the result that the region formerly occupied by an island arc evolved into a Middle Jurassic and younger continental arc located within an orogen akin to the Andes.

Terrane Accretion

The process in which a continental plate collides with a subduction zone ((3) above) provides a simple explanation for the accretion of offshore magmatic arc terranes to the western margin of the North American craton. As noted earlier, the accreted terranes appear to be *tectonic flakes,* with each preserving an upper crustal section ≤10 km thick that apparently was delaminated from its underlying lower crustal and upper mantle lithosphere. As the relatively strong wedge of cold, attenuated North American continental lithosphere converged with hotter, weaker island arc and marginal basin lithosphere that was anchored by the subduction zone, it drove into and separated the upper crustal arc rocks from their lower crustal and mantle lithosphere roots, and eventually overrode the subduction zone. During the advance of the wedge, the delaminated lower crustal and mantle lithosphere roots evidently became entrained in the underlying subduction zone. The upper crustal rocks are the tectonic flakes that form the collage of accreted terranes. Eventually, the Cordilleran foreland fold and thrust belt formed as the growing collage of tectonic flakes advanced over the underthrusting North American continental margin, and weak supracrustal (miogeoclinal) rocks, deposited on North American continental lithosphere, were scraped off and accreted to the overriding collage.

Oblique Subduction and Its Bearing on the Five Morphogeological Belts

The configuration of the Cordilleran orogenic collage is made far more complex because the direction of the long-lived plate convergence and subduction was probably mostly oblique to the convergent plate boundary. *Oblique subduction* results in the formation of arc-parallel strike-slip faults in weak arc lithosphere (Fitch, 1972; Jarrard, 1986; Oldow et al., 1990).

The widely distributed Late Cretaceous and Tertiary orogen-parallel dextral strike-slip faults in the Canadian Cordillera reflect the dextral motions of the oceanic plates relative to the North American plate as reconstructed by Engebretson et al. (1985). Fragmentary evidence of earlier orogen-parallel movements is provided by locally preserved sinistral or dextral faults that range in age from Devonian(?) to Early Cretaceous. In addition, the part of the Cordillera underlain by the accreted terranes features Permian through Early Jurassic faunas, and magnetic paleopoles ranging in age from Paleozoic to Late Cretaceous, that seemingly were displaced latitudinally relative to coeval faunas and paleopoles on the craton (Fig. 9-3).

With the exception of pre-Jurassic arc rocks of Alexander terrane and Wrangellia, the arc terranes in the Canadian Cordillera probably are fragments of the same North American plate margin arc. These were dispersed or duplicated by intra-arc strike-slip faults, rather than originating as discrete arcs that migrated across ocean basins (Samson and Patchett, 1991; Monger and Nokleberg, 1996). The Middle Jurassic-Early Cretaceous Gravina-Gambier island arc (emplaced in and on Alexander terrane and

Wrangellia; Fig. 8) may have been the offshore continuation of the coeval continental arc preserved in the southern half of British Columbia and in the conterminous western United States. The arc possibly was cut by a sinistral strike-slip fault and its northern part displaced southwards in Early Cretaceous time to lie oceanward of the continental arc in southern British Columbia (Monger et al., 1994). Lithological, stratigraphic, and faunal similarities suggest that Stikine terrane is a fragment of the same late Paleozoic-early Mesozoic island arc as Quesnel terrane, with Cache Creek terrane as the accretionary complex associated with both arc terranes. Possibly Stikine terrane was emplaced oceanward of the Cache Creek accretionary complex and Quesnel terrane by latest Triassic-Early Jurassic sinistral strike-slip faulting (C.J.R. Hart, pers. comm. 1999), rather than by oroclinal bending of the arc as suggested earlier by Mihalynuk et al. (1994).

Intraplate arc- (and orogen-) parallel displacements thus may be responsible for positioning large offshore arc terranes (Stikine, and Alexander, Wrangellia and the overlapping Gravina-Gambier arc) prior to their accretion to the former continental margin. During the Jurassic and Cretaceous advance of the North American plate towards the trench and the accompanying terrane accretion, the displaced arc terranes may have acted as stronger parts of the weak arc lithosphere and focused compressional deformation in the regions that eventually became the Omineca and Coast belts. Thus, oblique subduction may have played the key role in differentiation of the Canadian Cordillera into its five morphogeological belts.

Trip Outline

The road log is divided into four sections corresponding to: (1) Foreland Belt; (2) Omineca Belt; (3) Intermontane Belt; and (4) Coast Belt. The distances given are mostly calculated from the boundaries of each belt. Unfortunately, logistical considerations do not allow us to coordinate days with sections. On Day 1 of the field trip, we cross the displaced miogeoclinal (or craton margin) deposits of the Foreland Belt (Rocky Mountains), and the eastern parts of the Omineca Belt (Selkirk Mountains). We emphasize the relatively orderly transition from shelf to proximal to distal miogeoclinal facies, even though these rocks are now disposed in a fold and thrust belt on the ancient continental margin. On Day 2, we cross the Omineca Belt (Monashee Mountains, Shuswap Highlands) and much of the Intermontane Belt (Interior Plateaus). We examine parautochthonous old continental crust exhumed by early Tertiary extension in the Monashee Complex, the pericratonic Kootenay terrane containing Devonian arc magmatic rocks, the accreted Quesnel arc terrane and part of its accompanying accretionary complex, the Cache Creek terrane. On Day 3, we traverse the Cache Creek terrane in the westernmost Intermontane Belt, and then the Coast Belt, which is dominated by Middle Jurassic to early Tertiary granitic rocks. The granitic intrusions obscure a Mesozoic basin floored by late Paleozoic oceanic crust (Bridge River and Methow terranes) in southeastern Coast Belt that separates Wrangellia and Gravina-Gambier arc rocks in southwestern Coast Belt from the previously accreted terranes to the east. Our general intention is to provide an impression of the evolution of the entire mountain belt, whilst not concealing the current problems.

Acknowledgments

The guidebook incorporates material from several earlier guidebooks on the southern Canadian Cordillera (e.g. Price et al., 1985; Monger and Journeay, 1994a; Smith, 1998). JoAnne Nelson (British Columbia Geological Survey Branch, BCGSB) read and commented on the entire manuscript, Bob Thompson (Geological Survey of Canada, Sidney) and Ken Daughtry (Discovery Consultants of Vernon, B.C.) provided new information and comments on the western Omineca Belt, and Paul Schiarizza (BCGSB) reviewed the Coast Belt section. Ron Clowes (University of British Columbia) provided Figure 5. In preparing the guidebook, we received help from David Kisilevsky, Tom Ullrich, and Ela Rusak at Queen's University, who scanned figures and modified and prepared them for publication, and from Bev Vanlier of the Geological Survey of Canada in Vancouver, who completed the reference list.

1. FORELAND BELT

CALGARY TO LAKE LOUISE

INTRODUCTION

This segment of the trip crosses the eastern part of the foreland thrust and fold belt. It begins in the autochthonous Paleocene strata of the foreland basin along the northeast margin of the foreland thrust and fold belt at Calgary, and extends through the Foothills and Front Ranges, into the thick carbonate facies of the Cordilleran miogeocline in the Eastern Main Ranges of the Rocky Mountains near Lake Louise (Figs. 1-1, 1-2). The focus is on the nature and tectonic significance of the thrust and fold structures (Figs. 1-3, 1-4), and on the tectonic implications of the stratigraphy of the Upper Jurassic to Paleocene foreland basin succession, the underlying Cambrian to Jurassic platformal strata of the North American craton, and the transition into the Cordilleran miogeocline.

This route follows in the footsteps of R.G. McConnell of the Geological Survey of Canada (McConnell, 1887). During the field season of 1886, one year after the completion of the trans-continental Canadian Pacific Railroad, McConnell prepared a remarkably perceptive structure section, at a scale of 2 miles to 1 inch (Fig. 1-5), extending from the western edge of the Foothills to the Columbia River valley (Rocky Mountain trench). This section, published at a time when the very existence of large overthrust faults was controversial and vigorously disputed, outlined most of the basic features of the structure and stratigraphy of the southern Canadian Rockies, including the large thrust fault that now bears his name (McConnell thrust), and the fact that this fault is folded. As the following quotation shows, McConnell realized what he had accomplished:

> In regard to the completeness of the work, it may be stated that while much remains to be done in the way of filling in local details, yet the general features of the section, as presented, are believed to be fairly accurate, and will not be much modified by future investigations. In a mountainous and mostly wooded region like the one examined, where the possession of any geological fact implies a long and laborious climb of from 2000 to 6000 feet, the construction of a section showing all the structural minutiae, becomes an arduous undertaking, and would require much longer time than the four months or so at my disposal. [McConnell, 1887, p. 6D].

More than a century later, we are still filling in the details.

The Cordilleran foreland thrust and fold belt is an accretionary wedge. It consists of supracrustal rocks that have been scraped off the under-riding North American plate and accreted to the over-riding Intermontane terrane (Price, 1994). The structure of the foreland thrust and fold belt is dominated by thrust faults. Most of the thrust faults are southwest dipping and concave upward in profile. They flatten at depth and merge with each other and with a regional décollement, which, where outlined by seismic reflection imaging, generally lies close to the autochthonous Paleoproterozoic basement complex. The deformation is "thin-skinned." The basement slopes under the thrust and fold belt without disruption, from beneath the western Alberta Plains, at least as far southwest as the Rocky Mountain Trench (Bally, 1966; Price and Mountjoy, 1970; Cook et al., 1988; Cook and van der Velden, 1995; van der Velden and Cook, 1996). The faults cut upsection through the stratigraphic layering northeastward, but en route, they commonly follow the layering over large areas. The mechanical anisotropy and heterogeneity associated with the stratigraphic layering has exerted a fundamental control on the style of the thrusting and the folding.

The supracrustal rocks that have been scraped off the under-riding North American plate include two distinct stratigraphic successions: (1) an Upper Jurassic to Paleocene synorogenic clastic wedge succession that was eroded from the evolving thrust and fold belt, deposited in the isostatic lithospheric flexure produced by the evolving thrust and fold belt, and partly recycled through the evolving thrust and fold belt (Price, 1973, 1994; Beaumont, 1981; Beaumont et al., 1993); and (2) a Neoproterozoic to Middle Jurassic pre-orogenic continental terrace wedge that formed along the western margin of the Neoproterozoic Laurentian craton (Bally et al., 1966; Price and Mountjoy, 1970; Price, 1994). The continental terrace wedge had a long and protracted history, and several different components are discernible within it including: (a) the late Neoproterozoic-early Paleozoic Cordilleran miogeocline, its outboard deep-water equivalents, and its inboard equivalents which form

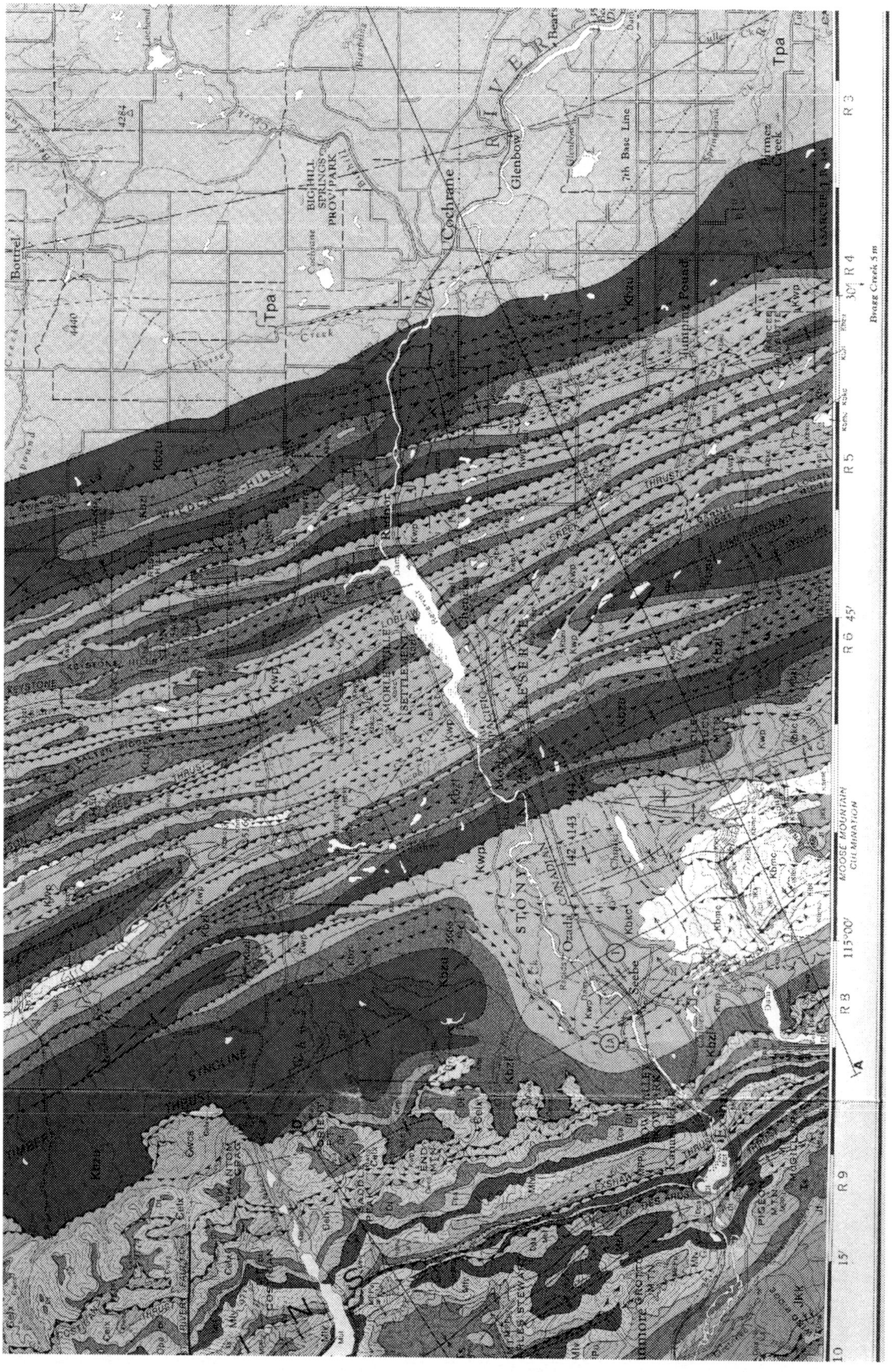

Figure 1-1. Geological map of the southern Canadian Rocky Mountains along the Bow Valley from the Alberta Syncline (east of Cochrane) to Lac des Arcs. From Ollerenshaw (1978).

Figure 1-2. Geological map of the southern Canadian Rocky Mountains along the Bow Valley from Lac des Arcs to Castle Mountain (Mount Eisenhower). From Ollerenshaw (1978).

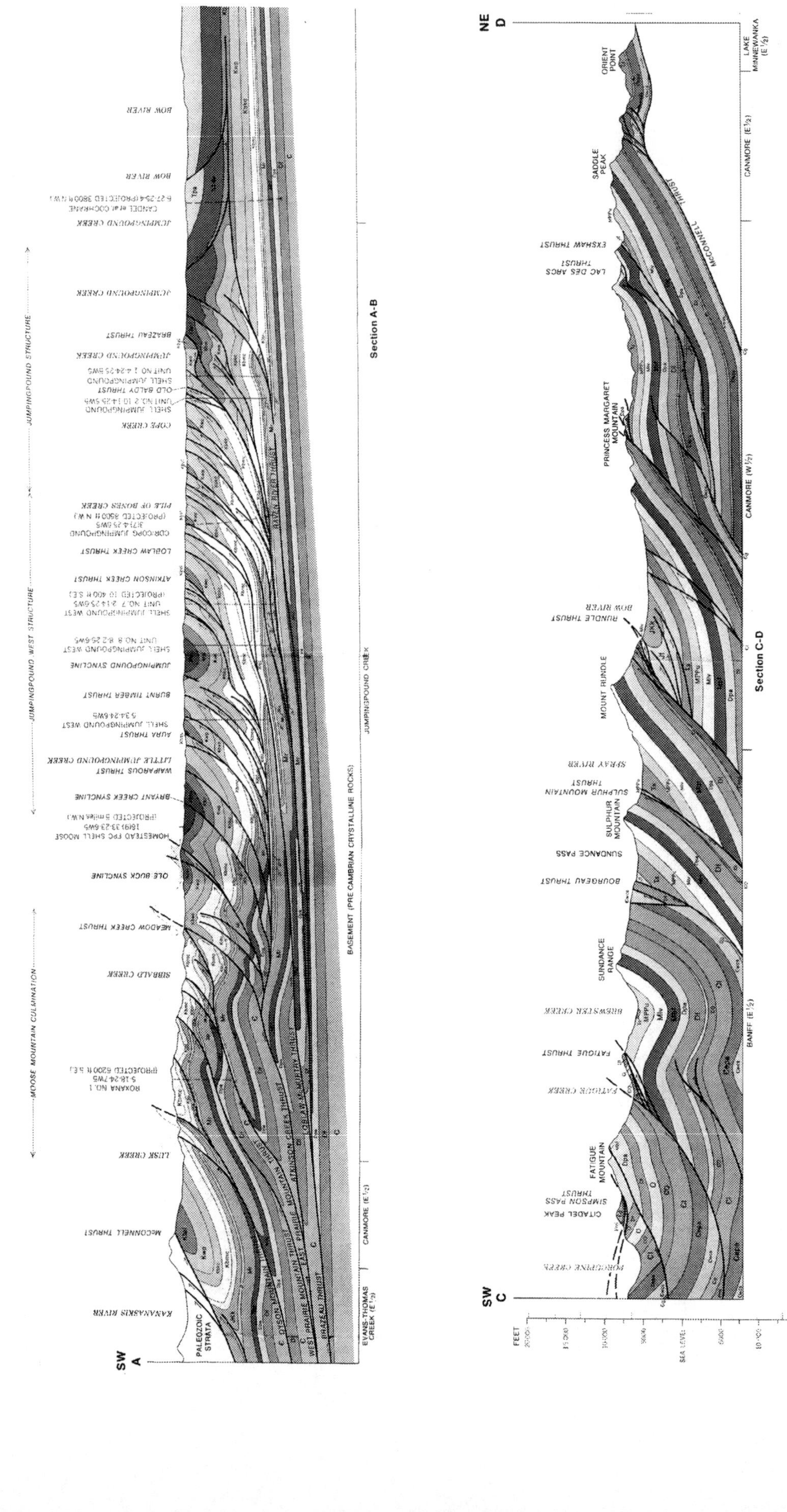

Figure 1-3. Structure sections: A-B—cross the Foothills south of the Bow Valley (see Fig. 1-1 for the location of the line of section); and C-D—cross the Front Ranges in the vicinity of Canmore. From Ollerenshaw (1978); after Price and Mountjoy (1970a, b, 1972a, b). See Figs. 1-1 and 1-2 for the location of the lines of section.

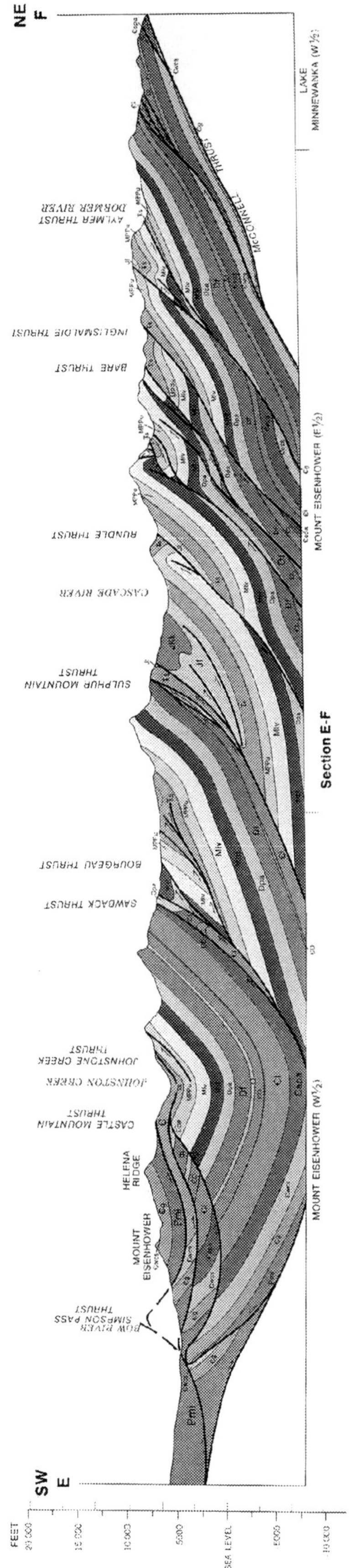

Figure 1-4. Structure section across the Front Ranges to Castle Mountain (Mount Eisenhower); the line labelled "E" on Figure 1-2 is location of the section. From Ollerenshaw (1978); after Price and Mountjoy (1972c, d).

the platformal cover along the margin of the craton (Fritz et al., 1991); (b) the Upper Devonian and lower Carboniferous distal foreland basin deposits associated with the mid-Paleozoic deformation and magmatism that occurred outboard of the continental margin in the southern Canadian Cordillera, and along the continental margin in Innuitian belt of Arctic Canada; (c) the upper Carboniferous and Permian shallow-water platformal carbonate and fine clastic deposits that thicken toward the edge of the late Paleozoic continental shelf; and (d) the shallow marine Triassic to Middle Jurassic siliciclastics of cratonal origin, and the subordinate carbonate rocks, that also thickened southwestward to the shelf edge (Gordey et al., 1991). Prior to the onset of the thrusting and folding, these rocks formed a northeastward tapering supracrustal wedge comprising up to 12 km of Neoproterozoic to Middle Jurassic mainly miogeoclinal rocks, and about 2.5 km of laterally equivalent Middle Cambrian to Middle Jurassic platformal rocks, both of which were overlain by up to 4 km of Late Jurassic to Paleocene synorogenic foreland basin fill (Fig. 1-6). A generalized stratigraphic table for this segment of the trip is given in Fig. 1-7.

Conspicuous variations in physiography within the thrust and fold belt are associated with abrupt variations in structural style, and can be related to variations in the thickness and in the composition and ductility of the supracrustal rocks (North and Henderson, 1954). In the Foothills belt, where the supracrustal wedge consists of a very thin Lower Paleozoic to Middle Jurassic platformal succession overlain by a thicker late Jurassic to Paleocene foreland basin succession, the structure is dominated by closely spaced, steeply dipping, imbricate, listric thrust slices, which give rise to relatively subdued topography dominated by ridges that are held up by sandstones and valleys that are eroded in shales (Fig. 1-3, Section AB). In the Front Ranges Subprovince of the Rocky Mountains, where the Lower Paleozoic to Middle Jurassic platform succession grades into the eastern edge of the miogeocline and thrust displacement is greater, the structure is dominated by three to six major listric thrust sheets, of which some are homoclinal, but others involve several imbrications and related fault-propagation folding (Fig. 1-3, Section CD). The steeply inclined resistant Paleozoic carbonate rocks, which are etched into relief by erosion of the overlying Mesozoic shaly rocks, form conspicuous linear mountain ranges (Figs. 1-1, 1-2) that are commonly homoclinal (e.g. Figs. 1-14, 1-17). The transition from the cratonic platform to the miogeocline involves a dramatic increase in the thickness and competency of the Lower Paleozoic to Middle

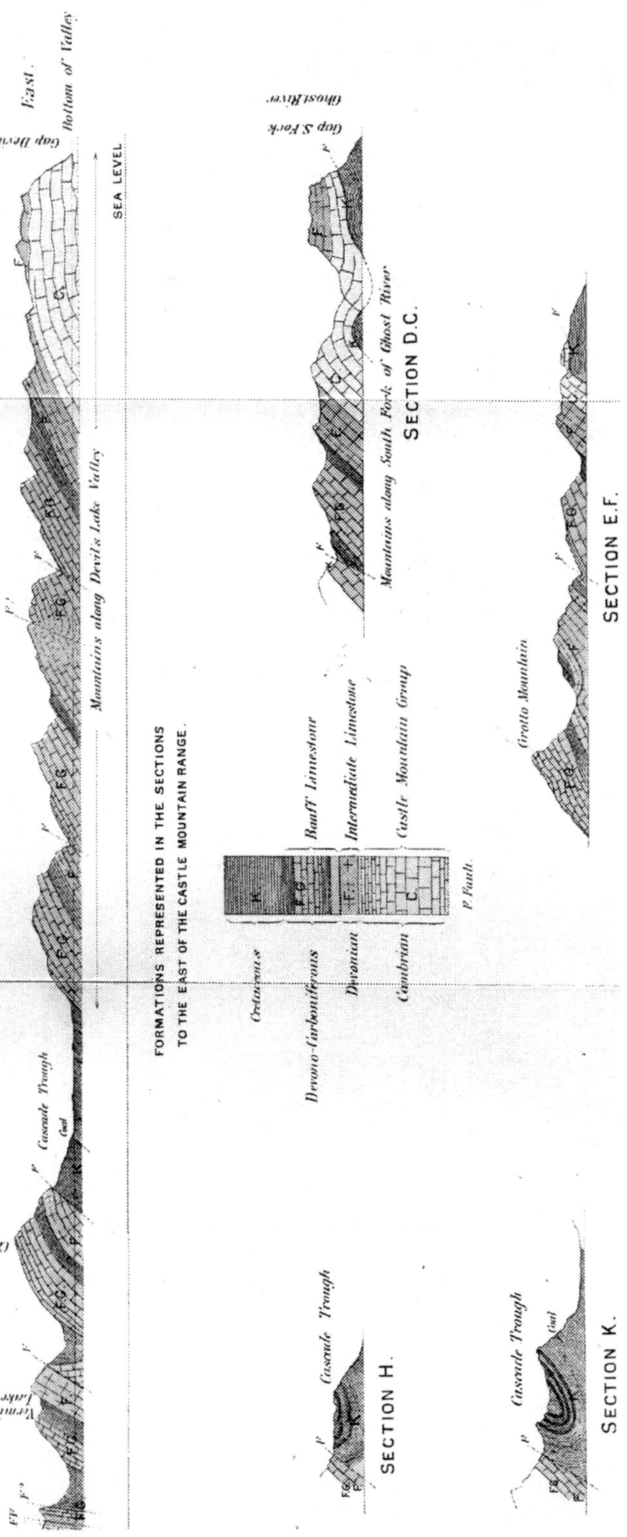

Figure 1-5. R.G. McConnell's structure sections though the eastern Front Ranges near the Bow Valley (from McConnell, 1887); section E.F. includes the Mount Yamnuska massif which is illustrated in the sketch in Figure 1-10.

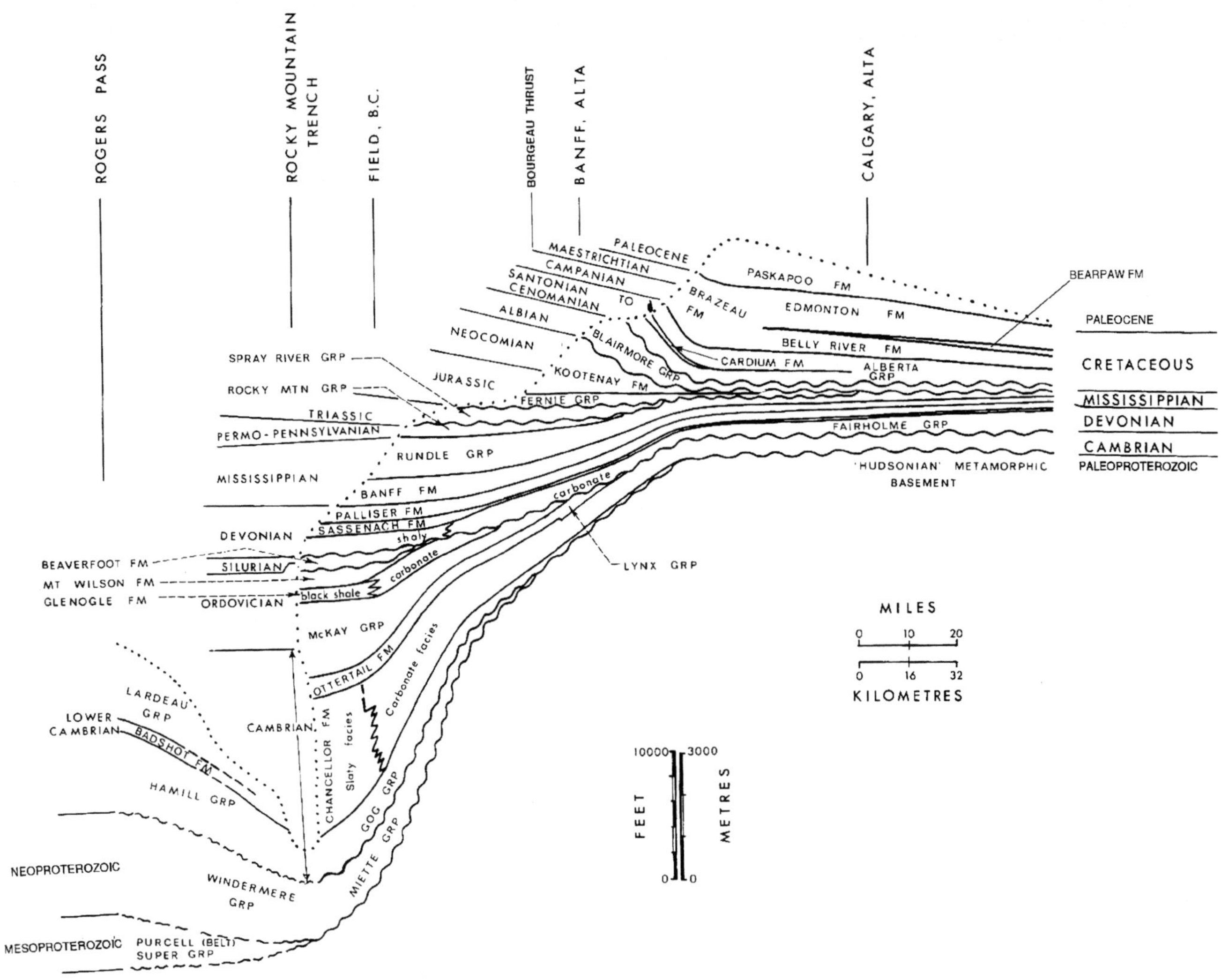

Figure 1-6. Schematic east-west stratigraphic section of the northeastward tapering wedge of Neoproterozoic to Paleocene supracrustal rocks in which the Cordilleran foreland thrust and fold belt developed. The horizontal datum is the transition form marine to non-marine Jurassic rocks, a sea-level surface that corresponds approximately with the transition between Neoproterozoic to Middle Jurassic miogeoclinal and platformal strata and the overlying Late Jurassic to Paleocene foreland basin strata (modified after Price et al., 1972).

Jurassic succession, and a change to more widely spaced thrust faults involving relatively gently dipping thicker sequences of carbonate rocks that form the distinctive castellated peaks of the Eastern Main Ranges Subprovince (Figs. 1-4, 1-22).

ROAD LOG

0.0 km Junction of Crowchild Trail and Nose Hill Drive.

Westward from the City of Calgary to beyond Cochrane the bedrock comprises autochthonous fluvial deposits of the Paleocene Paskapoo Formation, the youngest unit of the Cordilleran foreland basin succession in this region. The depositional slope for these rocks was toward the northeast, but bedding now dips very gently to the southwest (Figs. 1-1, 1-3) because of isostatic flexure of the underlying continental lithosphere in response to the weight of the tectonically thickened supracrustal rocks in the foreland thrust and fold belt (Price 1973, 1994).

15.0 km Approximate location of the axial surface of the Alberta syncline.

The Cordilleran foreland thrust and fold belt ends in the subsurface at the tip of a northeastward-tapering tectonic wedge of thrust faulted and folded

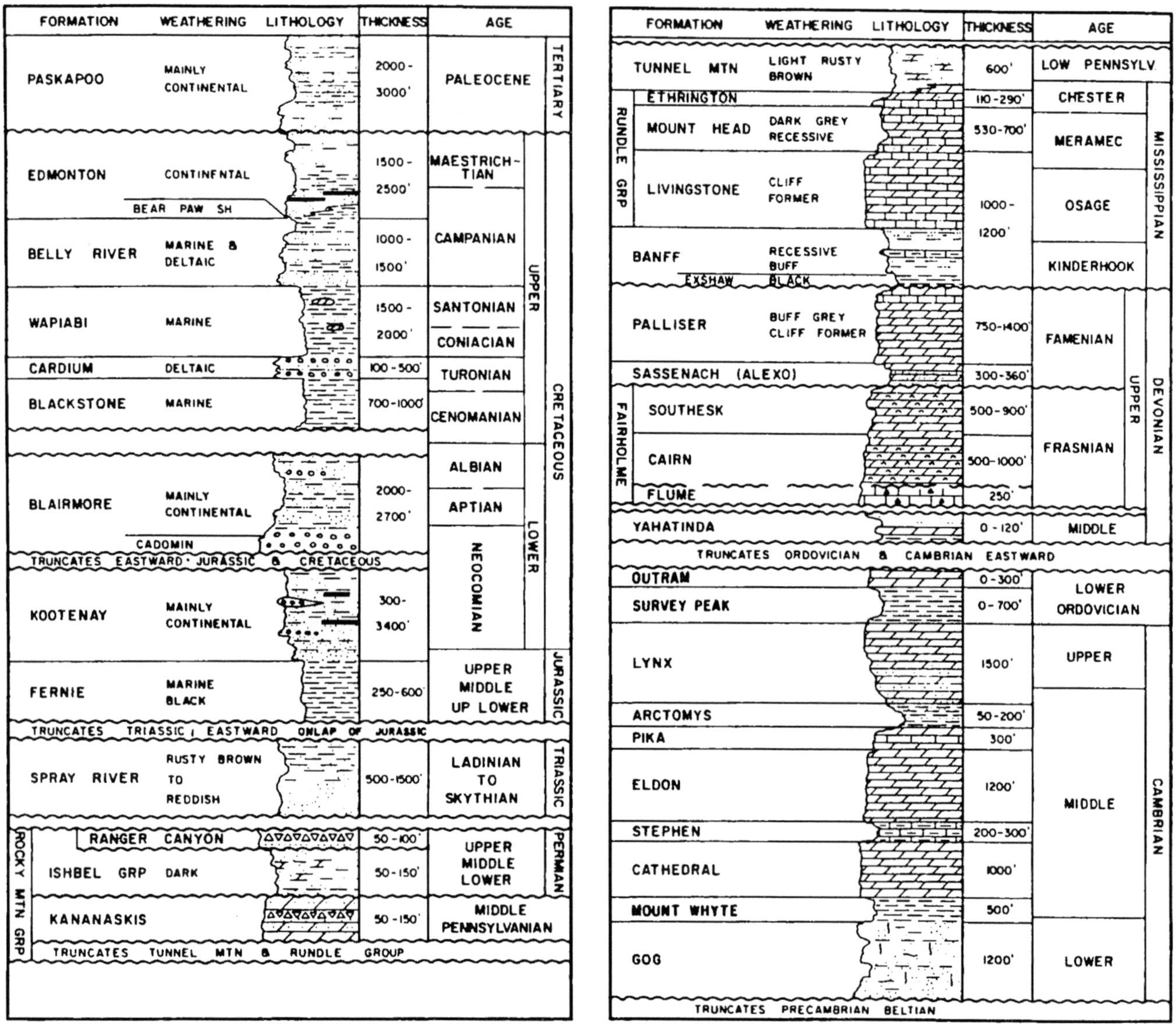

Figure 1-7. Generalized stratigraphic table for the region between Calgary and Lake Louise (from Bruce et al., 1995).

Upper Cretaceous rocks (Fig. 1-3). The wedge is bounded above and below by oppositely verging thrust faults that merge along the tip of the wedge (Price, 1986). The floor of the wedge is the northeast-verging basal detachment of the fold and thrust belt. The roof of the wedge is a southwest-verging back-thrust, above which detached, but autochthonous Paleocene and Upper Cretaceous strata have been tilted to the northeast to form the southwest limb of the Alberta syncline.

The Alberta syncline is a fault-propagation fold. Its northeast limb comprises the gently inclined strata that slope toward the Rocky Mountains with the rest of the foreland basin. The axial surface of the syncline emanates from the tip-line of the tectonic wedge, and it obviously must have migrated northeastward through the autochthonous Upper Cretaceous and Paleocene strata as the tip of the underlying tectonic wedge propagated northeastward. With the advance of the evolving tectonic wedge, the Paleocene and uppermost Cretaceous strata of the foreland basin succession have been progressively tectonically delaminated from the underlying parts of the foreland basin succession. The triangular "structure" that is defined by the roof thrust and the floor thrust, together with the nearest southwest-dipping thrust fault exposed at the surface to the southwest, is widely known as "the triangle zone."

Stop 1-1 Cochrane Hill viewpoint, Foothills Belt

18.2 km Pull off on south side of highway at top of hill northeast of Cochrane. Cochrane Hill viewpoint provides an overview across the low-lying mainly Cretaceous rocks of the Foothills belt to the grey cliff-forming Paleozoic limestones of the Front Ranges. The structure of the Foothills belt is dominated by northeast-verging listric thrust faults (Figs. 1-1, 1-3). At the surface the thrust faults are closely spaced and generally dip steeply (>45°) to the southwest; but at depth they flatten and merge with each other, with a regional detachment near the top of the Paleozoic rocks, and eventually, with the basal detachment of the foreland thrust and fold belt, below which the Paleoproterozoic crystalline basement and overlying supracrustal rocks are undeformed (Fig. 1-3).

The Foothills belt typically consists of long ridges, held up by resistant Cretaceous sandstone units, and linear valleys eroded in the weaker shale units; but in the vicinity of the Bow Valley these features have been partly obscured by glacial erosion and deposition associated with Pleistocene ice that flowed out of the Rocky Mountains along the Bow Valley. In the eastern part of the Foothills belt, some thrust sheets carrying Paleozoic strata emerge locally at the surface to form resistant homoclinal ridges or fault-propagation anticlinoria; but there are none of these in the vicinity of the Bow Valley. Locally, in the western part of the Foothills belt, imbricate thrust sheets involving substantial tectonic overlap of the Paleozoic carbonate rocks form antiformal stacks that produce broad structural culminations. The Moose Mountain culmination, which occurs in the western Foothills belt, south of the Bow Valley (Figs. 1-1, 1-3) is an example that can be seen from this viewpoint. It forms the higher, mainly forest-covered hills that are situated in front of the Front Range and south of the Bow Valley.

Fluvial sandstones and mudstones of the Paskapoo Formation (Paleocene) exposed in hill slope northwest of Cochrane have a shallow northeast dip, and form part of the southwest limb of the Alberta syncline. The axis of the Alberta syncline is situated about 3 km northeast of this locality.

Hydrocarbon Reservoirs

Several productive gas fields occur in the Foothills near the Bow Valley (Fig. 1-8). The gas is trapped in porous carbonate rocks of the Mississippian Rundle Group in structural culminations above thrust faults. In the eastern Foothills, the culminations occur where the reservoir rocks have been raised above the regional southwest slope of the autochthonous Mississippian along footwall ramps on which the sole fault changes from a bedding-parallel detachment to a cross-cutting southwest-dipping thrust fault. The Jumping Pound field (Fig. 1-9) is an example of this kind of trap. In the western Foothills belt reservoirs are associated with broad culminations formed by thrust duplex structures or by imbricate thrusts that have folded overlying thrust sheets of Upper Paleozoic carbonate rocks. The Moose Mountain culmination (Figs. 1-3, 1-9) is an example of this kind of trapping structure.

The petroleum resources of the Foothills belt occur in thrust-fault related structural traps. With one important exception, the Turner Valley oil field, they consist mainly of natural gas. The main reservoir horizon in the Turner Valley field and elsewhere is the Turner Valley member of the Livingstone Formation (Mississippian, Rundle Group), but commercial gas reserves have also been established in the Upper Devonian Palliser Formation, and in the Peechee Member of the Southesk Formation (Upper Devonian, Fairholme Group), which is a lateral equivalent of the prolific Upper Devonian Leduc reef oil and gas reservoirs of the central Alberta Plains. Petroleum production in the vicinity of Calgary, outside of the deformed belt, is from stratigraphic traps that involve several reservoir zones within the very gently southwest-dipping Upper Paleozoic cratonic platform succession, and the unconformably overlapping Mesozoic foreland basin succession (Bruce et al., 1995).

From this locality to our next stop, which is at the foot of the Front Range, our route traverses a part of the Foothills in which the exposed rocks consist mainly of Upper Cretaceous marine shale (Alberta Group) and overlying fluvial (Brazeau Formation) foreland basin deposits that occur in steeply dipping, closely spaced thrust slices (Figs. 1-1, 1-3).

20.0 km Cochrane.

21.9 km Paskapoo sandstones on hillside to the north dip to the northeast. They are in the southwest limb of Alberta syncline.

23.7 km Paskapoo sandstones on hillside to the north dip about 20° northeast.

24.3 km Horse Creek.

26.3 km Approximate location of the base of the Paleocene Paskapoo Formation.

29.2 km Junction of Highway 1A and Grand Valley Road.

29.6 km Grand Valley Creek. A small southwest-dipping thrust fault along the valley of Grand Valley

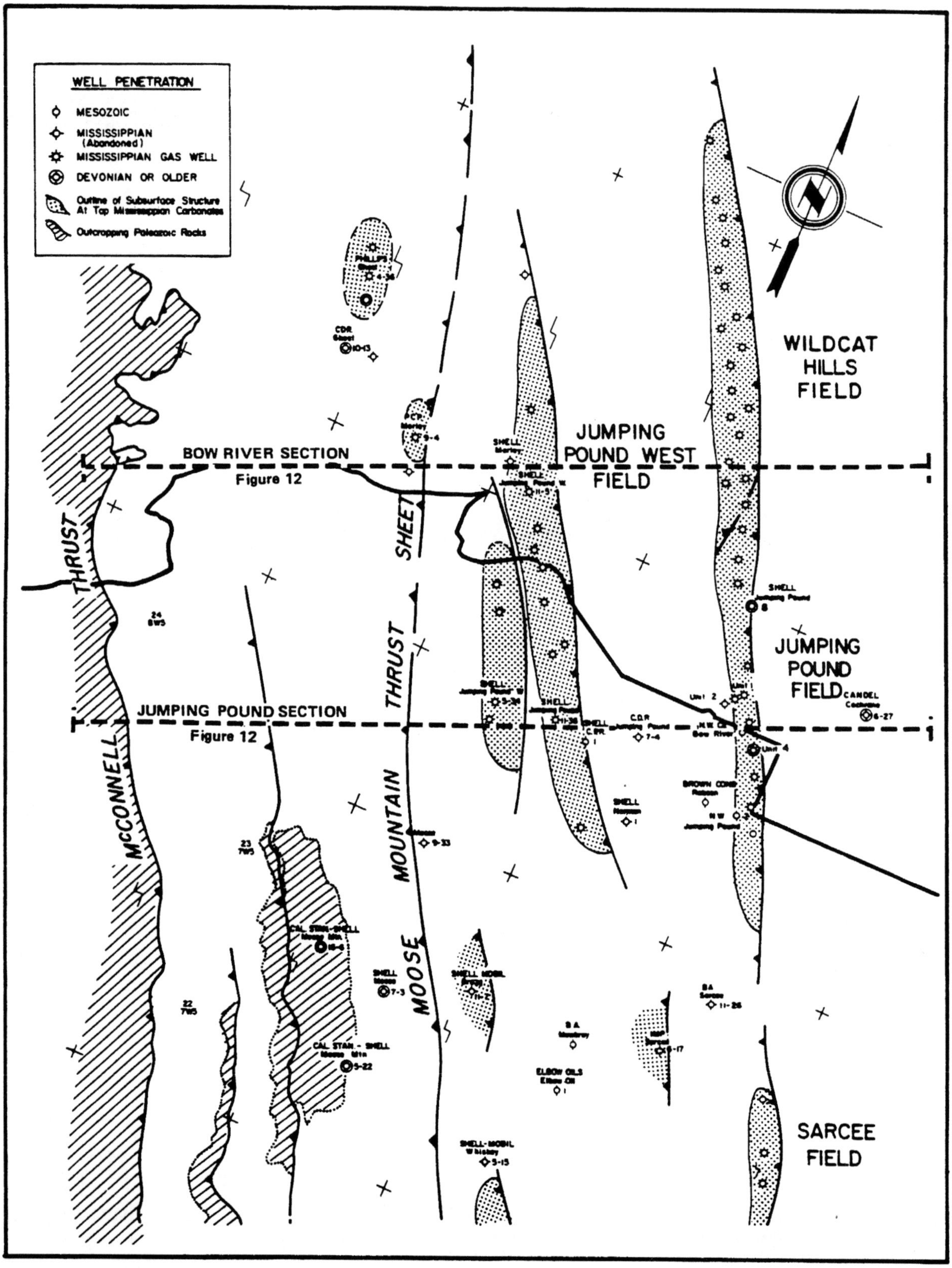

Figure 1-8. Map of well locations and gas fields in Foothills belt near Bow River (Bruce, et al., 1995).

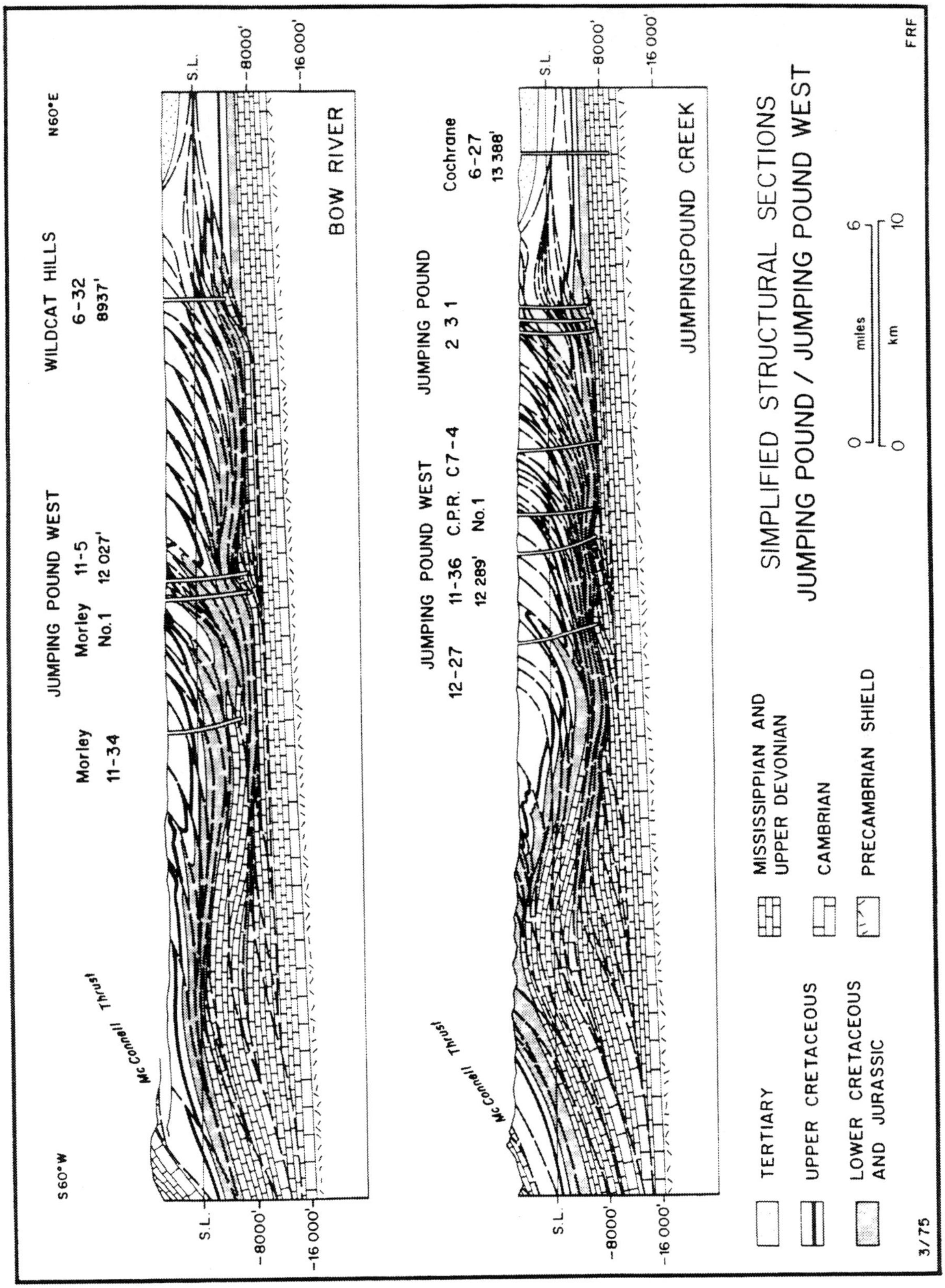

Figure 1-9. Simplified structural sections: Jumping Pound, and Jumping Pound West gas fields (Bruce, et al., 1995); see Fig. 1-8 for locations of sections.

Creek (the Grand Valley fault) marks the west side of "the triangle zone."

33.4 km Petro-Canada Wildcat Hills gas processing plant to the south at 11:00 o'clock.

39.2 km Ghost River; hydro-power dam and reservoir on Bow River.

46.4 km Roadcut in Upper Cretaceous fluvial deposits of the Brazeau Formation.

49.0 km Dark grey Upper Cretaceous marine shales (Wapiabi Formation/Alberta Group).

49.9 km Historic site to the south: McDougall Memorial Church. Reverends George and John McDougall, father and son, Methodist Church missionaries, ministered to the Stony Nation for 77 years. They established the settlement of Morleysville in 1864.

51.4 km Junction with the road to Morley.

53.2–55.3 km Roadcuts and stream cuts (Bow River) in Upper Cretaceous fluvial deposits (Brazeau Formation).

67.8 km Oldfort Creek.

Stop 1-2 Panoramic view of the Front Range—McConnell Thrust Sheet

74.9 km Junction of Highway 1A & Highway 1X.

The McConnell thrust fault, which marks the boundary between the Foothills and Front Ranges in the vicinity of the Bow River, juxtaposes the light-grey, cliff-forming limestones of the Middle Cambrian Eldon Formation over recessive-weathering fluvial sandstones and mudstones of the Upper Cretaceous Brazeau Formation (Fig. 1-10). The knife-sharp fault contact is exposed in gullies on the south slope of Mount Yamnuska. The fault is relatively flat under Mount Yamnuska, but it is folded and dips moderately to the southwest under the Front Range. The basic characteristics and significance of the fault were identified by R.G. McConnell of the Geological Survey of Canada, in 1886 (Fig. 1-5), at a time when the very existence of large overthrusts was a controversial idea. The McConnell thrust is one of the major thrust faults of the southern Canadian Rockies. It follows a broad arc about 480 km long within the area between the Crowsnest and Yellowhead Passes, and the maximum displacement on it is

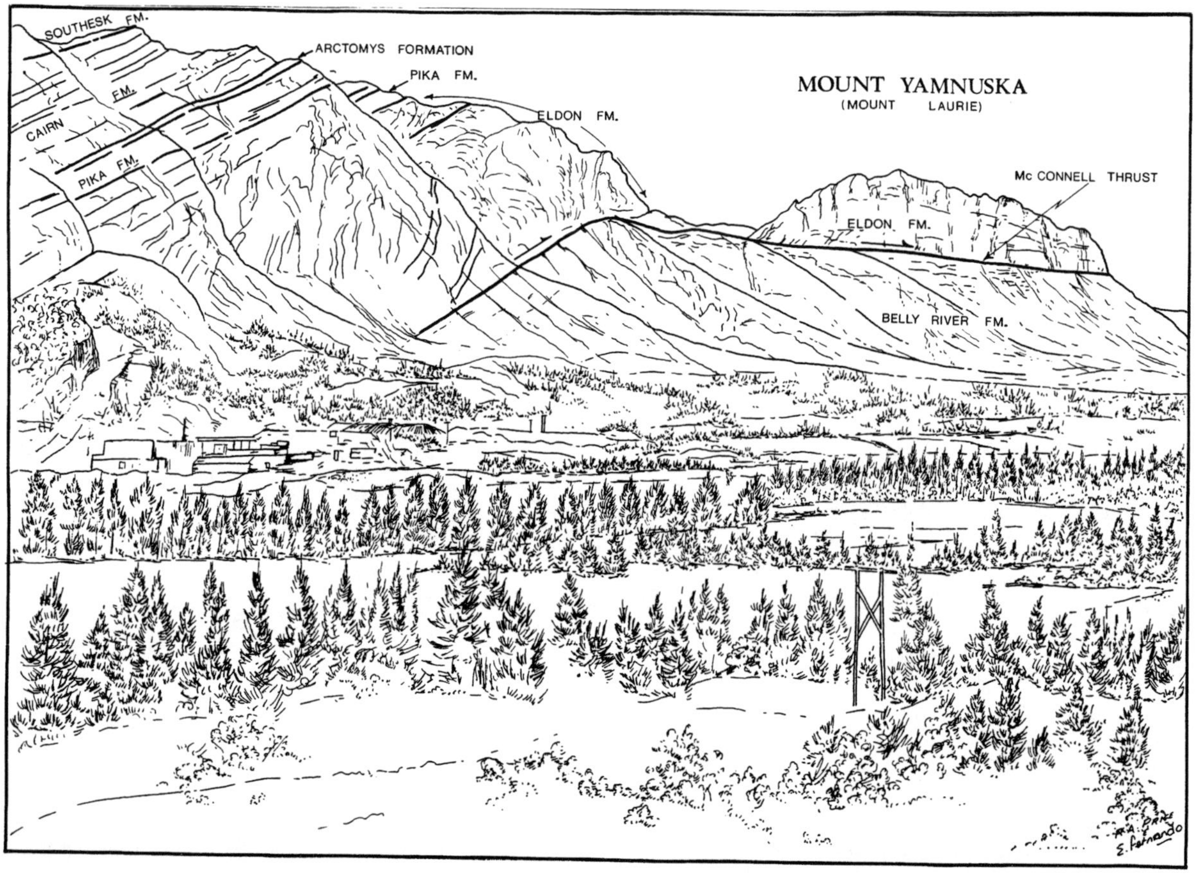

Figure 1-10. Sketch of the McConnell thrust sheet at Mount Yamnuska; view northwest across the Bow Valley from the Trans-Canada Highway (Price et al., 1972).

greater than 35 km (Price and Mountjoy, 1970; Price and Fermor, 1985). At this locality, the Cambrian carbonate rocks in the hanging wall of the McConnell thrust are deformed by a complex array of minor imbricate thrust faults and related folds, some of which are discernible in the cliffs southwest of Mount Yamnuska; in contrast, the Upper Cretaceous strata in the footwall of the fault are relatively undisturbed.

Platformal Cover

The southwest-dipping homoclinal panel of resistant Paleozoic rocks that forms the Front Range southwest of Mount Yamnuska is the thinnest and most incomplete Paleozoic succession in this transect of the Rocky Mountains. It is part of the thin, condensed, platformal cover succession that was deposited on the North American craton; and it forms part of a broad, northwest-plunging epeirogenic structure that was uplifted and eroded in Late Middle Devonian time and has been called the West Alberta Ridge. The Middle Cambrian Arctomys, Pika, Eldon, Stephen, and Cathedral Formations, all of which are represented either here or in the mountains immediately to the northwest, have an aggregate thickness of less than 600 m; and the remainder of the Lower Paleozoic succession has been removed by erosion that preceded the deposition of unconformably overlying Devonian rocks. The Upper Devonian (Frasnian) Fairholme Group (darker weathering carbonate rocks in Fig. 1-10) either lie directly on the Middle Cambrian Arctomys, Pika or Eldon Formations, or else are separated from them by the yellow and orange-brown weathering Middle Devonian (Upper Givetian) fluvial deposits of the Yahatinda Formation that fill channels eroded into the Middle Cambrian strata (Aitken, 1966a). In this area the Fairholme Group consists of massive dolomites of the Cairn and Southesk Formations, which comprise the carbonate or "reef" facies of area the Fairholme Group. To the southwest this facies is replaced by shaly or "off-reef" facies, comprising the Flume, Perdrix, and Mount Hawk Formations.

The northwest-plunging Moose Mountain culmination forms the higher hills in the western Foothills to the southwest. Carboniferous carbonate rocks (Rundle Group) are exposed in the core of the culmination (Fig. 1-1).

Proceed south on Highway 1X to Seebe Dam (Kananaskis Falls).

77.1 km Seebe Dam on Bow River. The Kananaskis Falls dam and hydro-electric power plant have been built on southwest-dipping marine sandstones of the Cardium Formation (Upper Cretaceous-Turonian to Coniacian) which form the middle part of the marine Upper Cretaceous succession of the Alberta Group. These southwest-dipping beds are in the southwest limb of one of a series of northwest plunging anticlines that form the northwest end of the northwest-plunging Moose Mountain culmination. Small, outcrop-scale thrust faults and fault-related folds are exposed in the bed of the Bow River below the dam. Upstream from the dam, on the south side of the river, quarries in the river terraces provide shale from the Upper Cretaceous Wapiabi Formation to the cement plants that are located close to the limestone quarries, farther up the valley at town of Exshaw.

79.5 km Junction of Highway 1X and Highway 1. Proceed west on Highway 1.

84.0 km Cross the trace of the McConnell thrust fault, which places Middle Cambrian Eldon Formation over Upper Cretaceous Belly River (Brazeau) Formation.

Stop 1-3 Lac des Arcs viewpoint

88.6 km Several southwest-dipping thrust faults that are splays from the McConnell thrust produce repetitions in the distinctive sequence of Paleozoic formations in the McConnell thrust sheet along the Bow Valley. One of these faults is well displayed in the view to the north across Lac des Arcs (Fig. 1-11). The Lac des Arcs thrust juxtaposes Middle Cambrian strata of the Eldon Formation over Mississippian rocks of the Livingstone Formation. A smaller, unnamed thrust is responsible for the emergence of the Palliser Formation beyond the west end of Lac des Arcs. The Exshaw thrust, which crosses the Bow Valley east of Lac des Arcs, juxtaposes the Upper Devonian Fairholme Group over Mississippian to Triassic rocks (Fig. 1-1).

89.6–90.8 km Roadcuts in the Upper Devonian Fairholme Group (carbonate facies).

93.8 km Contact between Banff Formation and Rundle Group on the south.

102.6 km Overpass.

Stop 1-4 Canmore viewpoint, Rundle Thrust Sheet

106.6 km The Bow Valley follows the weak Mesozoic rocks in the upper part of the McConnell thrust sheet for about 25 km in the vicinity of Canmore (Figs. 1-1, 1-2). Throughout this interval the northeast side of the valley, in the Fairholme Range, is a

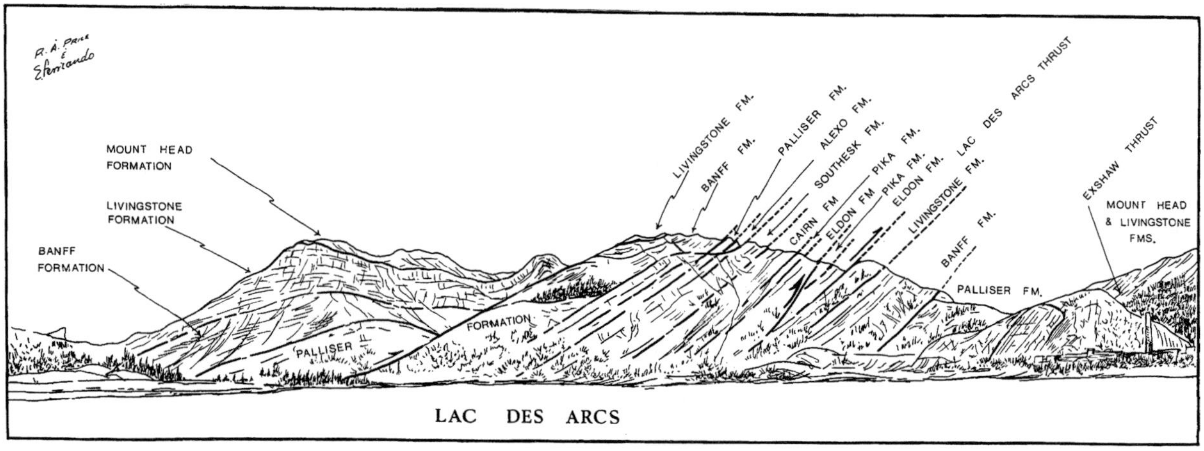

Figure 1-11. Sketch of Lac des Arcs thrust and related structures in the McConnell thrust sheet; view northwest across Lac des Arcs from the Trans-Canada Highway (after Price et al., 1972).

Figure 1-12. Sketch of Rundle thrust sheet and Mount Allan syncline near Mount Lougheed, view from southwest the Trans-Canada Highway south of Canmore (after Price et al., 1972).

series of dip slopes cut on the resistant beds of the upper part of the Rundle Group. The southwest side is rimmed with the steep cliffs of the Rundle Range (the second of the Front Ranges), which are formed by the southwest-dipping homoclinal panel of Fairholme, Palliser, Banff and Rundle strata that occurs in the hanging wall of the Rundle thrust sheet.

Structural relationships along the Rundle thrust can be studied in the view to southwest from the Trans-Canada Highway south of Canmore. To the south, between Mount Allan and Mount Lougheed (Fig. 1-12), the distinctive triplet of upper Paleozoic units (Palliser, Banff, Rundle) outlines a series of folds along the frontal edge of the Rundle thrust sheet; and resistant conglomerate in the Cadomin Formation, the basal unit of the Blairmore Group (Lower Cretaceous; Albian), outlines the northeastward overturned, Mount Allan syncline, which has developed in the footwall of the Rundle thrust. Northwest of Mount Lougheed, at The Three Sisters (Fig. 1-13), erosion has etched out a syncline-anticline fold pair in the hanging wall of the Rundle thrust. These folds terminate farther northwest against a transverse, northeast-striking steep fault that crosses the Rundle Range near White Man Gap (Fig. 1-14). This transverse fault abuts the underlying

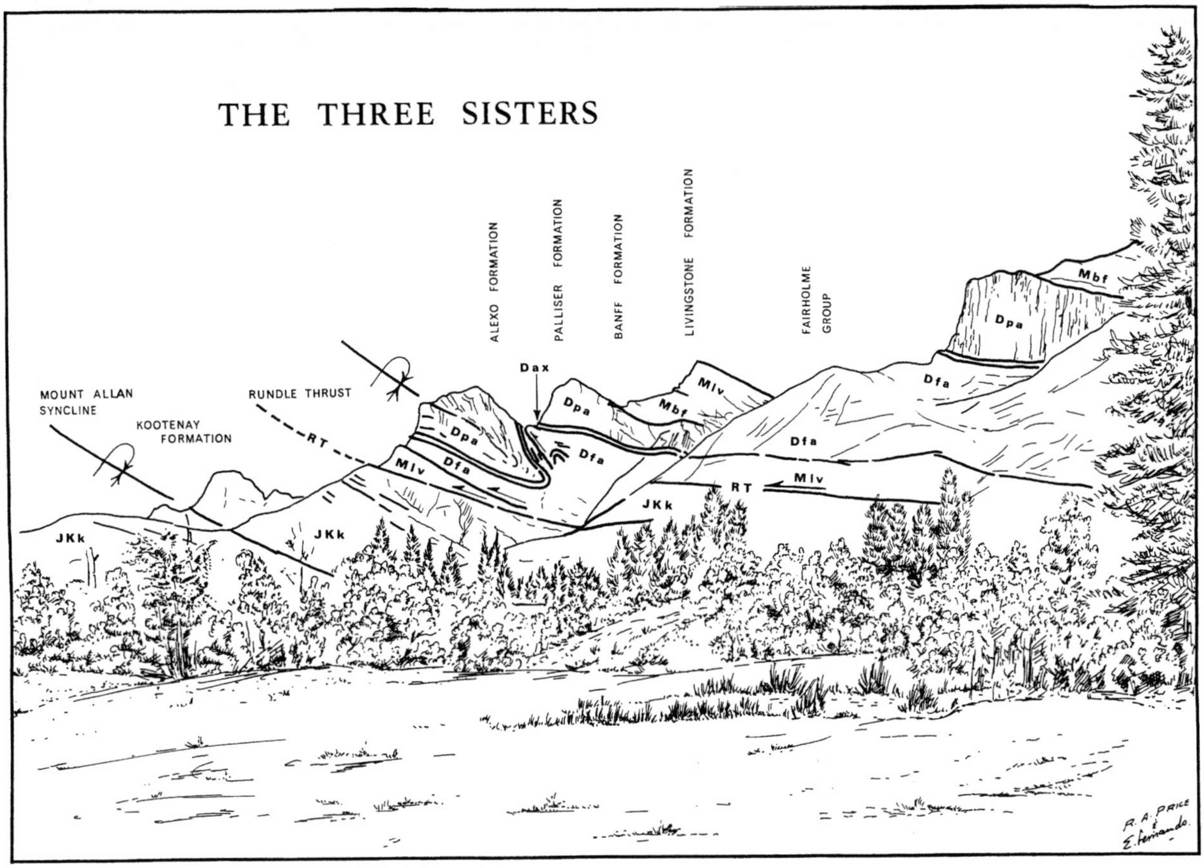

Figure 1-13. Sketch of The Three Sisters, Rundle thrust sheet, view southwest from the Trans-Canada Highway south of Canmore (Price et al., 1972, Fig. 9). JKk = Kootenay Group; Dfa = Fairholme Group; Dax = Alexo Formation; Dpa = Palliser Formation; Mbf = Banff Formation; Mlv = Livingstone Formation, Rundle Group.

Rundle thrust; but it offsets the overlying Sulphur Mountain thrust (Fig. 1-2).

The Upper Jurassic and Lower Cretaceous Kootenay Group and the Lower Cretaceous Blairmore Group strata comprise non-marine (fluvial) foreland basin deposits that record the emergence and erosion of thrust sheets in the western Rocky Mountains and adjacent Purcell-Dogtooth Mountains (Price and Mountjoy, 1970). These foreland basin deposits are substantially thicker and more coarse-grained here, within the McConnell thrust sheet, than they are northeast of the McConnell thrust in the Moose Mountain culmination. Coal seams in the lower part of the Kootenay Group, some of which reach semi-anthracite rank, were responsible for the establishment of Canmore as a major coal mining community at the close of the 19th century. Note the locality (former town) of Anthracite between Canmore and Banff (Fig. 1-2).

114.1 km Banff Park gate.

Stop 1-5 Overturned Kootenay-Fernie section (west limb of Mount Allan syncline) in the footwall of the Rundle thrust fault, and viewpoint for Cascade Mountain

125.2 km At Cascade Mountain (Figs. 1-15, 1-16) the Rundle thrust juxtaposes southwest-dipping rocks of the upper part of the Fairholme Group (Upper Devonian) over steeply dipping and overturned northeast-facing strata of the Spray River Group (Triassic) and Fernie Group (Jurassic). The steep southeast face of Cascade Mountain is carved in the sequence of three distinctive "marker" units that facilitate the delineation of the structures in the Front Ranges of the Southern Canadian Rockies: (1) lowermost, the cliff-forming Palliser Formation (Upper Devonian limestone); (2) the more recessive Banff Formation (Lower Mississippian shale); and (3) uppermost, resistant Rundle Group (Upper Mississippian limestone). The conspicuous fold in the Palliser Formation on Cascade Mountain is part of a

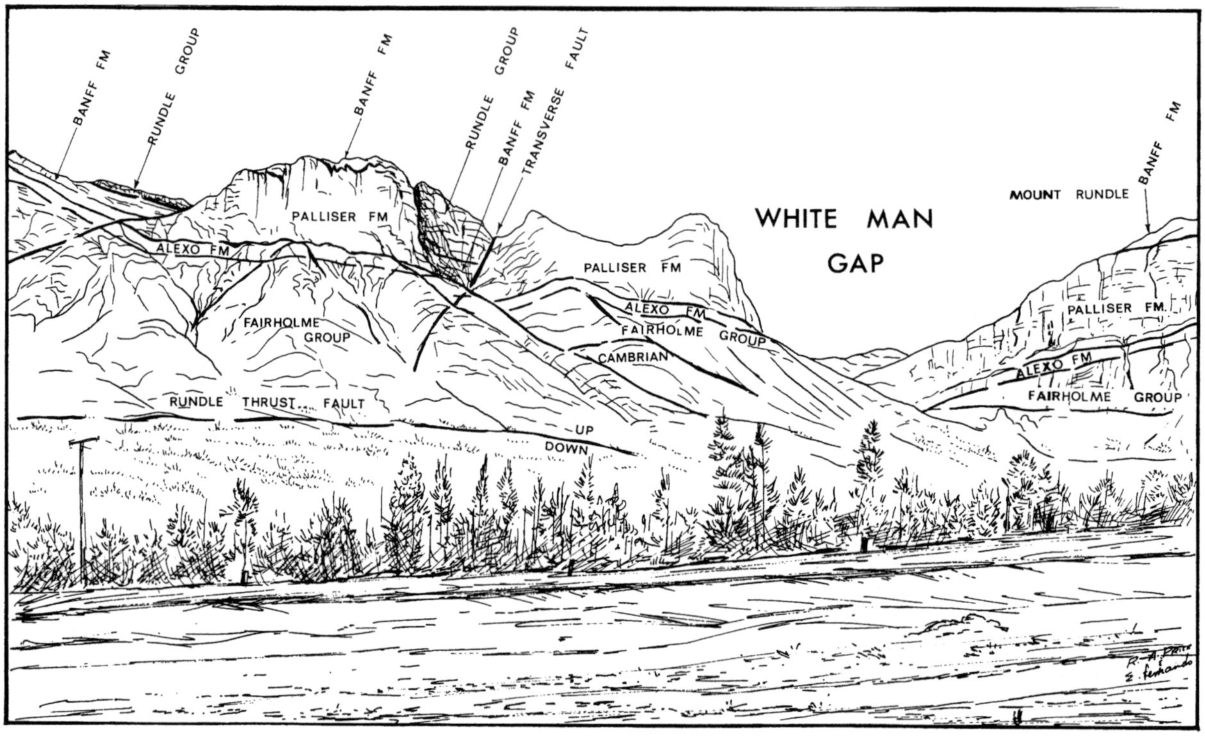

Figure 1-14. Sketch of Rundle thrust sheet at White Man Gap, view northwest from the Trans-Canada Highway south of Canmore (Price et al., 1972).

Figure 1-15. Sketch of Cascade Mountain, view west-northwest from the Trans-Canada Highway (Price et al., 1972).

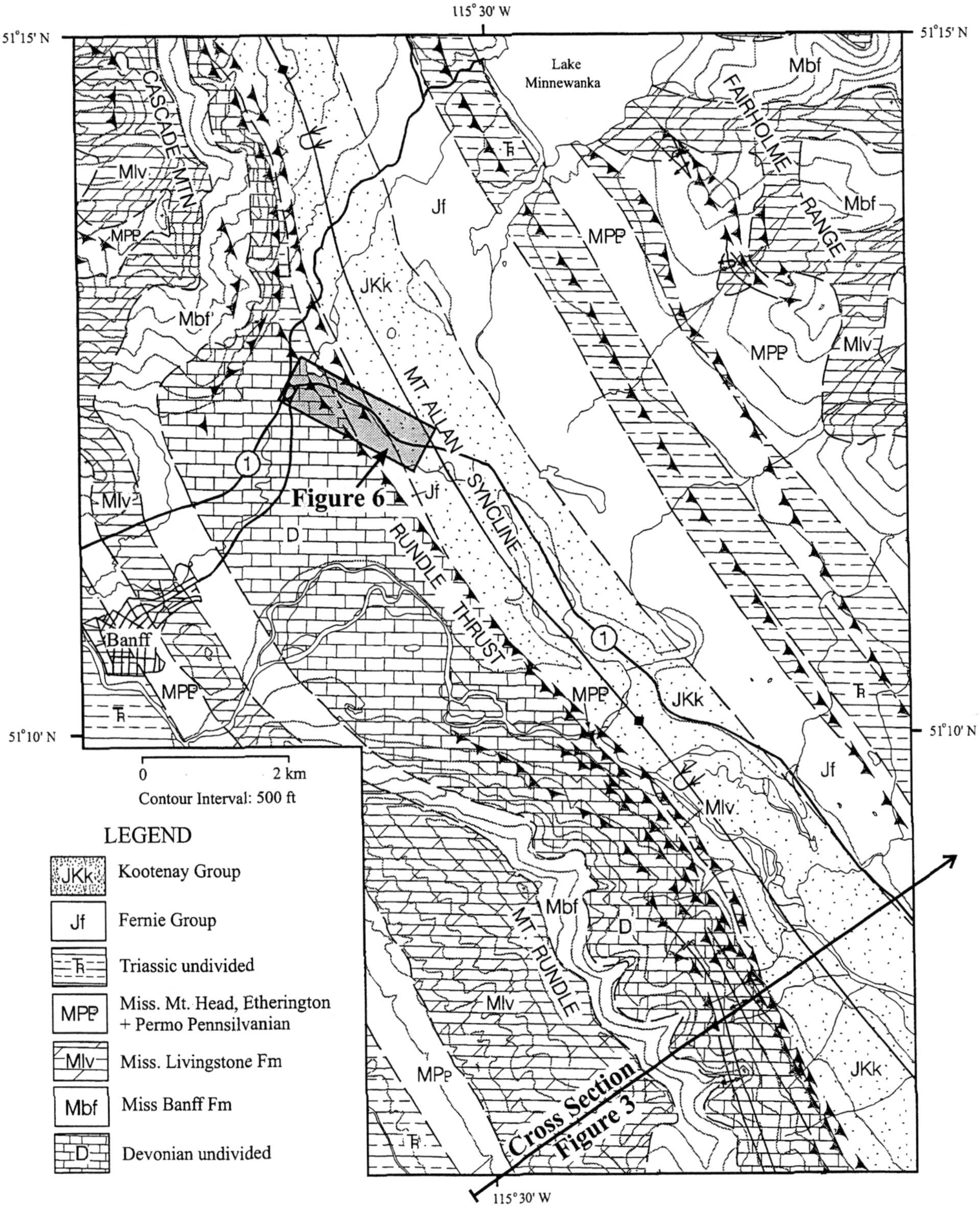

Figure 1-16. Geological map of the Rundle thrust sheet and Mount Allan syncline in the vicinity of Banff, Alberta (from Cant and Stockmal, 1999; simplified after Price and Mountjoy, 1970b and 1972a), showing locations of Figures 1-17 (labelled "Figure 3") and 1-18 (labelled "Figure 6").

transverse monocline that plunges obliquely down the southwest-dipping structural panel in the hanging wall of the Rundle thrust. This transverse monocline marks a transverse step along the hanging wall of the Rundle thrust, along which the Rundle thrust cuts up-section to the northwest from a bedding-glide zone in the Middle Cambrian section under the Rundle Range to the upper part of the Fairholme Group in the Cascade Range, from which it cuts up-section northwestward very gradually through the Palliser Formation. The cross-cutting relationships between this transverse monocline and structures in the overlying Sulphur Mountain and Bourgeau-Sawback) thrust sheets provide the basis for demonstrating that although the thrust faults developed sequentially from southwest to northeast, displacement on the older thrust fault continued while the younger ones were developing (Price, in review, Journal of Structural Geology).

The well exposed Jurassic section (marine Fernie Group and lower part of non-marine Kootenay Group) exposed along the north side of the highway at this locality (Fig. 1-16) provides opportunities to study the stratigraphy and sedimentology of the lowest part of the Rocky Mountain foreland basin succession, and also to study details of the deformation in the overturned southwest limb of the Mount Allan syncline, which developed along a footwall ramp beneath the Rundle thrust sheet (Fig. 1-17). A recently published detailed description and interpretation of these rocks by Cant and Stockmal (1999) provides much new information, only some of which can be considered during our brief traverse through the section from here to where we re-board the bus at the other end of the section.

The section consists of three distinct parts (Fig. 1-18). The upper (northeastern) part, which is exposed in new roadcuts close to the highway, comprises the lower 48 m of the Upper Jurassic-Lower Cretaceous Kootenay Group. The beds are overturned and dip to the southwest. A fine- to medium-grained sandstone unit, 8 m thick, at the stratigraphic base of the this section has been interpreted by Cant and Stockmal (1999) as a shoreface deposit.

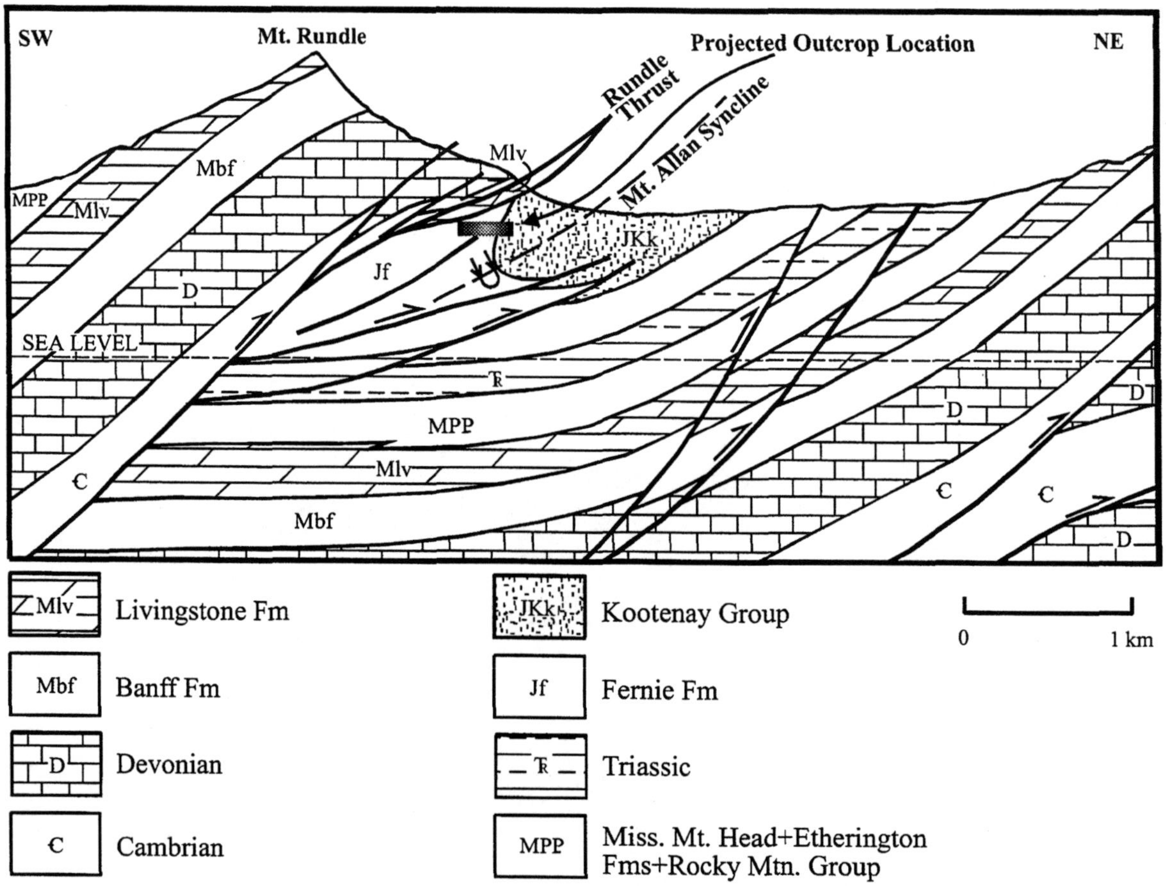

Figure 1-17. Cross section of Rundle thrust sheet and Mount Allan syncline (from Cant and Stockmal, 1999; simplified after Price and Mountjoy, 1970b, 1972a). In Figure 1-16, the location of this cross section is labelled "Figure 3."

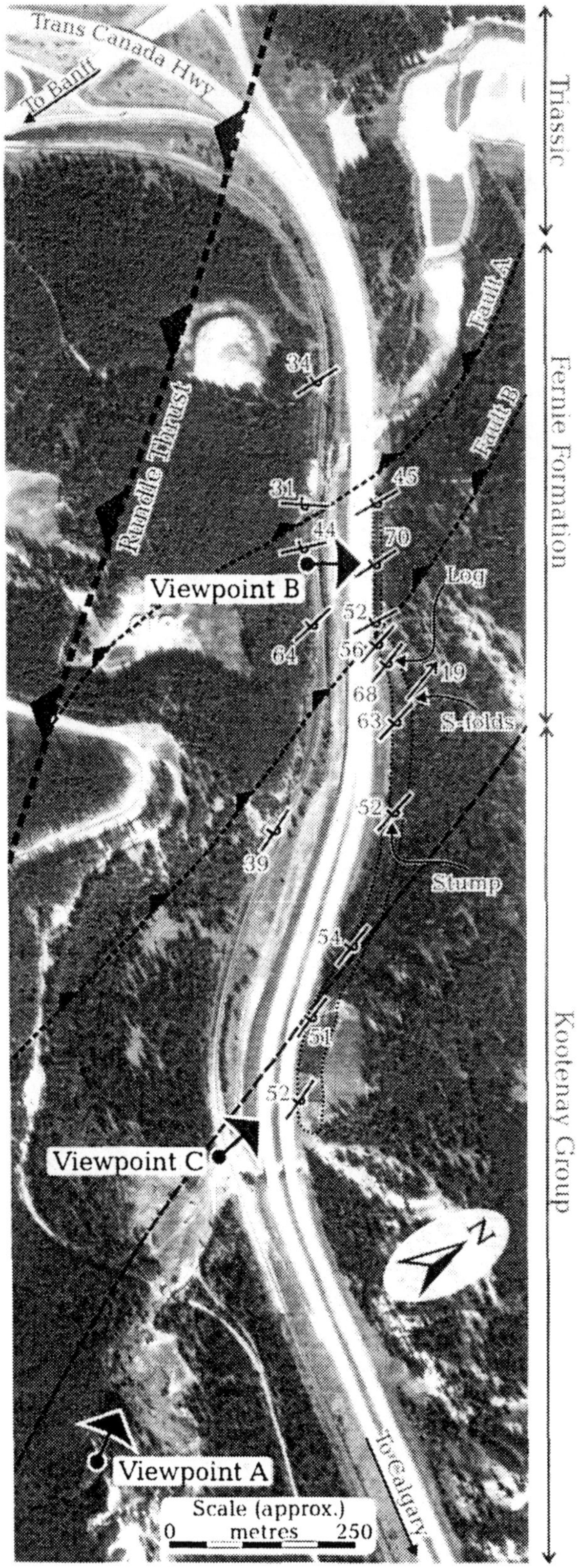

It is succeeded by about 15 m of mainly shaly strata with some interbedded sandstone and with abundant plant debris including local "log conglomerates," all of which has been interpreted as representing a lagoon or floodplain environment that developed behind the beach. The sandstones in the upper part of this segment have been interpreted as fluvial deposits, but could be interpreted as shoreface deposits (Cant and Stockmal, 1999).

The middle part of the section, which is exposed in natural stream cuts, is set back from the highway, and is largely covered with vegetation. It comprises strata of the Upper Jurassic (Oxfordian and Kimmeridgian) Passage Beds of the Fernie Group, and it forms a relatively simple homoclinal panel. It records a transition from marine shales with thin interbeds of "turbidite"-like siltstone and fine sandstone that were deposited below storm-wave base, through hummocky cross-stratified sandstones that were deposited above storm-wave base, to the shoreface deposits at the base of the first part of the section. A fossilized tree stump occurs in the hummocky cross-stratified sandstones in the upper part of this interval, and a fossilized "driftwood" log is embedded in the finer grained sediments near the base (see Fig. 1-18 for exact locations). The middle part of the section is separated from the lower part by a fault that is hidden by overburden, but marked by an abrupt change in the orientation of the bedding (see Fig. 1-18).

The lower part of the section consists of much older roadcuts, but these provide excellent exposures of the lower levels within the Fernie Group. This part of the section is closer to the Rundle thrust fault, and is cut by a large number of small faults that have been described by Cant and Stockmal (1999). Some of the faults, which dip less steeply than the beds and appear to cut out parts of the stratigraphic sequence, evidently are thrust faults that developed after the beds in the southwest limb of the Mount Allan syncline had been rotated to a nearly vertical orientation. Other faults are southwest-verging thrusts that formed in the southwest limb of the Mount Allan syncline, presumably in conjunction with flexural-slip folding. This part of the section is structurally complex in detail. Among the noteworthy details are cm-spaced small faults that offset the surfaces of fine sandstone and siltstone beds, and that appear to record "penetrative" horizontal compression and vertical extension of steeply dipping strata. The lower one-third to one-half of the lower part of the section

Figure 1-18. Portion of Natural Resources Canada aerial photograph A27851-173, 1:20,000 nominal scale (June 22, 1992), with annotations after Cant and Stockmal (1999), showing the Fernie-Kootenay section along the Trans-Canada Highway east of Banff. In Figure 1-16, the location of this photograph is labelled "Figure 6."

consists of the Bathonian Grey Beds (Frebold, 1962), which unconformably underlie the Green Beds (Oxfordian) and the overlying Passage Beds (Kimmeridgian) of the Fernie Group and pre-date the onset of foreland basin subsidence. The lowest part of the Fernie Group appears to have been cut out by a concealed northeast-verging thrust fault (Fig. 1-18) that carries dolomitic siltstones of the Triassic Whitehorse Formation in its hanging wall (Cant and Stockmal, 1999).

In the view to the northwest from the bridge at the northwest end of the section, the overturned Triassic strata can be seen in a gully on the lower slope of Cascade Mountain, sitting discordantly below the southwest-dipping Devonian strata of the Rundle thrust sheet.

126.3 km Board bus at west end of exposure.

126.4 km Cascade River bridge.

127.1 km Junction with road to Banff.

131.8 km Cross trace of Sulphur Mountain thrust.

135.4 km Cross trace of Bourgeau thrust, with northeast-facing Devonian Fairholme Group over southwest-facing Triassic Sulphur Mountain Formation.

Bourgeau Thrust

The Bourgeau thrust is one of the major thrust faults of the southern Canadian Rocky Mountains. It has been traced southward more than 400 km, from Maligne Lake, near Jasper, Alberta (~52°45′N) to the Crowsnest Pass ~49°30′N), and may continue farther south into northern Montana as the MacDonald thrust (Price and Mountjoy, 1970; Price, 1981). Over most of the interval from Maligne Lake to Crowsnest Pass it juxtaposes the eastern part of the Cordilleran miogeocline over platformal and foreland basin strata. In addition, it is the northeasternmost thrust that brings Neoproterozoic Windermere Group strata to the surface. Near the Bow Valley, the horizontal separation across the Bourgeau thrust is about 30 km (Price and Fermor, 1985).

Figure 1-19. Sketch of Mount Cory, view northwest at south end of Sawback Range from Trans-Canada Highway (Price et al., 1972).

Stop 1-6 Viewpoint for Mount Cory, south end of Sawback Range: Platform to miogeoclinal transition

138.9 km The south end of the Sawback Range (Fig. 1-19) is a natural cross-section through the Bourgeau-Sawback thrust sheet, exposing a lower Paleozoic succession that is transitional between the cratonic platform succession and the Cordilleran miogeocline. The lower Paleozoic succession, which is typical of the eastern edge of the miogeocline, is thicker and more complete that in the more easterly Front Ranges. The typical weathering profile of alternating resistant carbonate units and recessive shale units is well displayed. The Arctomys and Waterfowl Formations, each of which is less than 30 m thick where they are preserved from pre-Upper Devonian erosion in the McConnell thrust sheet, are both more than 60 m thick in the Sawback Range. The remaining Upper Cambrian and Lower Ordovician formations, which are missing beneath the sub-Upper Devonian unconformity in the McConnell Range, have an aggregate thickness of 825 m here.

The Middle Cambrian to Middle Ordovician strata of the Cordilleran miogeocline and adjacent cratonic platform occur in three *facies belts* termed: (1) inner detrital, (2) middle carbonate, and (3) outer detrital (Robinson, 1960). Intermittent expansion of the inner detrital belt, which occasionally led to the extinction of the middle carbonate belt, resulted to the development of *Grand Cycles* (Fig. 1-20) that approximate chronostratigraphic units (Aitken, 1966b, 1978). The Grand Cycles each span from one to three trilobite zones and are marked by an abrupt change from a unit of middle-carbonate character to a shaly unit of inner detrital character that is succeeded gradationally by another unit of middle-carbonate character. The alternation of resistant carbonate and less resistant shaly units has provided the basis for dividing the succession into the distinctive lithostratigraphic units that have been very helpful in mapping the structure of this part of the thrust and fold belt.

Several small northeast-trending, near-vertical transverse faults, which cross the south end of the Sawback Range (Fig. 1-21), disrupt the continuity of

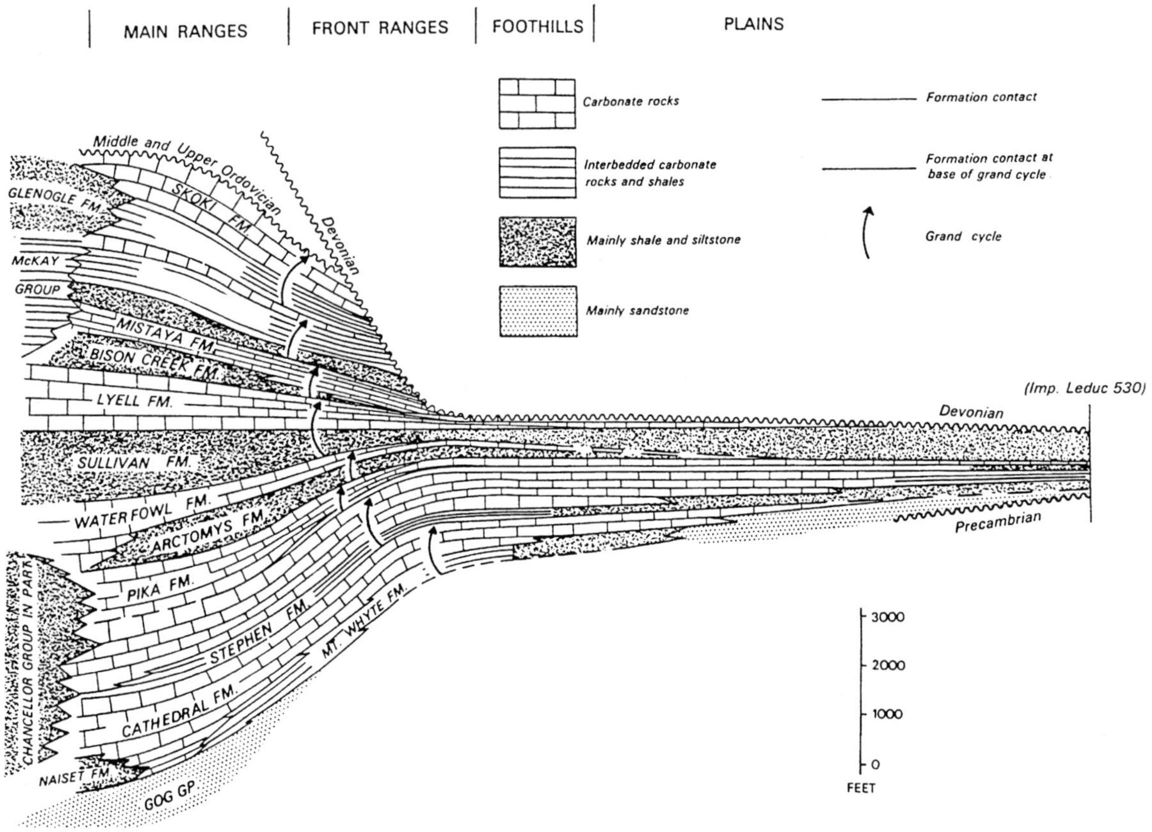

Figure 1-20. Schematic, palinspastically restored stratigraphic section of Cambrian and Ordovician grand cycles between the Western Main Ranges shaly succession and the thin cratonic platform succession of the Interior Plains (from Douglas et al., 1972; modified after Aitken, 1966b).

the stratigraphic units as seen from this viewpoint. These faults terminate to the northeast along the Sulphur Mountain thrust, in a zone within which the Sulphur Mountain thrust has been folded over a transverse monocline that marks a transverse hanging wall ramp on the underlying Rundle thrust (Fig. 1-21).

The overall structure of this part of the Bourgeau thrust sheet is homoclinal, but a very large syncline-anticline fold pair occurs immediately to the west, adjacent to the edge of the Main Ranges (Fig. 1-2). The Bow River valley cuts across these structures at a low angle where it crosses the boundary between the Front Ranges and the Main Ranges. This boundary is marked by the Johnston Creek thrust (not labelled in Fig. 1-2), the lowest thrust fault on the southeast slope of Castle Mountain (Mount Eisenhower). It carries Lower Paleozoic rocks over the Triassic rocks in the Brewster Creek syncline. The Johnston Creek thrust is the lowest of several faults that branch northward from the Simpson Pass thrust. The Simpson Pass thrust is one of the major thrust faults of the southern Canadian Rocky Mountains. It extends ~275 km northward from Mount Assiniboine (~50°45′N) to near Jasper, Alberta (~53°N), and it carries the thick, carbonate-dominated, shallow-water facies of the miogeocline and the underlying thick, Neoproterozoic succession of the Windermere Group over the thinner eastern part of the miogeocline.

143.5 km Roadcuts in northeast-dipping Rundle Group strata (Mount Head Formation, Upper Carboniferous).

145.6 km Roadcuts in northeast-dipping Rundle Group strata (Livingstone Formation, Lower Carboniferous).

149.3 km Redearth Creek.

153.9 km Viewpoint on north side of highway.

Stop 1-7 Castle Mountain (Mount Eisenhower) viewpoint: Eastern miogeoclinal succession

154.2 km Pull off on shoulder east of guard rail. The broad, open folds, and the castellated peaks in

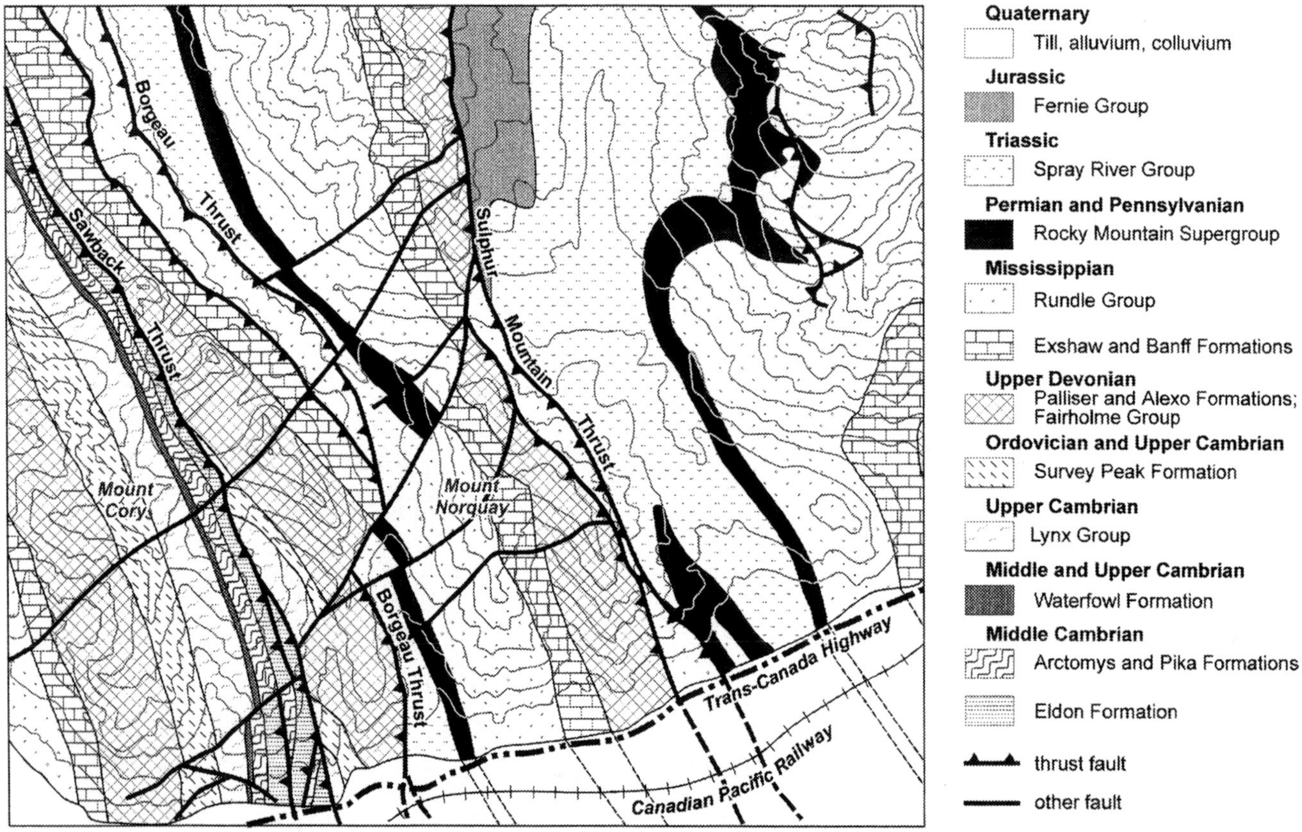

Figure 1-21. Geological map of transverse faults in the Rocky Mountain Front Ranges on the north side of the Trans-Canada Highway between Cascade Mountain (Banff) and Mount Cory (southern Sawback Range) (simplified after Price and Mountjoy, 1972a).

relatively flat-lying strata in Castle Mountain (Fig. 1-22) are typical of the Eastern Main Ranges Subprovince of the Canadian Rockies, which coincides with the thick, carbonate-dominated succession in the eastern part of the Cordilleran miogeocline. The changes in structural style and physiographic expression from the Front Ranges Subprovince reflect the lateral changes in the supracrustal succession from the relatively thin platformal facies to the thick miogeoclinal facies of the lower Paleozoic succession. This change also coincides with the first appearance, beneath the miogeoclinal lower Paleozoic rocks, of the rifting related Neoproterozoic rocks of the Windermere Supergroup.

All of the Cambrian formations exposed in Castle Mountain are *substantially thicker* than their lateral equivalents in the Front Ranges; but these units become even thicker southwest of the Bow Valley in the vicinity of Lake Louise. The quartz arenites of the Lower Cambrian Gog Group, which are about 350 m thick at Castle Mountain, are about 1000 m thick at Lake Louise. The succession of Middle Cambrian carbonate units, which is about 950 m thick at Castle Mountain, is about 1700 m thick at Lake Louise.

The Castle Mountain thrust merges southeastward with the Simpson Pass thrust. It is underlain by the Johnston Creek thrust, which also merges southeastward with the Simpson Pass thrust (Fig. 1-2).

The hummocky terrain along the Bow River valley southeast of Castle Mountain is a landslide mass comprising siltstones of the Triassic Spray River Group that came from the scarp upslope, in the core of the syncline under the Johnston Creek thrust.

156.2 km Copper Mountain. Roadside exposures of Middle Cambrian carbonate rocks; Castle Mountain (Mount Eisenhower) viewpoint for eastbound traffic.

158.9 km Junction of Highway 1 and Highway 93 (to Radium Hot Springs and Cranbrook via Vermilion Pass). Switch to Figure 1-23 (geological map, Lake Louise to Golden).

180.4 km Bow River bridge.

180.5 km Junction with Johnston Canyon Road.

180.6 km Beginning of roadcuts in Miette Group (Neoproterozoic) slate, sandstone and feldspathic pebble conglomerate.

182.6 km End of roadcuts in Miette Group.

183.7 km Road to Lake Louise.

185.7 km Junction Highway 1 and Highway 93. Proceed north 1 km on Highway 93 toward Jasper.

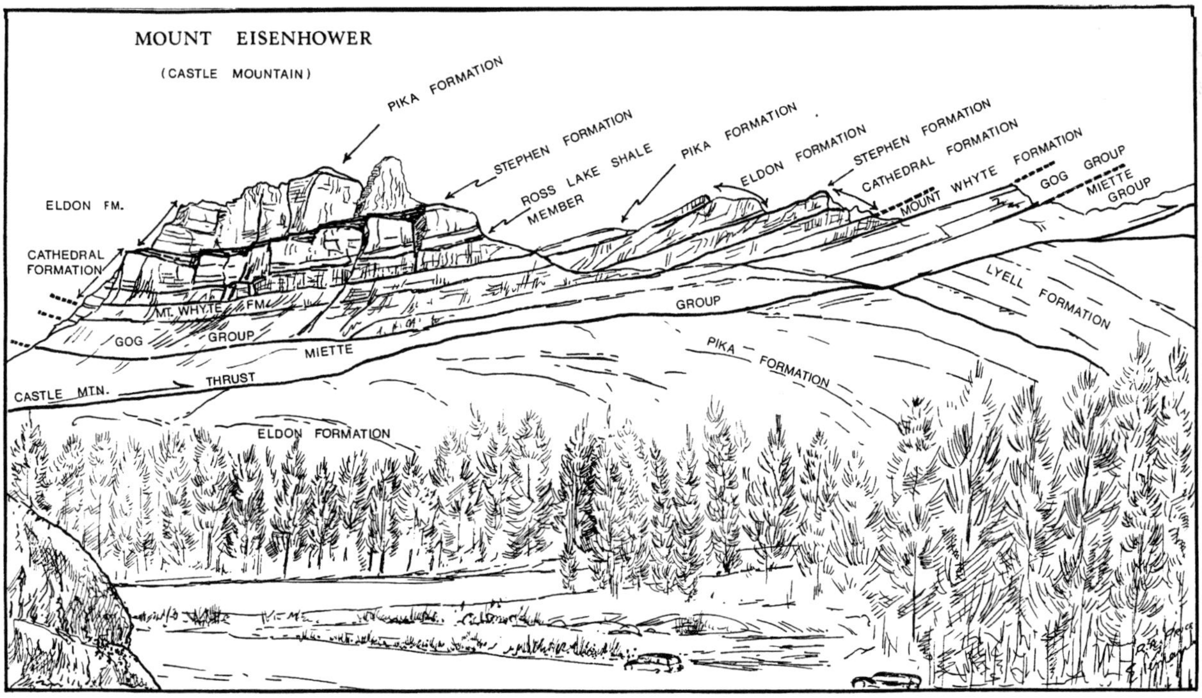

Figure 1-22. Sketch of Castle Mountain (Mount Eisenhower) from the Trans-Canada Highway at Copper Mountain (Price et al., 1972).

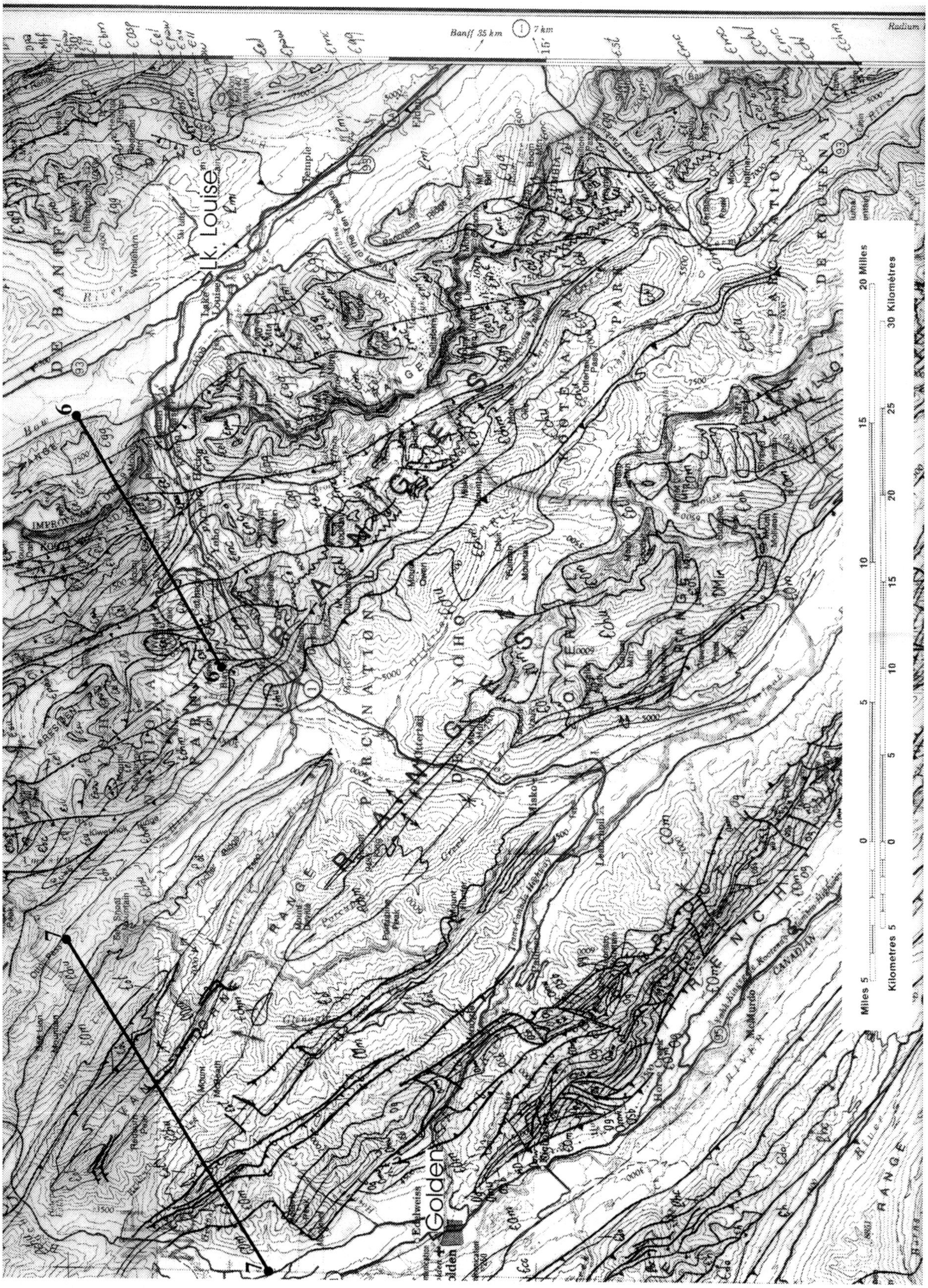

Figure 1-23. Geological map, Lake Louise to Golden (simplified after Price and Mountjoy, 1978a, b; Price et al., 1979a, b; Price et al., 1980, 1980a, b; and Balkwill et al., 1980a, b).

Stop 1-8 Exposures of Miette Group (Neoproterozoic) submarine fan deposits

186.7 km Park near National Park permit check point. Proceed with care across Highway 93 to examine the Miette Group exposures in nearby railway cuts.

These rocks are typical of the lower and thicker part of the Miette Group, which is the equivalent, in the southern Canadian Rockies, of the Windermere Supergroup farther west in the Purcell, Selkirk, and Cariboo mountains.

In the southern and central Cordillera the stratigraphy of the Windermere Supergroup is dominated by thick-bedded, coarse-grained, arkosic sandstones and granule conglomerates ("grits") that are intercalated with sandstones, and siltstones, and varying amounts of slate or shale. These rocks form a large submarine fan system in which the flow was mainly southeast to northwest (Ross et al., 1995), parallel with the axis of the overlying lower Paleozoic miogeocline. Detrital zircon geochronometry indicates that the coarse clastic sediments were derived from adjacent Paleoproterozoic and Archean basement rocks that form the basement of the Western Canada Sedimentary Basin (Ross and Bowring, 1990).

The sedimentology and stratigraphy of the Miette rocks in the Lake Louise area have been studied by Arnott and Hein (1986). They have described a 135 m-thick channel-fill sequence containing feldspathic pebbly sandstone and conglomerate and chaotic conglomerate/slate mixtures, with intercalated turbiditic sandstones and multiple scour zones.

The Old Fort Point Formation, a distinctive thin marker unit, which overlies the strata exposed at this location, has been followed along the Eastern Main Ranges from Lake Louise to Jasper, and is widespread farther west. It consists of purple and green mudstones (slates), limestone flat-pebble conglomerates, and limestone-mudstone couplets that culminate in a carbonaceous sulphide-rich interval. It has been interpreted as marking a major sea-level rise that stopped the influx of the coarse clastic sediments into the Windermere basin (Ross and Murphy, 1988).

Time of Rift-Drift Transition

Although there is general agreement that the Windermere Supergroup is associated with the rifting and ensuing sea-floor spreading that gave birth to the Cordilleran miogeocline, there has been disagreement about the time of the rift-drift transition and the birth of the miogeocline. The eastern limit of the Windermere Supergroup and correlative strata coincides with the eastern edge of the Cordilleran miogeocline from east-central Alaska to southern California. Thus, the Windermere Supergroup and correlative strata have been interpreted as the initial rift-related deposits of Cordilleran miogeocline (Stewart, 1972). However, the rifting that marked the onset of Windermere deposition occurred at about 750 Ma, more than 150 Ma prior to the onset of the deposition of the thick, carbonate-dominated, Cambro-Ordovician passive margin succession that is the hallmark of the Cordilleran miogeocline. Glacial diamictite and associated volcanic rocks at the base of the Windermere Supergroup provide stratigraphic evidence of syn-depositional rifting (Gabrielse and Campbell, 1991; Ross et al., 1995). The associated volcanism has been dated at ~750 Ma in western Yukon (Roots and Parrish, 1988), and ~760 Ma in northeastern Washington (Devlin et al., 1988).

Bond and Kominz (1984), using an algorithm that compensates for the effects of sediment compaction and isostatic subsidence due to sediment loads, showed that the changing rates of subsidence in the Cambro-Ordovician succession of the southern Canadian Rockies imply that the onset of passive margin subsidence, and therefore the transition from rifting to sea-floor-spreading and continental drift, must have begun in earliest Cambrian or latest Neoproterozoic time. Stratigraphic relationships in the northern Purcell and Selkirk Mountains and in the western Rocky Mountains demonstrate that deposition of the distinctive, laterally extensive, mature quartz arenites of the lower part of the Gog Group and the overlying archeocyathid-bearing limestones of the Mural Formation (and their correlatives) occurred after an episode of latest Neoproterozoic and(?) earliest Cambrian rifting that was accompanied by local accumulation of feldspathic quartz sandstones and associated mafic volcanic rocks (Devlin and Bond, 1988; Lickorish and Simony, 1995; Warren, 1997).

Return to junction of Highway 1 and Highway 95.

LAKE LOUISE TO GOLDEN

INTRODUCTION

Between Lake Louise and Golden (Fig. 1-23), the Trans-Canada highway crosses an abrupt, long-lived (lower Middle Cambrian to Middle Ordovician) carbonate bank margin known as the Kicking Horse Rim (Aitken, 1971). The Kicking Horse River separates the thick, shallow-water, Cambrian-Ordovician carbonate facies of the miogeocline that is characteristic of the Eastern Main Ranges, from an equivalent, thicker, deeper water, shale and limestone facies that is characteristic of the Western Main Ranges.

A conspicuous change in style of deformation is associated with the facies change (Cook, 1970, 1975). The thick, competent, carbonate-dominated succession in the Eastern Main Ranges forms massive, coherent thrust sheets, in which bedding is generally gently dipping. Tthere is relatively little internal compressional deformation these rocks. In the thicker, incompetent shale and argillaceous limestone facies of the Western Main Ranges, penetrative deformation and cleavage are pervasive, complex disharmonic folds and "polyphase" thrusting and folding are common, and the influence of structure on topography is much less obvious than in the Eastern Main Ranges.

The *Porcupine Creek anticlinorium,* a large asymmetric fan-fold (Fig. 1-33), is the dominant structure of the western part of the southern Canadian Rockies (Balkwill, 1972; Gardner, 1977; Lickorish, 1993). The dip of the cleavage and the vergence of smaller folds and thrust faults changes across the anticlinorium from southwest-dipping and northeast-verging on the northeast side, to northeast-dipping and southwest-verging on the southwest side.

On the southwest side of the anticlinorium, early northeast-verging thrust faults that were kinematically congruent with the developing fold, were rotated during the evolution of the anticlinorium, along with the beds that they cut, and now are commonly overturned, and easily mistaken for normal faults. Locally, these older faults and related folds have been deformed by younger, southwest-verging structures that developed as the core of the evolving anticlinorium was thrust over the southwestern flank. The younger, southwest-verging structures extend across the floor of the Rocky Mountain trench into the eastern Dogtooth Mountains, where they are truncated along the footwall of the Purcell fault, a major, out-of-sequence, northeast-verging thrust fault.

The youngest strata preserved in the western Rockies along this transect occupy a synclinorium on the west flank of the Porcupine creek anticlinorium. They range in age from Middle Ordovician to Middle Devonian, and they fill part of the stratigraphic gap that occurs between the Upper Devonian Fairholme Group and the unconformably underlying Lower Ordovician strata in the Sawback Range, along the eastern edge of the miogeocline, and between the Fairholme Group and the Middle Cambrian strata in the platformal cover of the West Alberta Arch, which is preserved in the McConnell thrust sheet at the mountain front. In contrast to the underlying Cambrian and Lower Ordovician shaly rocks, the Middle Ordovician graptolitic shales (Glenogle Formation) and the overlying Upper Ordovician quartz arenites (Mount Wilson Formation) and dolomites (Beaverfoot Formation) are not penetratively deformed. However, they are deformed by early northeast-verging thrusts and related folds, which have been deformed by younger, southwest-verging thrusts and related folds, which have, in turn, been deformed by still younger northeast-verging thrusts related to the Purcell thrust fault. This record of three distinct phases of deformation has been attributed to the progressive co-evolution of the Porcupine Creek anticlinorium and the Dogtooth thrust duplex, which is exposed in the northern Purcell Mountains, southwest of the Rocky Mountain Trench (Balkwill, 1972; Gardner, 1977; Price, 1986; Kubli and Simony, 1994). These relationships (Figs. 2-1, 2-2, 2-4) will be a focus of the Omineca Belt segment of the field trip.

Reset odometer to 0 km.

0.0 km Junction of Trans-Canada Highway and Highway 93.

6.0 km Abandoned quarry on south side of highway in Miette Group arkosic grit and pebble conglomerate. Purple slates of Old Fort Point Formation are exposed in the railway cut north of the Highway. The Old Fort Point Formation is an important marker unit within the Neoproterozoic Windermere Supergroup.

7.2 km Kicking Horse Pass: the Continental Divide and the boundary between Banff and Yoho National Parks.

8.1 km Lower Cambrian Gog Group quartz arenites exposed along the north side of the highway.

10.3 km Carbonate rocks of the Middle Cambrian Cathedral Formation exposed along the north side of the highway.

12.3 km Cross, at west end of lake, the trace of the west-dipping Cataract Brook normal fault, which juxtaposes east-dipping Middle Cambrian Eldon Formation over west-dipping Cathedral Formation.

Stop 1-9 Sherbrooke Creek

13.6 km Begin a traverse through the thick Middle Cambrian section of the carbonate facies of the miogeocline (Aitken, 1997), comprising the lower part of Eldon Formation (500 m of massive micritic limestone and grey mottled dolostones), the Stephen Formation (50 m of cyclically interbedded green-grey shale and thinly bedded limestone with minor siltstone and oolitic limestone), and the Cathedral

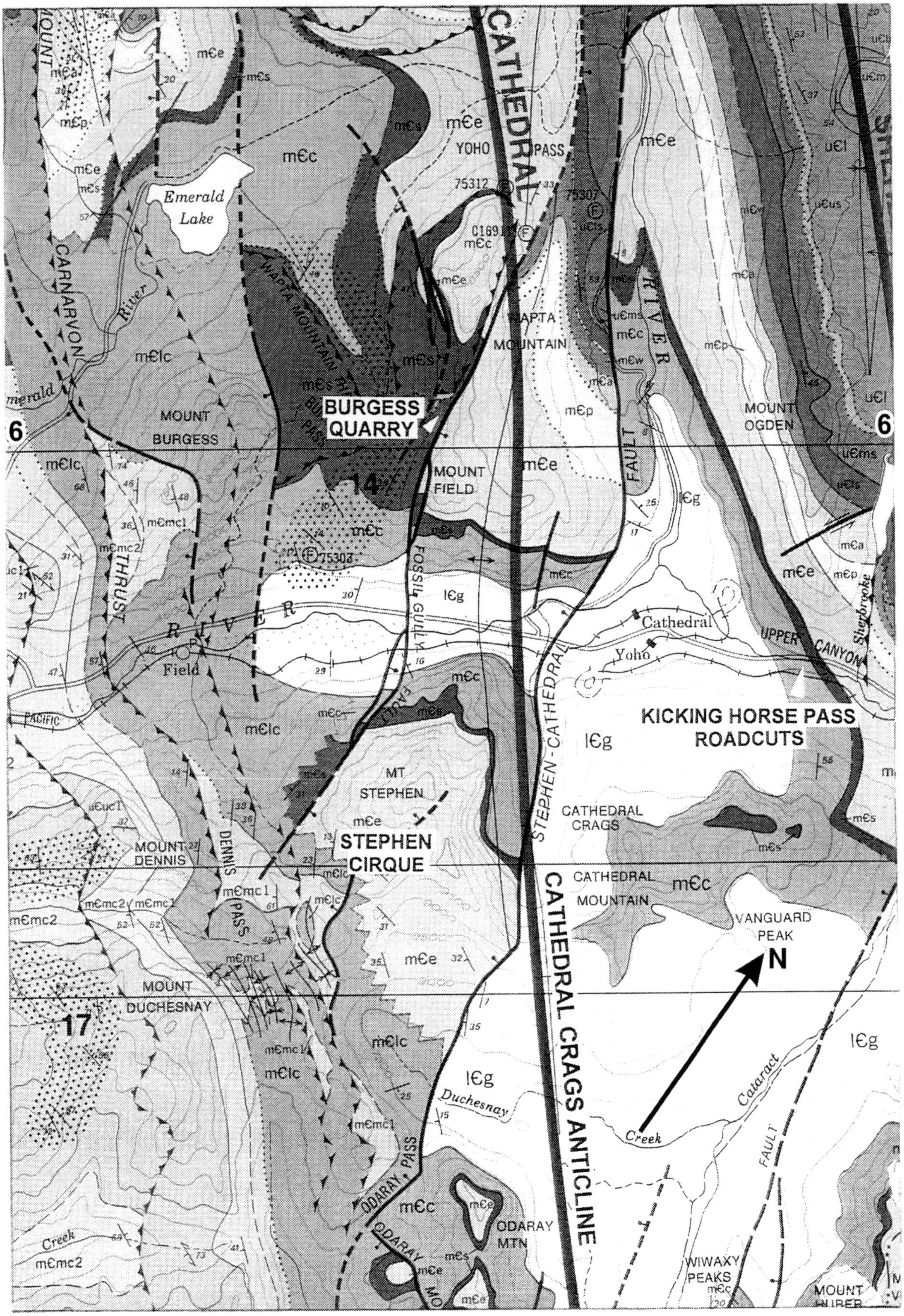

Figure 1-24. Geological map of the area along the Trans-Canada Highway between Sherbrooke Creek and Field, British Columbia (from Cook, 1975, with modifications by Stewart, 1997).

Formation (limestone and dolomite), all of which are exposed in roadcuts along the north side of the highway between Sherbrooke Creek and the bridge that carries the highway to the south side of Kicking Horse River (Fig. 1-24). The top of the Stephen Formation is gradational into the Eldon Formation; but the base is relatively sharp, and marks the base of a Grand Cycle. It coincides with the boundary between the *Bathryuriscus-Elrathina* and the *Glossopleura* trilobite zones (see Fig. 1-28) and is considered to be chronostratigraphic marker. The Cathedral Formation consists of about 600 m of medium to dark grey and light brown, massive, micritic limestone with dolomitic mottling, and of completely dolomitized equivalents, which become dominant southwestward near the Kicking Horse Rim carbonate-bank margin. Conspicuous cross-cutting zones of yellowish weathering, coarse sparry dolomite are common near the carbonate bank margin, and may be much younger than the penecontemporaneous dolomitic mottling. The contact with the Mount Whyte Formation is transitional.

14.6 km Kicking Horse River Bridge. For the lower part of the Cathedral Formation, the Mount Whyte Formation, and the Gog Group, shift to roadcuts along south side of highway. The lower part of the Cathedral Formation and the Mount Whyte Formation are partly exposed in old roadcuts upstream from the bridge. The Mount Whyte Formation consists of about 40 m of light green and grey shale interbedded with siltstones, fine-grained sandstones and dark grey fragmental to oolitic limestones. The Gog Group, ~2000 m thick, is exposed in roadcuts along the south side of highway from the bridge to the Spiral Tunnels viewpoint and beyond. It consists of quartz arenites, commonly cross-bedded, with intercalated grey, green and red shales and siltstones. A thin (~5 m) limestone unit (Peyto Limestone Member) at the top con-

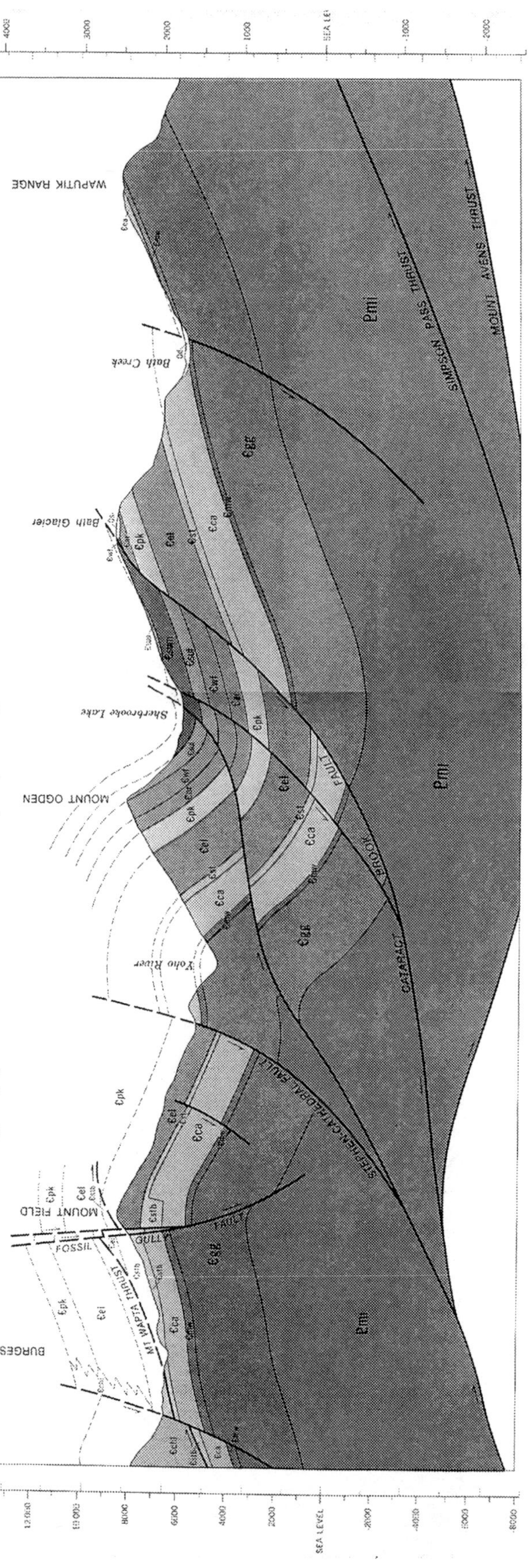

Figure 1-25. Structure section across the Eastern Main Ranges on the north side of the Trans-Canada Highway (Section 6 from Price et al., 1980b); see Figure 1-23 for location of line of section.

tains the late Early Cambrian *Bonnia-Olenellus* trilobite fauna.

15.8 km Spiral Tunnels Viewpoint. The panoramic view northward up Yoho River valley from this viewpoint (see Figs. 1-24 and 1-25) extends from Mount Ogden on the east (Sullivan, Waterfowl, Arctomys, Pika, Eldon, Stephen, Cathedral, and Mount Whyte Formations), across the lower Yoho Valley (Gog Group), and across the west-dipping Stephen-Cathedral normal fault, which passes west of the low ridge west of Yoho River, to Mount Field (Eldon, Stephen, and Cathedral Formations). Cathedral Mountain (to the south) is in the footwall of the normal fault, and on it the Gog Group is capped by the Cathedral Formation and the Stephen Formation (the Stephen Formation is on the summit and out of the line of view). Mount Stephen (to the southwest) comprises the Eldon, Stephen, Cathedral, and Mount Whyte Formations, and it is in the hanging wall of the west-dipping Stephen-Cathedral normal fault.

The spiral tunnels were constructed by the Canadian Pacific Railway Company after coping, for many years, with the hazards and expense of the formerly much steeper railway grade between Kicking Horse Pass and Field, British Columbia.

17.7 km Cross Monarch Creek and the trace of west-dipping Stephen-Cathedral normal fault.

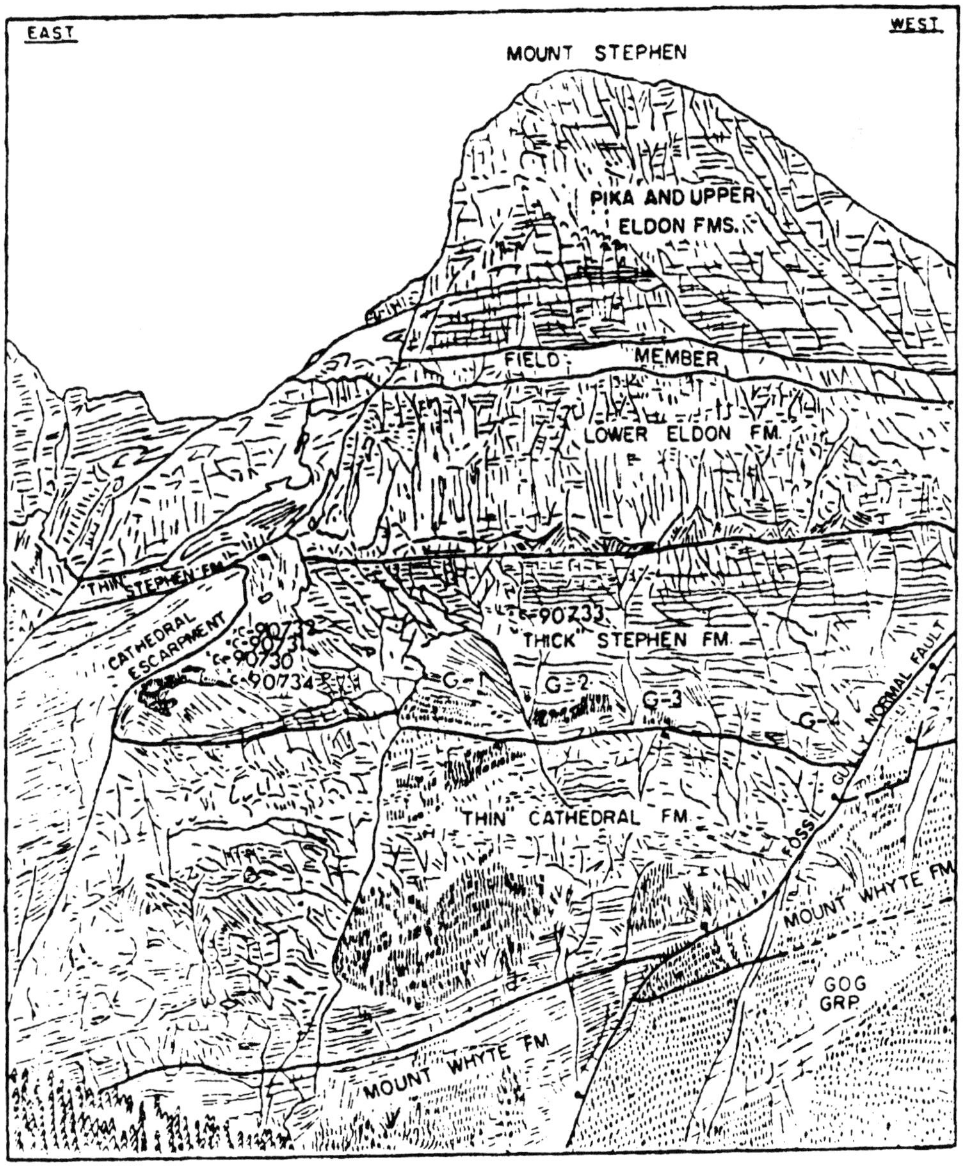

Figure 1-26. Sketch of Mount Stephen, view to the southeast from near Field, showing the Cathedral escarpment separating the "thin" and "thick" facies of the Stephen Formation. Sketch by I.A. McIlreath, from Bruce et al. (1995).

Lead-Zinc-Silver Deposits

Note the abandoned mine site on the north side of the valley (adits and cones of waste rock) in the lower part of the Cathedral Formation on Mount Field. The Kicking Horse Mine on Mount Field and the Monarch Mine on Mount Stephen, which were developed on tabular bodies of sphalerite-galena-pyrite-chalcopyrite-dolomite, operated until 1946, and produced about 850,000 tons of ore grading 7% lead, 10% zinc, and 1.2 ounces of silver per ton (Ney, 1954). The ore bodies were located in the lower part of the Cathedral Formation, about 120 m above the Mount Whyte Formation, along the contact between thin-bedded black limestone and an extensive overlying mass of light coloured, coarsely crystalline dolomite. The dolomite cross-cuts the stratification and contains local diffuse relict patches of limestone. The mineralization has been attributed to hydrothermal activity (Ney, 1954).

19.6 km Cross the trace of east-dipping Fossil Gully normal fault, which is antithetic to the Stephen-Cathedral fault and forms the west side of a small graben.

Stop 1-10 Middle Cambrian carbonate bank margin and the Burgess Shale

23.5 km Park at the roadcut on the north side of the highway. At the Kicking Horse Rim, east of Field, British Columbia, the Cathedral carbonate platform ends in a near vertical algal-reef escarpment (McIlreath, 1977). The Cathedral escarpment extends across the field trip route, from Mount Stephen on the south side of the Trans-Canada Highway (Fig. 1-26) to Mount Field on the north side (Fig. 1-27). Southwest of the escarpment the Cathedral Formation consists of a thin succession of deep-water lime mudstones or calcareous slates (the "thin" Cathedral Formation of Fig. 1-28), and it is overlain by a "thick" Stephen Formation. The configuration of the boundary between the *Glossopleura* Zone and *Bathyuriscus-Elrathina* Zone (G and B-E respectively in Fig. 1-28) shows that at the onset of deposition the shales of the Stephen Formation, the Cathedral escarpment was about 200 m high. The famous Burgess Shale, with its exquisitely preserved soft-body fossils, is located adjacent to the Cathedral escarpment, north of the Trans-Canada Highway, on the west slope of Mount Field.

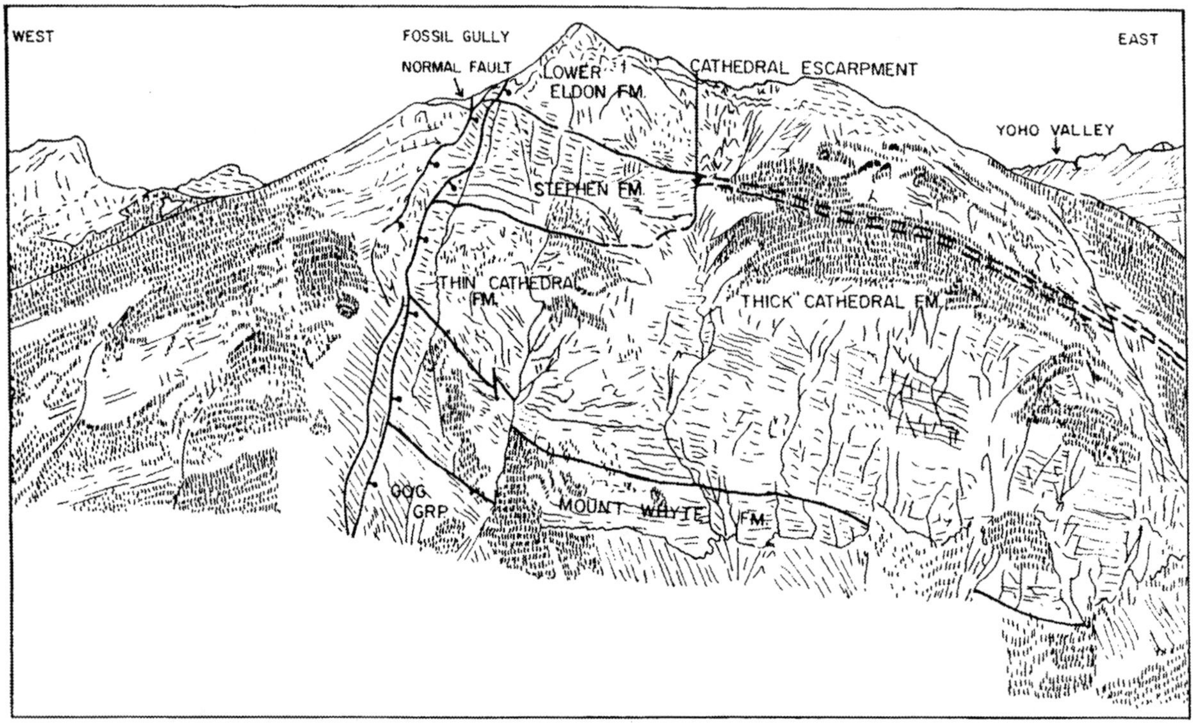

Figure 1-27. Sketch of Mount Field, view to the northeast from near Field, showing the east-dipping Fossil Gully normal fault, and the juxtaposition at the Cathedral escarpment of the thick Cathedral Formation and overlying "thin" Stephen Formation with the "thin" Cathedral Formation and overlying "thick" Stephen Formation. The famous Burgess shale fossil quarry is located out of the field of view, on the west flank of Mount Field, in the footwall of the Fossil Gully fault. Sketch by I.A. McIlreath, from Bruce et al. (1995).

At the Kicking Horse Rim, the Mount Whyte, Cathedral, Stephen, Eldon, and Pika Formations are replaced southwestward by the lower part of the Chancellor Formation, the Arctomys and Waterfowl Formations are replaced by the middle part of the Chancellor Formation, and the Sullivan Formation by the upper part of the Chancellor Formation (Figs. 1-29, 1-30).

Above the level of the Cathedral escarpment, the Kicking Horse Rim appears to have been a steep, gravitationally unstable carbonate ramp. At the level of the Eldon, Pika, Arctomys, and Waterfowl Formations, the transition to the Chancellor Formation is marked by areally extensive megatruncation surfaces (gravity slide scars) and by olistostromes and carbonate debris flows (Fig. 1-30).

The roadcut at this stop is in Takakkaw Tongue (Fig. 1-30), the lower part of the Chancellor Formation. It consists of the deep-water, basinal, lime mudstone facies equivalent of the Cathedral Formation, which characteristically is cleaved and provides evidence of high levels of penetrative compressional strain.

The younger Cambrian and overlying Lower Ordovician strata, which are not exposed near the Highway, also have been mapped and correlated across the Kicking Horse Rim (Fig. 1-31). All of the limestone formations, with the exception of the mid-Upper Cambrian Lyell Formation (Ottertail Formation in the basinal facies) change facies across the Kicking Horse Rim to shaly rocks, and all of the shaly formations are thicker on the southwest side of the Kicking Horse Rim. The Ottertail Formation differs from the shallow water deposits of the Lyell Formation in that it consists mainly of deeper-water lime mudstones.

31.3 km Ottertail River bridge. View to northwest (Fig. 1-32) at the south end of trough of the Split Creek synclinorium, which is outlined by the resistant limestones of the Ottertail Formation (medial Upper Cambrian).

See also the structure section of Figure 1-33, which crosses this structure down-plunge, about 20 km to the northwest of the highway.

Stop 1-11 Kicking Horse River viewpoint; structures in upper part of Chancellor Formation

34.0 km Park in picnic area on northwest side of highway. **Check for traffic before crossing the highway!** The roadcuts on the east side of the highway expose folded, boudinaged dykes and deformed cleavage in the upper part of the Chancellor Formation, in the core of the Porcupine Creek anticlinorial fan structure. Note the gently folded, strongly boudinaged, chloritized mafic dykes. The dykes are associated with the Ice River Complex, a funnel-shaped, composite alkaline igneous complex that occurs within the core of the Porcupine Creek anticlinorium about 7 km south of this roadcut (unit DMir in the south-central part of Fig. 1-23). The Ice River Complex consists of a tabular, layered mafic alkalic complex, a central, cross-cutting alkali-syenite and

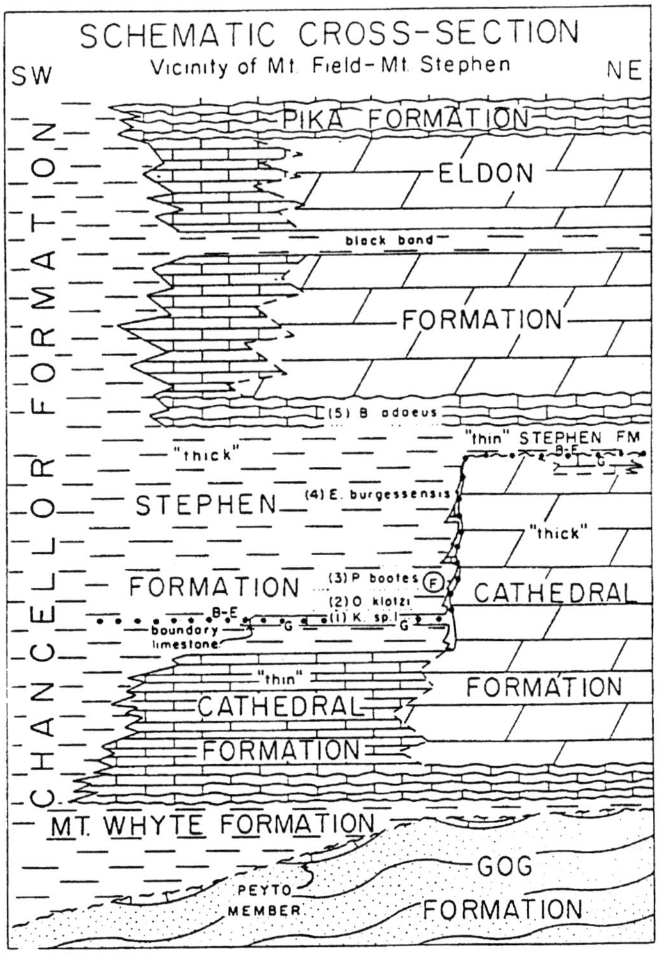

Figure 1-28. Schematic cross-section showing stratigraphic relationships exposed adjacent to the Trans-Canada Highway between Mount Field and Mount Stephen. Thicknesses are greatly exaggerated relative to horizontal distance. The approximate location of the Burgess Shale quarry is at the circled "F". A heavy dotted line marks the boundary between the *Glossopleura* Zone (G) and the *Bathyuriscus-Elrathina* Zone (B-E), which lies close the base of the Stephen Formation both at the base and at the top of the escarpment. From Fritz (1971).

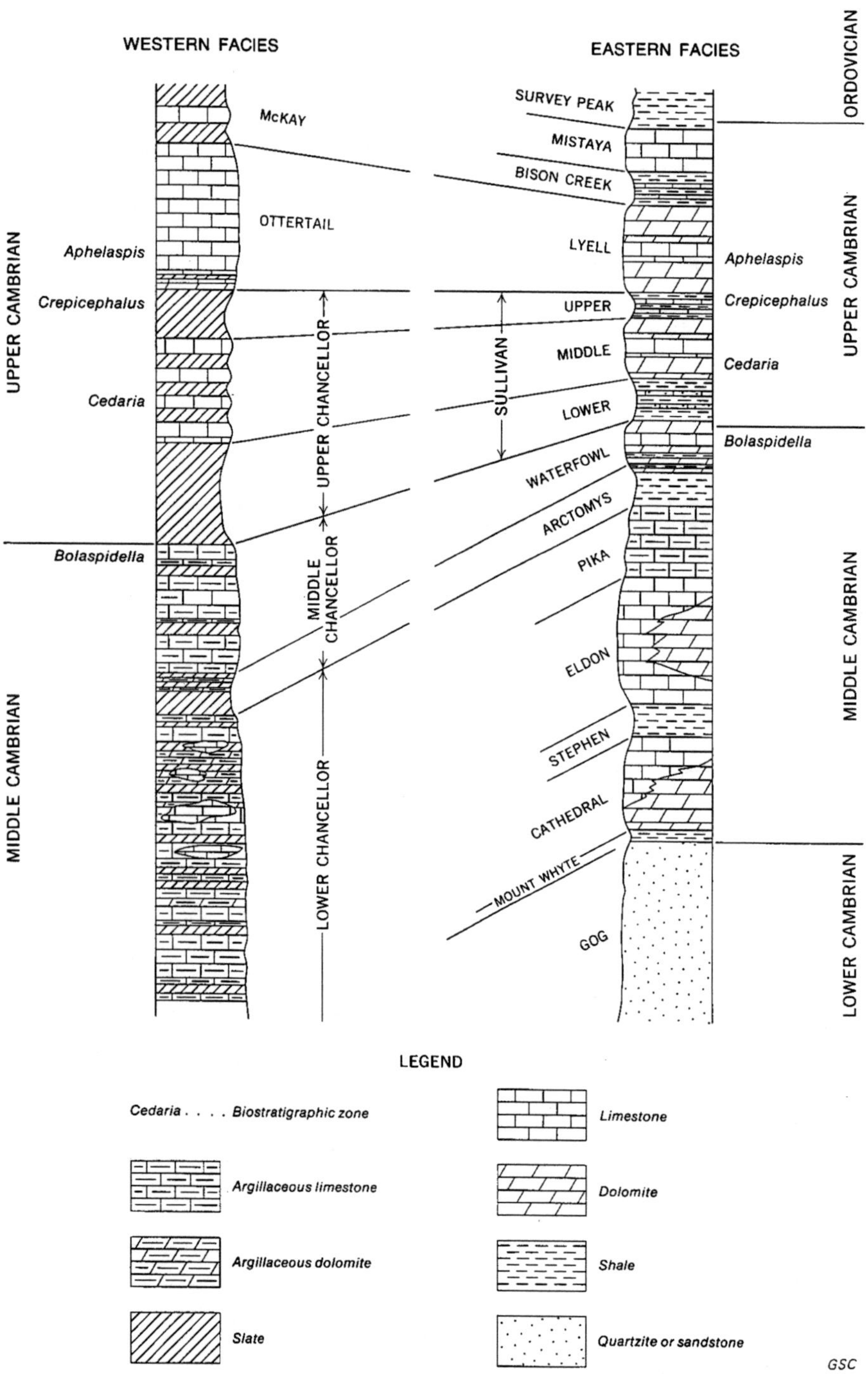

Figure 1-29. Cambrian-Lower Ordovician correlations across Kicking Horse Rim (Cook, 1975).

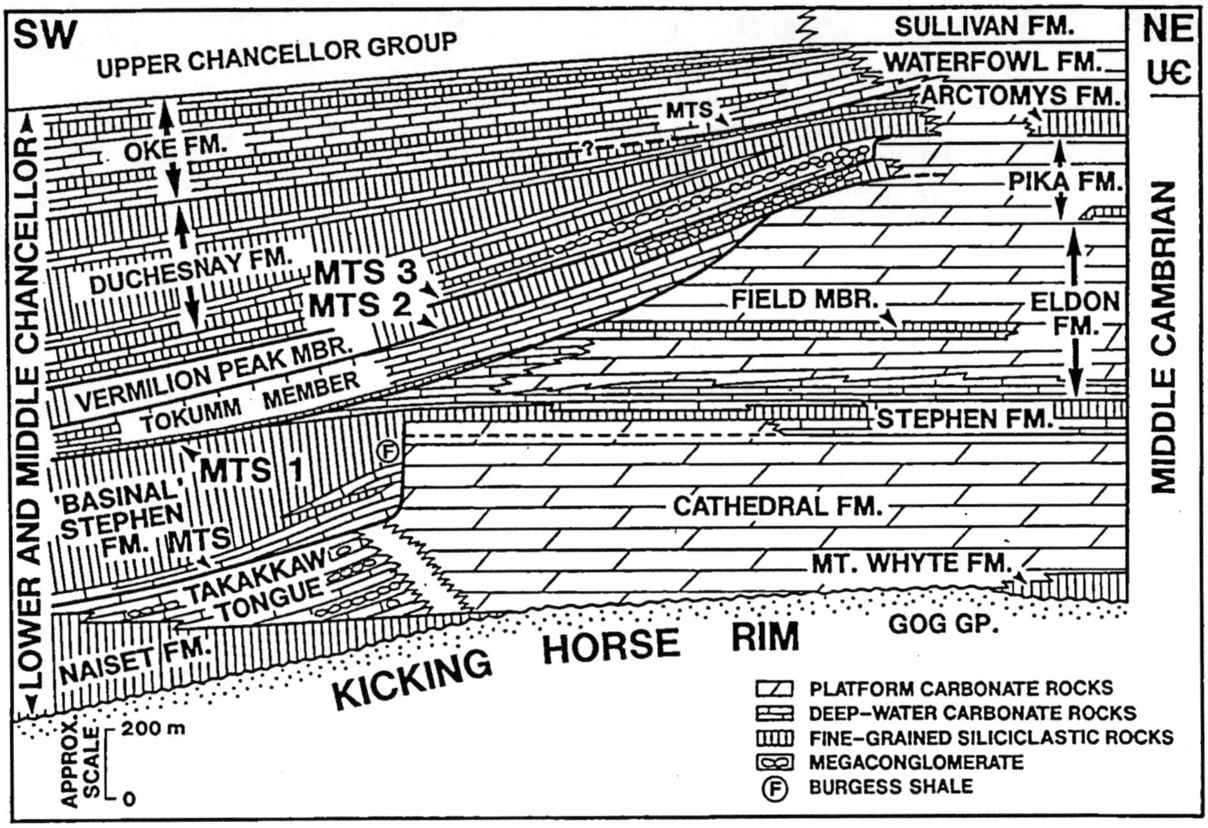

Figure 1-30. Schematic regional stratigraphic model for the Middle and Upper Cambrian composite carbonate bank margin at the Kicking Horse Rim (Stewart, 1991; Stewart et al., 1993). MTS indicates a megatruncation surface.

carbonatite body, and associated cross-cutting melanocratic lamprophyre dykes (Currie, 1975). It has been dated at 368 ± 4 Ma (Parrish et al., 1987), and is broadly coeval with the magmatic arc rocks that occur far to the west in the outboard part of the Cordilleran miogeocline and in the pericratonic terranes. Chloritized dykes and widespread pre-folding hornfels metamorphism are common in the core of the Porcupine Creek anticlinorium northwest of these roadcuts (Balkwill, 1972; Balkwill, et al., 1980a).

The shallowly west-dipping slaty cleavage seen in these roadcuts is a local phenomenon. The younger, steeply east-dipping to vertical crenulation cleavage is congruent with the slaty cleavage that elsewhere in the region defines the downward-converging cleavage fan of the Porcupine Creek fan structure.

The rocks with the shallowly west-dipping slaty cleavage have been interpreted as a local high-strain zone associated with the detachment folding of the high ductility shaly rocks (Middle Cambrian Chancellor Formation) above the low ductility sandstones (Lower Cambrian Gog Group). This local high strain zone is thought to have developed early in the evolution of the Porcupine Creek anticlinorium, before the back-folding that produced the southwestward overturning and southwest-verging thrusting in the southwest flank of the anticlinorium (Fig. 1-34).

The bedding in these roadcut is sub-horizontal and is outlined by thin lime mudstone laminae in calcareous slates. The dykes are inclined at <20° to the slaty cleavage, and are very strongly boudinaged, with coarse-crystalline calcite and quartz filling the gaps between boudins. The crenulation cleavage parallels axial surface of the core of the Porcupine Creek anticlinorium, and also the axial surfaces of the small-scale, open folds that have deformed the boudinaged dykes.

Stop 1-12 (optional) Leanchoil: View of west flank of the Porcupine Creek anticlinorium and the Ice River Complex

45.7 km Park at the west side of the Kicking Horse River bridge, near the abandoned quarry. In the view

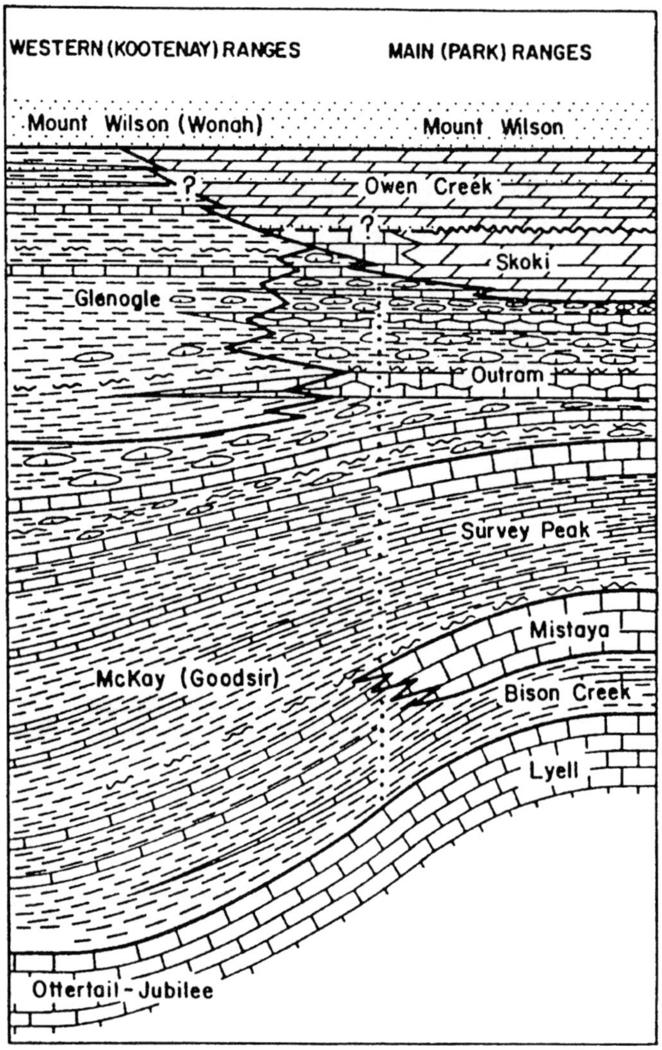

Figure 1-31. Upper Cambrian to Middle Ordovician correlation across Kicking Horse Rim (Aitken and Norford, 1967).

Figure 1-32. Sketch of Split Creek Synclinorium from Trans-Canada Highway at Ottertail River bridge (Price et al., 1972).

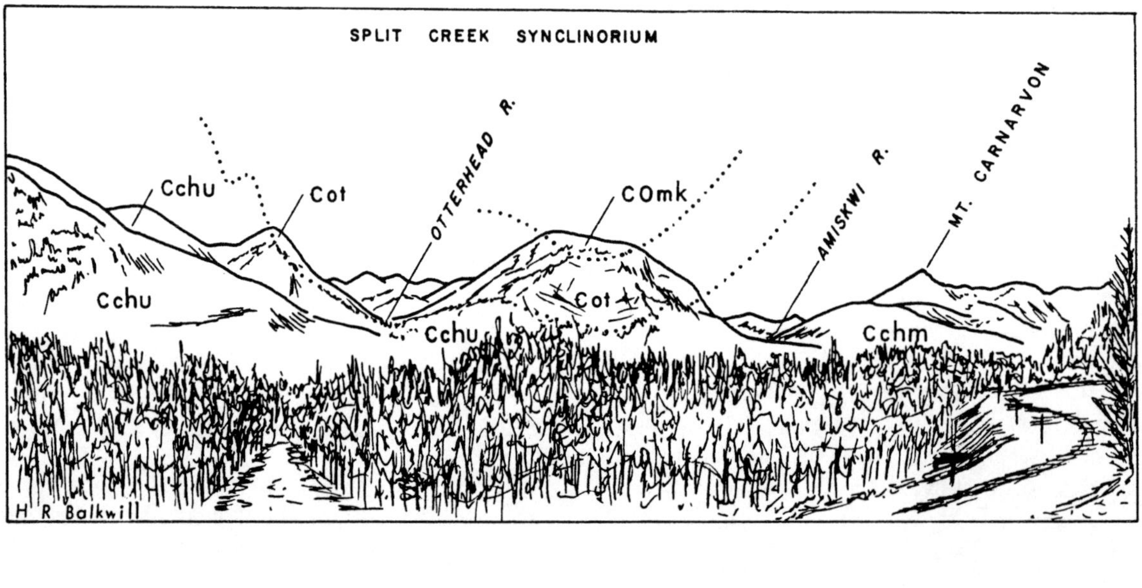

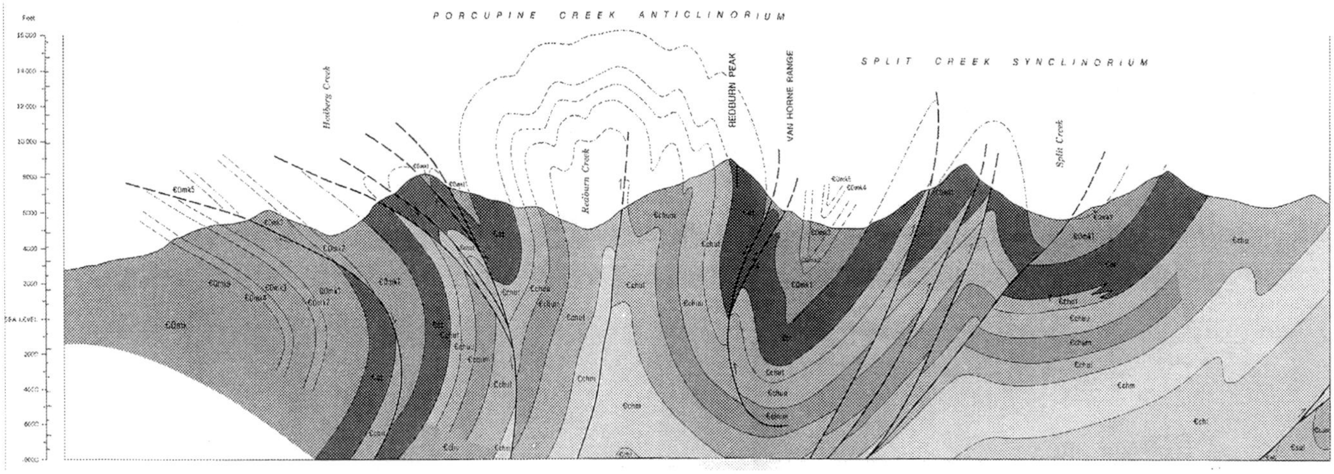

Figure 1-33. Structure section across the Split Creek synclinorium and Porcupine Creek anticlinorium between Kicking Horse and Blaeberry Rivers (Balkwill et al., 1980b; part of Section 7 accompanying GSC Map 1497A). Symbols: Cchl = lower Chancellor, Cchm = middle Chancellor, Cchu = upper Chancellor (4 sub-units), Cot = Ottertail, COmk = McKay (5 sub-units).

to the east (Fig. 1-35), toward Mount Hurd, Mount Vaux and Chancellor Peak and up Hoodoo Creek, the Ottertail Limestone can be seen wrapping over the Chancellor Formation along crest of the Porcupine Creek anticlinorium, which here plunges northwest. The quarry northwest of the bridge is in the Ottertail Formation (medial Upper Cambrian), which here is in the overturned southwest flank of the Porcupine Creek anticlinorium/fan structure. The beds dip about 50° to the northeast, but face to the southwest. The basal beds of the McKay Group (upper Upper Cambrian, *Elvina* zone) occur immediately to the southwest in railway and highway cuts.

52.1 km Yoho National Park boundary.

Stop 1-13 Upper Kicking Horse Canyon bridge: Overturned thrust in upper McKay Group

64.0 km Pull off in parking area at west end of bridge (Fig. 1-36).

The upper part of the Upper Cambrian to Lower Ordovician McKay Group and the lower part of the Glenogle Formation (Middle Ordovician, Arenig to very early Caradoc) are well exposed along the highway in the vicinity of the upper Kicking Horse Canyon bridge. The upper unit of the McKay Group (unit COmk6 in Figs. 1-37, 1-38), comprising interbedded lime mudstone and shale, faces to the southwest and is repeated by an overturned thrust fault. This fault is a "normal fault" (hanging wall down relative to footwall). However, regional relationships, outlined by structural mapping, including cross-cutting relationships with southwest-verging thrust faults, show that it formed as a northeast-verging thrust fault (a compressional or contractional fault) when these strata, which now form the overturned southwest limb of the Porcupine Creek anticlinorium, were horizontal or dipped much less steeply to the southwest. In the lower part of the exposure, the fault is parallel with the bedding on both sides. In the middle part of the exposure, it cuts up-section at about 30° to the bedding on the northeast side (now the hanging wall, but formerly the footwall), but is parallel with the bedding on the southwest side. At the top of the exposure, it is parallel with the bedding on both sides. Fault-bend folds occur in the strata on the left side of the fault, at each end of the ramp that occurs on the right side of the fault. This fault lies within unit COmk6 where it crosses the Kicking Horse River in Figure 1-37; and it is the fault that lies directly below the words "Kicking Horse River" in the structure section of Figure 1-38.

The thinly bedded McKay shales and limestones are deformed on a small scale by two conjugate sets of kink bands that record a minor amount of vertical compression and horizontal extension. The kink bands are larger and more conspicuous in the older, more shaly McKay strata that occur upstream from the bridge on the north side of the canyon.

Glenogle Formation black carbonaceous shales containing graptolites (zone of *Paraglossograptus*

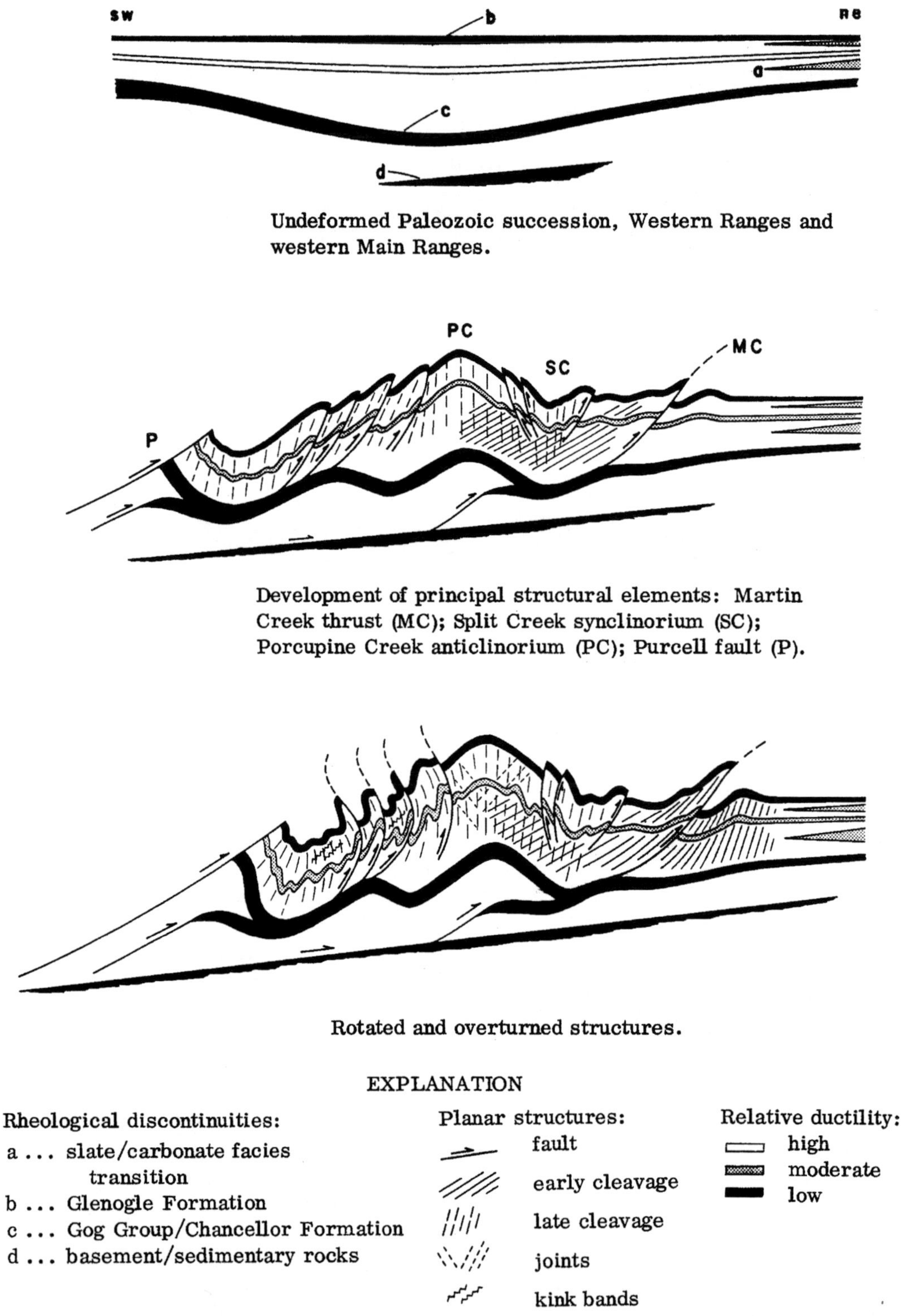

Figure 1-34. Stages of Porcupine Creek anticlinorium outlined by H.R. Balkwill (in Price et al., 1972).

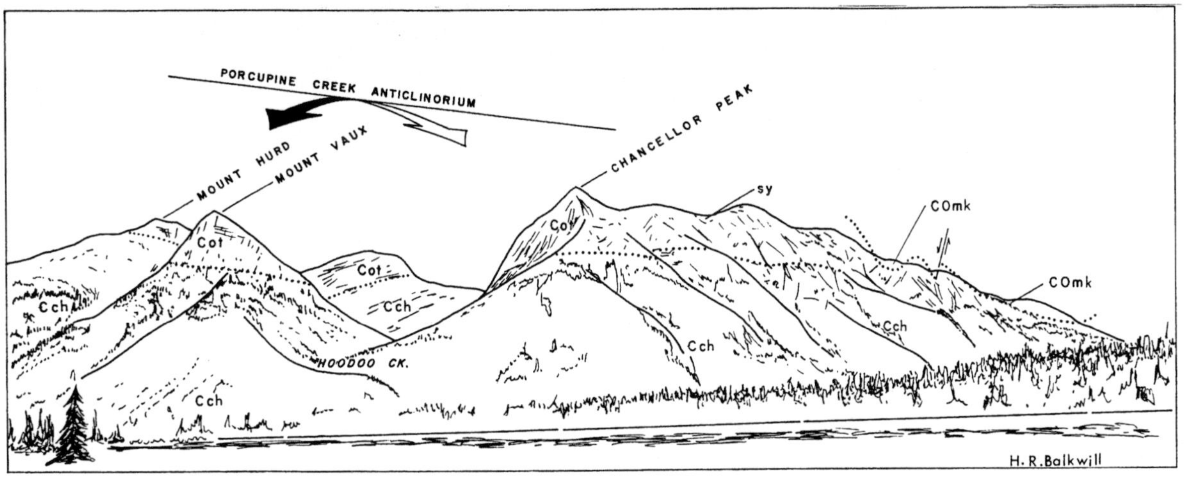

Figure 1-35. Sketch of the crest of the Porcupine Creek anticlinorium near Mount Vaux and Chancellor Peak from the Trans-Canada Highway at Leanchoil. The Ottertail Formation (Cot) outlines the relatively flat top of the anticlinorium; southwest-verging folds are outlined by the contact with the McKay Group (COmk) on the southwest slope of Chancellor Peak. The edge of the Ice River complex, (sy) is visible on the skyline south of Chancellor Peak. Price et al. (1972).

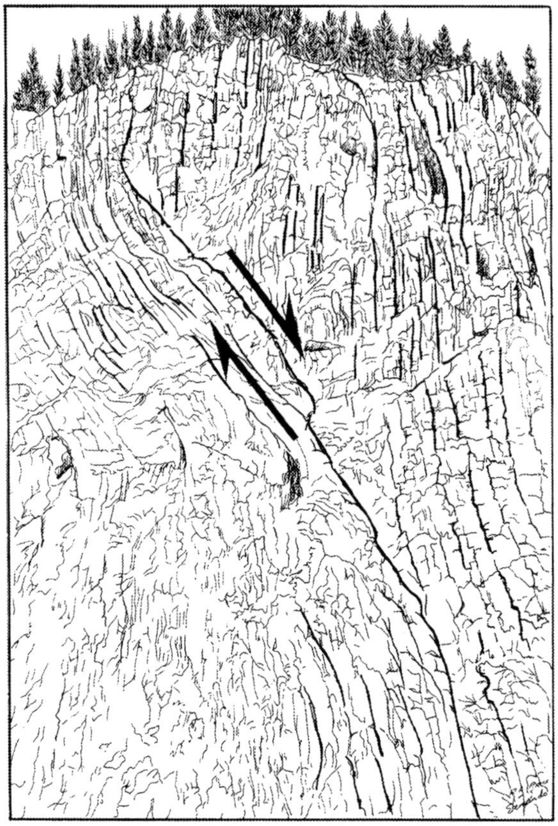

Figure 1-36. Sketch of overturned thrust fault in a limestone unit in the upper part of the McKay Group at upper Kicking Horse canyon bridge; view to the northwest from south side of the bridge (Price et al., 1972).

etheridgei) occur above the McKay Group in the roadcuts on the south side of the highway. The amount of penetrative deformation decreases rapidly up-section toward the overlying competent quartz arenites of the Mount Wilson Formation (Middle Ordovician) and carbonate strata of the Beaverfoot Formation (Upper Ordovician and Lower Silurian).

66.7 km Overturned, northeast-dipping Mount Wilson Formation (Middle Ordovician quartz arenite) in roadcuts on the south side of the highway.

67.7 km Overturned, northeast-dipping Beaverfoot Formation (Ordovician and Silurian limestone and dolomite) in roadcuts on the south side of the highway.

68.3 km Kicking Horse River bridge; lower Kicking Horse River canyon.

69.0 km Approximate location of trace of overturned northeast-dipping (formerly southwest-dipping and northeast-verging) thrust fault which is sub-parallel to the highway; the Beaverfoot Formation (Ordovician and Silurian), on the northeast side, overlies the Glenogle Formation on the southwest side.

69.5–71.7 km Roadcuts in Glenogle Formation; bedding dips gently east and faces east.

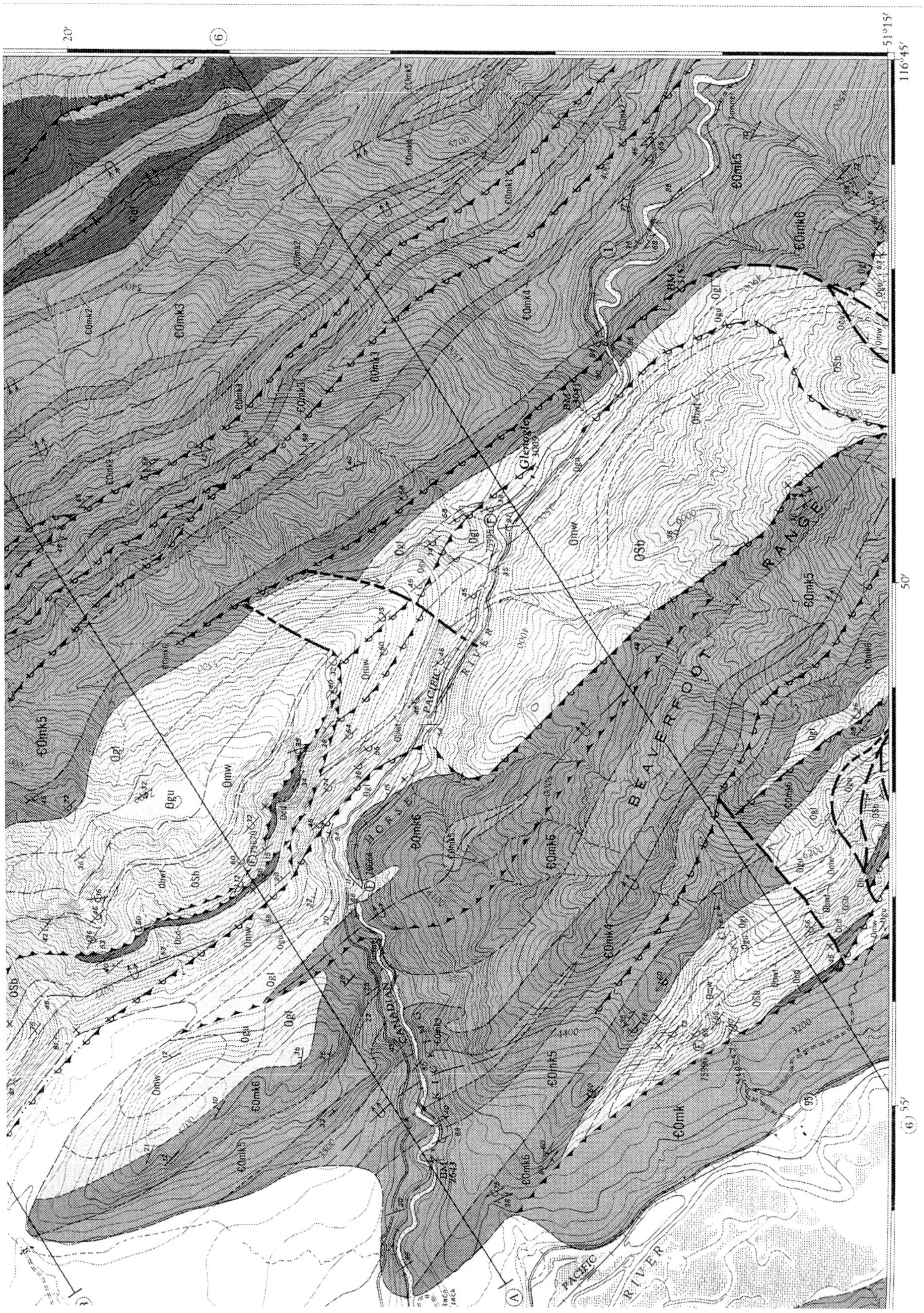
BEAVERFOOT
RANGE
Glenogle
CANADIAN
PACIFIC
RIVER

Figure 1-37 (facing page). Map of the southwest flank of the Porcupine Creek anticlinorium, adjacent to the Trans-Canada Highway, along the Lower Kicking Horse Canyon. east of Golden, British Columbia (Balkwill et al., 1980a; part of GSC Map 1497A). Symbols: Cchu = upper Chancellor, Cot = Ottertail, COmk = McKay (units 1 to 6), Ogl = lower Glenogle (shale), Ogu = upper Glenogle (siltstone), Omw = Mount Wilson, Obwt = Whiskey Trail Member of Beaverfoot Formation, Osb = Beaverfoot Formation, Dcd = Cedared Formation (Middle Devonian), Dh = Harrogate Formation (Middle Devonian).

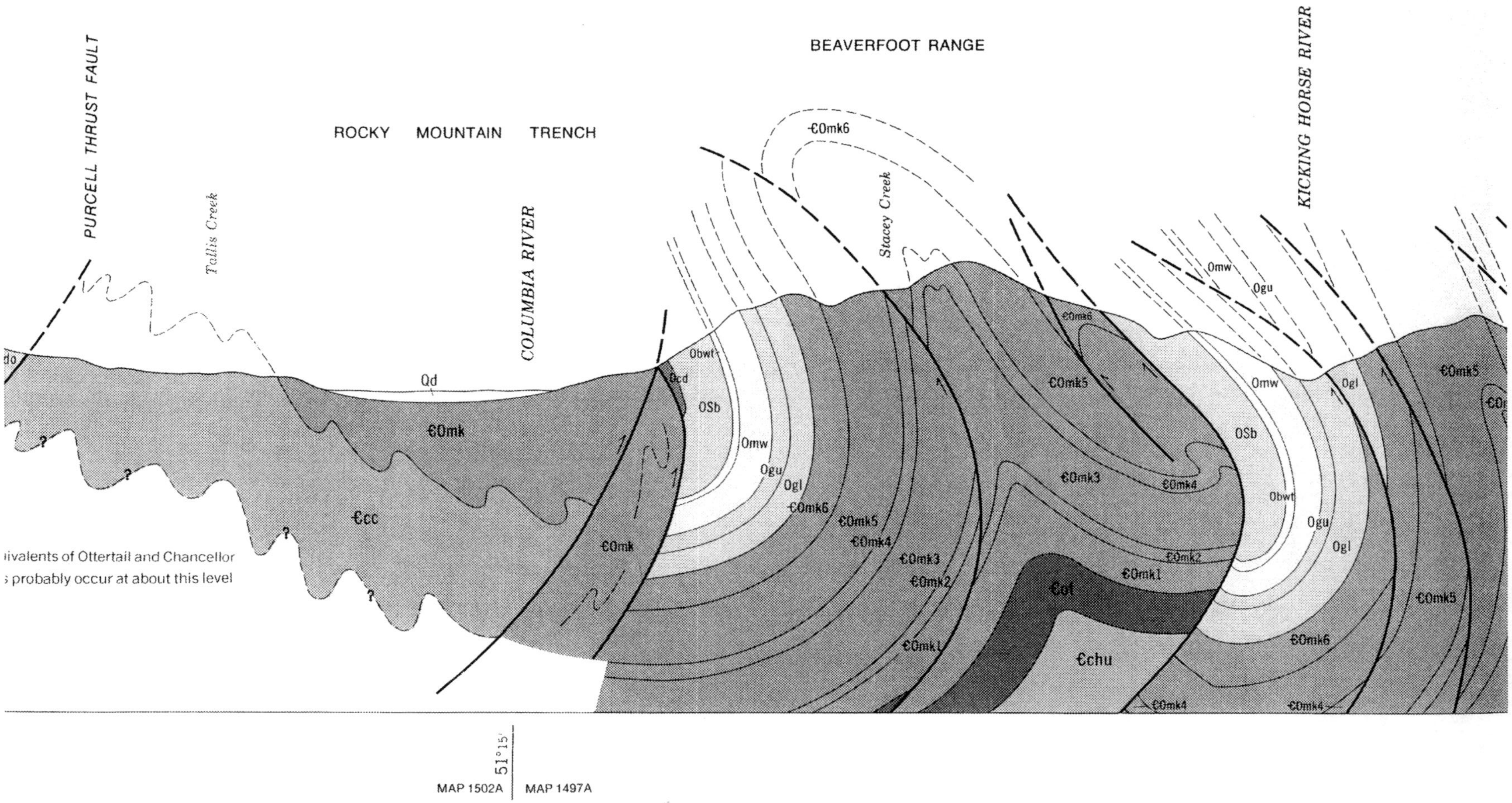

Figure 1-38. Section through the southwest flank of the Porcupine Creek anticlinorium, adjacent to the Trans-Canada Highway, along the Lower Kicking Horse Canyon east of Golden (Balkwill et al., 1980a; part of Section 6 accompanying GSC Map 1497A). Symbols as in Figure 1-37.

Stop 1-14 (optional) Roadcut in McKay Group and viewpoint for southwest overfolds

74.1 km The northeast-dipping folds in McKay Group strata on the south side of the of Kicking Horse canyon are part of a back-rotated, subsidiary, southwestward-overturned anticlinorium within the southwest flank of the main Porcupine Creek anticlinorium (Figs. 1-37, 1-38). The underlying, backrotated, overturned thrust fault is the fault that parallels the northwest-trending segment of the Trans-Canada Highway downstream form the lower Kicking Horse Canyon bridge in Figure 1-37. The McKay Group strata in the roadcut occupy the tightly appressed core of this fold and show evidence of intense compressional strain.

Stop 1-15 Columbia Valley-Rocky Mountain Trench viewpoint at Golden

77.8 km Stop at top of "the hill" east of Golden; the eastbound parking area is on south side of highway at the curve; the westbound parking area is on north side of highway, east of curve. **Look for oncoming traffic before attempting to cross the highway!**

The View Southwestward (Fig. 1-39)

The Columbia River valley is part of the southern Rocky Mountain Trench. In the vicinity of Golden it has the classical U-shaped profile of a glaciated valley. However, a narrow U-shaped valley, occupied by the Columbia River and associated wetlands has been incised in the base of the larger U-shaped valley. Above this smaller, sediment filled valley, the bottom of the large U-shaped valley consists of broad bedrock terraces that are covered with a thin veneer of glacial drift. The soft slaty rocks of the Canyon Creek Formation (an Upper Cambrian, non-calcareous shaly facies equivalent of the lower McKay Group) and the McKay Group are exposed in the beds of streams that cross the terraces. The northeast-dipping structures in these rocks show that the Porcupine Creek fan structure extends across the floor of the valley.

The trace of the *Purcell thrust fault* follows the break in slope at the top of the broad terrace. In the hanging wall of the Purcell thrust fault, rocks of the Neoproterozoic Horsethief Creek Group and the Lower Cambrian Hamill Group and Donald Formation form an imbricate stack of northeast-verging thrust slices. These thrust slices are part of a thrust duplex which is truncated by the Purcell thrust fault (Kubli and Simony, 1994).

The View Southeastward

The Beaverfoot Range, which forms the northeast side of the Rocky Mountain Trench southeast of Golden, is the northern end of the Western Ranges Sub-province of the southern Canadian Rocky Mountains (North and Henderson, 1954). It is characterized by overturned, southwest-verging folds and faults, and is held up by resistant carbonate rocks of the Upper Ordovician-Lower Silurian Beaverfoot Formation and by the quartz arenites of the Middle Ordovician Mount Wilson Formation (Figs. 1-37, 1-38). The penetrative deformation and slaty cleavage that is characteristic of the underlying McKay Group strata die out upward into the overlying Middle Ordovician graptolitic, black shales (Glenogle Formation), which still retain their bedding fissility.

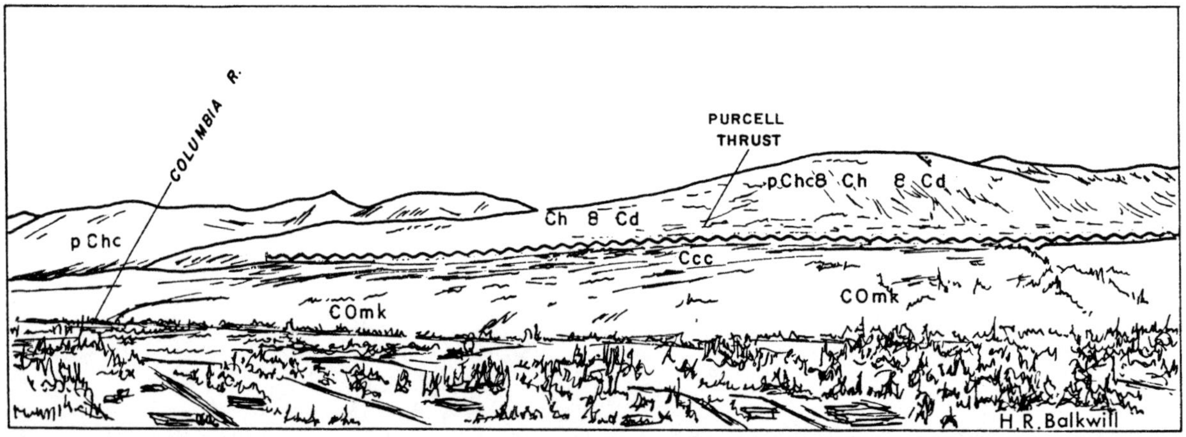

Figure 1-39. Sketch of the Rocky Mountain Trench and the Dogtooth Range, view southwestward from the viewpoint on the Trans-Canada Highway east of Golden, British Columbia (by H.R. Balkwill, from Price et al., 1972).

The Beaverfoot Formation is overlain by Middle Devonian limestone and gypsum (Harrogate and Cedared Formations).

The Middle Ordovician to Middle Devonian rocks occupy the core of a northwest-trending synclinorium with a complex internal structure that involves three distinct phases of thrust faulting and related folding. The oldest thrusts were northeast-verging structures that produced imbricate repetitions, shortening, and thickening of the stratigraphic succession. They were back-rotated during evolution of the Porcupine Creek anticlinorium, to form steeply dipping, vertical, and overturned structural panels, and were deformed by a younger set of southwest-verging thrust faults and related folds. The youngest thrust faults and related folds, which developed during the evolution of the Purcell thrust fault, deformed and offset both of the preceding groups of structures. These relationships can be seen in maps and cross sections of the Beaverfoot Range (Fig. 1-40).

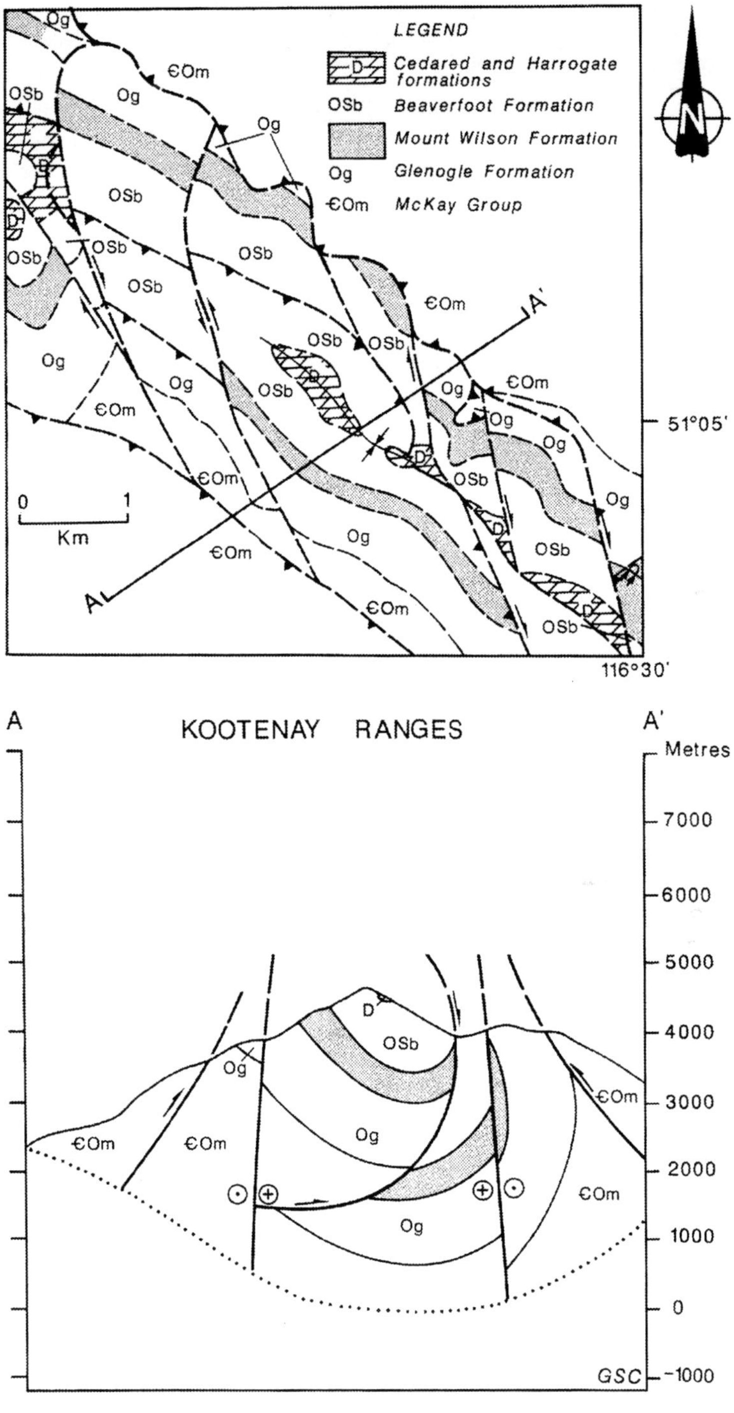

Figure 1-40. Map and structure section of part of the Beaverfoot Range southeast of Golden, showing three cross-cutting phases of thrusting in the southwest flank of Porcupine Creek anticlinorium (from McMechan and Thompson, 1989; after Price et al., 1980).

Southern Rocky Mountain Trench

The Rocky Mountain trench is a remarkable physiographic feature that has inspired many grand conjectures concerning Cordilleran tectonics. It is a relatively narrow, north- and northwest-trending valley that extends some 1600 km from Flathead Lake, Montana (48°N latitude) to the Liard Plain of northern British Columbia and southern Yukon Territory (60°N latitude). It is aligned with the Tintina Trench, which extends another 1000 km northwest into east-central Alaska. Many of the great rivers of the Cordillera have their headwaters in the Rocky Mountain Trench, with their mutual headwaters separated by anomalously low divides.

The northern part (Tintina and Northern Rocky Mountain trenches; extending southeastward from ~65°N latitude in east-central Alaska to ~53°N latitude) of this unusual physiographic feature is a right-hand strike-slip fault zone across which mid-Cretaceous and older rocks have been offset by up to 450 km (Roddick, 1967). The southern part (extending northward from about 48°N latitude, at Flathead Lake, Montana to about 50°N latitude, near Canal Flats, B.C.) is a half-graben that follows the hanging wall of a southwest-dipping listric normal fault (Leech, 1960; Bally et al., 1966; van der Velden and Cook, 1996). Older transverse structures, which can be matched across the graben without any significant strike separation, preclude the possibility of any significant strike slip along this part of the Trench (Leech, 1966; Benvenuto and Price, 1979; Price, 1981). The intervening segment is due mainly to erosion that was controlled by the presence of weaker rocks, but also in part to west-dipping normal faults. In the vicinity of Golden and Donald, B.C., the Rocky Mountain Trench is situated in the footwall of the Purcell thrust fault. It has been eroded in the less resistant shaly and slaty rocks of the Upper Cambrian to Lower Ordovician Canyon Creek Formation and McKay Group that occur in the footwall of the Purcell thrust. It is bounded on the northeast by the resistant carbonate rocks and quartz arenites of the Upper Ordovician-Lower Silurian Beaverfoot Formation and the Middle Ordovician Mount Wilson Formation, and on the southwest by the resistant siliciclastic rocks of the Lower Cambrian Hamill Group and the Neoproterozoic Horsethief Creek Group that occur in the hanging wall of the Purcell fault. North of Donald it follows the hanging wall of a southwest-dipping normal fault.

The Rocky Mountain Trench has been profoundly modified by Pleistocene glacial erosion and by the ensuing sedimentation and erosion during deglaciation. For example, the segment that extends southward from Donald to Canal Flats, which formerly was drained by a major southeast-flowing trunk stream, was dammed during deglaciation by ice and glacio-fluvial deposits that flowed into the Trench from the Rocky Mountains along the Kootenay River valley at Canal Flats. The sediment choked segment of the Trench is now occupied by Columbia Lake and Lake Windermere and by extensive wetlands through which the Columbia River meanders northward to Donald. Columbia Lake, the headwaters of this northwest-flowing segment of the Columbia River, is separated from the southeast-flowing Kootenay River, a major tributary of the Columbia River, by a very low divide formed by the glacio-fluvial and glacio-lacustrine deposits at Canal Flats.

The corollary to this description of what is known about the Southern Rocky Mountain Trench is that contrary to many inspired grand conjectures relating this dramatic topographic lineament to Cordilleran tectonics, the Southern Rocky Mountain trench is not a major through-going fault structure; it is not a terrane boundary; it does not mark the southwestern limit of the foreland thrust and fold belt; and it does not mark the northeastern limit of metamorphic and magmatic rocks. Based on physiography, and not on geology, it is somewhat arbitrarily taken as the eastern boundary of the Omineca Belt where crossed on Highway 1. The southern part is a Tertiary half-graben; the middle part is eroded in weak Lower Paleozoic shaly rocks, and the northern part is controlled by the Purcell thrust fault, and by younger west-dipping normal faults, some of which may be early Tertiary transtensional structures related to the dextral displacement on the Northern Rocky Mountain Trench-Tintina Trench fault system.

78.5 km Junction of Highway 1 and Highway 95 at Golden.

Reset odometer to 0 km.

2. OMINECA BELT

GOLDEN TO PRITCHARD

INTRODUCTION

Between Golden and Pritchard, the Trans-Canada Highway traverses from the western part of the Foreland thrust and fold belt, through the Omineca Belt, to the eastern edge of the Intermontane belt (Figs. 1, 4). The magmatism and regional metamorphism, which are the hallmarks of the Omineca Belt, are *superimposed* on the suture zone (interface) between the North American rocks of the Cordilleran miogeocline and the accreted rocks of the Intermontane superterrane (Fig. 4). Accordingly, the boundaries between the Omineca Belt and the Foreland and Intermontane Belts are gradational and somewhat arbitrary, except where they happen to coincide with a fault.

Three major regional fault zones are crossed on this segment of the trip. Each involves a profound change in the level of exposure within the orogen (Fig. 1a). The faults bound two broad structural domains: (1) the *Selkirk fan structure,* which straddles the boundary between the Foreland and Omineca belts, and (2) the *Shuswap metamorphic complex,* which lies within the Omineca belt. The changes that

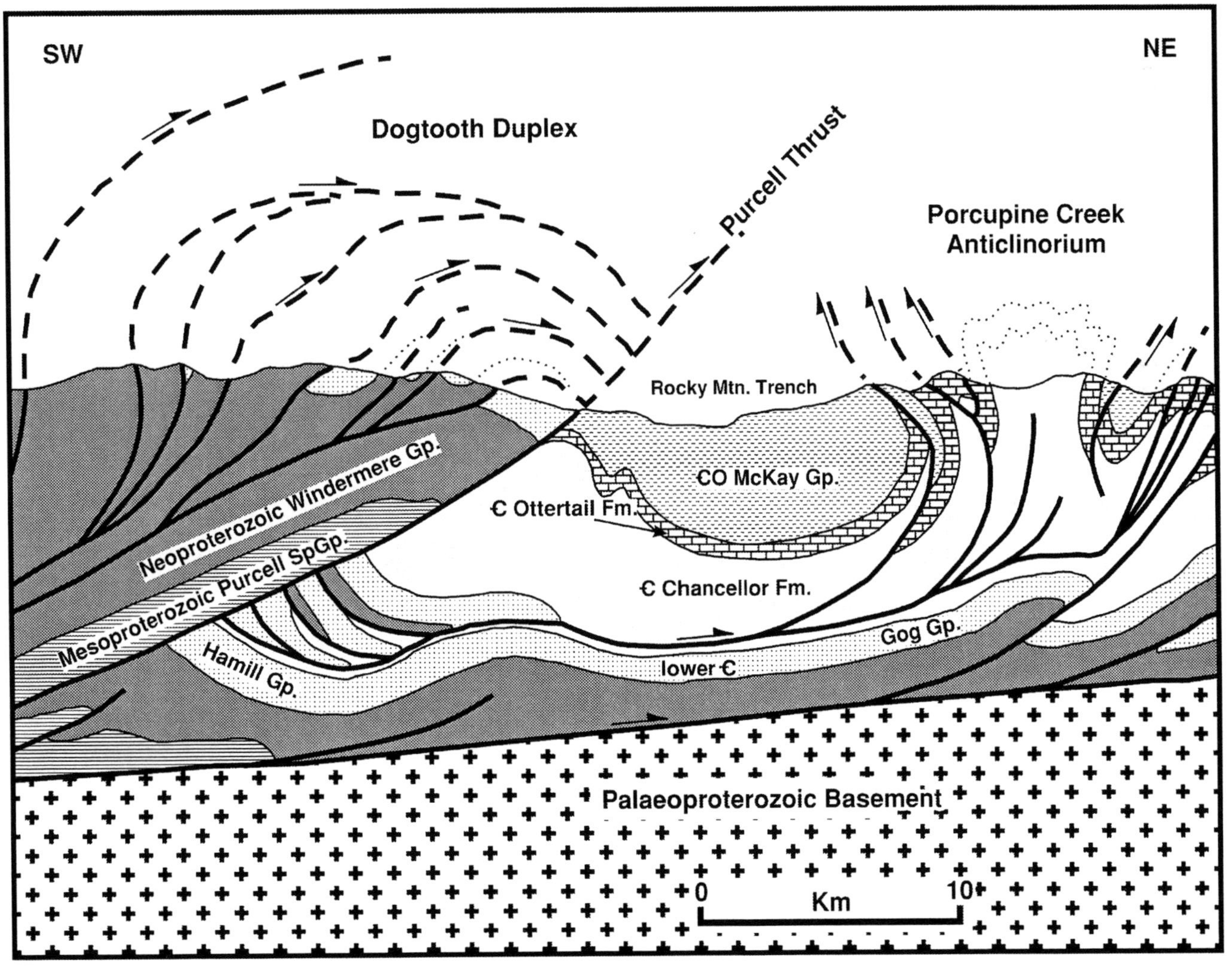

Figure 2-1. Schematic structural section of relationships between the Dogtooth thrust duplex and the Porcupine Creek anticlinorium (after Simony and Price, *in* Monger et al., 1985).

occur across these fault zones provide opportunities to examine variations in structural style and tectonic history that occur with increasing depth in the orogen, and to compare them with the changes that occur with increasing proximity to the Intermontane superterrane.

Purcell Thrust

The Purcell thrust is a major, regional, northeast-verging, pre-mid-Cretaceous, out-of-sequence thrust fault (Figs. 2-1, 2-2, 2-3, 2-4). In the vicinity of the Trans-Canada Highway, the Purcell thrust follows the western side of the Southern Rocky Mountain Trench; but farther north, it extends into the western Rocky Mountains, where it offsets older thrust and fold structures and regional metamorphic isograds (Craw, 1978). West of Golden, the Purcell thrust truncates underlying southwest-verging structures, and the older back-rotated northeast-verging thrusts and folds that form the southwest flank of the Porcupine Creek fan structure. Along its hanging wall, it truncates the overlying, northeast-verging *Dogtooth thrust duplex.* This implies that the matching counterpart of the truncated Dogtooth thrust duplex must lie in the footwall of the Purcell thrust, beneath the southwestern part of the Porcupine Creek fan

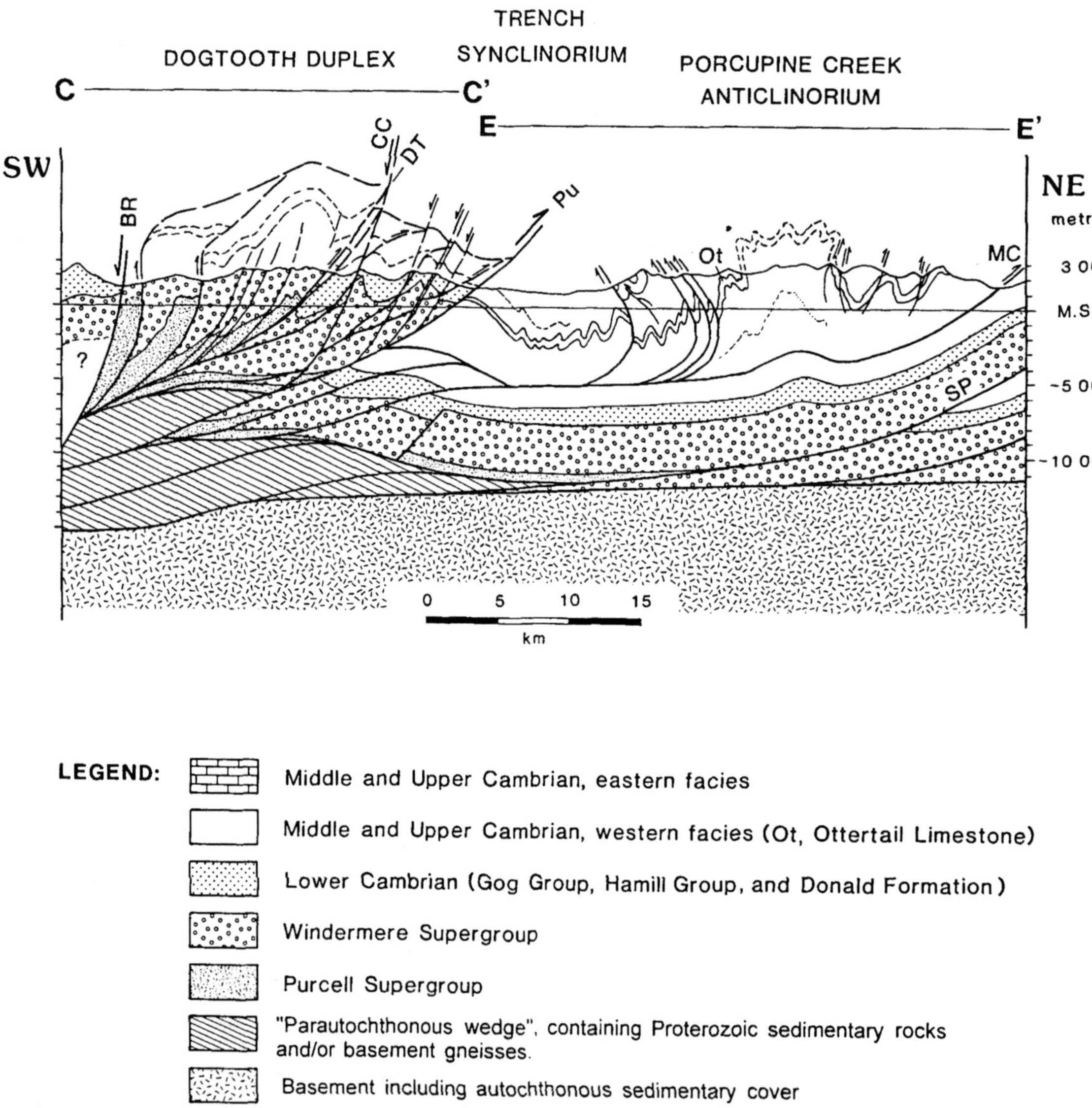

Figure 2-2. Composite structure section of the Dogtooth duplex. From Kubli and Simony (1994).

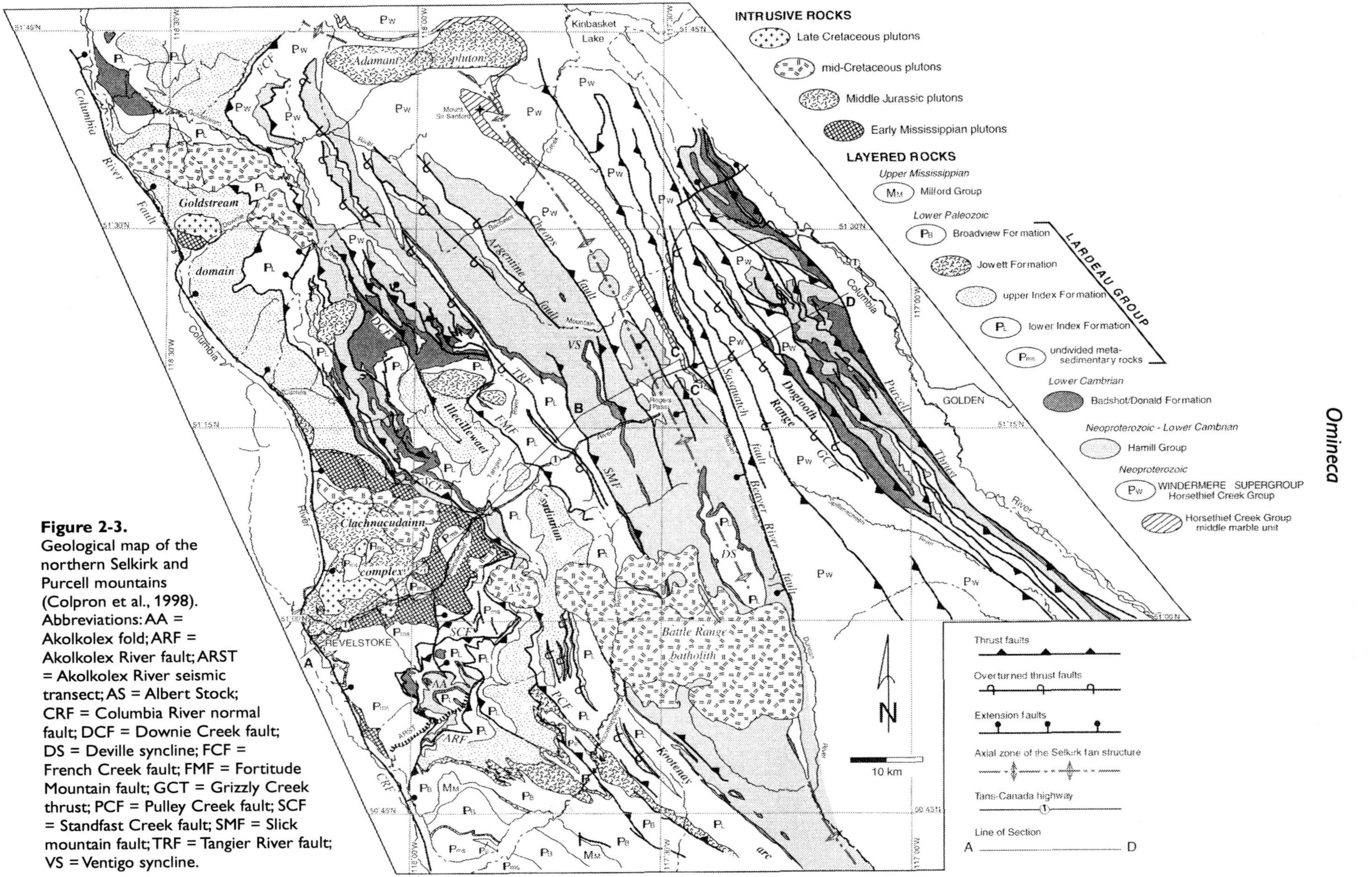

Figure 2-3.
Geological map of the northern Selkirk and Purcell mountains (Colpron et al., 1998). Abbreviations: AA = Akolkolex fold; ARF = Akolkolex River fault; ARST = Akolkolex River seismic transect; AS = Albert Stock; CRF = Columbia River normal fault; DCF = Downie Creek fault; DS = Deville syncline; FCF = French Creek fault; FMF = Fortitude Mountain fault; GCT = Grizzly Creek thrust; PCF = Pulley Creek fault; SCF = Standfast Creek fault; SMF = Slick mountain fault; TRF = Tangier River fault; VS = Ventigo syncline.

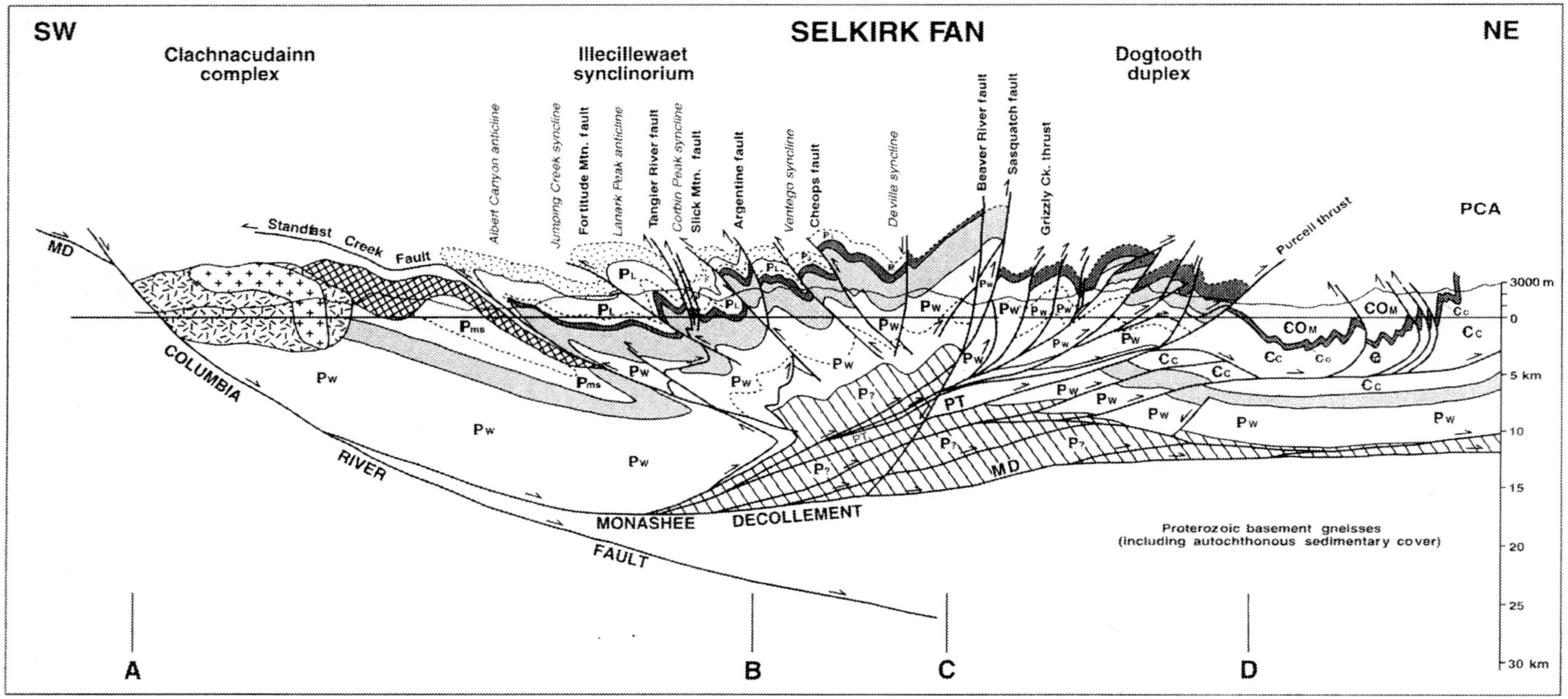

Figure 2-4. Regional structure section across the Selkirk fan structure from the Clachnacudainn complex to the Porcupine Creek anticlinorium (PCA) in the Rocky Mountains (Colpron et al., 1998). See Figure 2-3 for legend and location of section. Abbreviations: P? = Proterozoic sedimentary rocks (may include rocks of the Windermere and/or Purcell Supergroups, and/or basement gneisses?); Cc = Chancellor Formation; Co = Ottertail Formation; COm = McKay Group; PT = Purcell thrust; MD = Monashee décollement.

structure. On the basis of these relationships, horizontal shortening due to the imbricate thrust faulting of the Neoproterozoic Horsethief Creek Group and the Lower Cambrian Hamill Group within the Dogtooth thrust duplex has been interpreted to be an equivalent expression, at a deeper structural level, of the horizontal shortening that is expressed at a shallower structural level by the detachment folding that formed the Porcupine Creek anticlinorium (Monger et al., 1985; Kubli and Simony, 1994). This interpretation is illustrated in Figures 2-1 and 2-2. A sequence of schematic restored regional cross sections illustrating stages in the evolution of the Dogtooth duplex and the Porcupine Creek anticlinorium, as interpreted by Kubli and Simony (1994) in Figure 2-2, are presented in Figure 2-5.

The imbricate thrust slices of the Dogtooth thrust duplex are arranged in an asymmetric fan that has been called the *Purcell fan structure* (Price et al., 1972). There are significant contrasts between the Purcell fan structure and the Porcupine Creek fan structure (Fig. 2-2). The Porcupine Creek fan structure is a large disharmonic detachment fold. It is more symmetrical, and within its southwest flank, map-scale southwest-verging thrusts and folds are superimposed on older, back-rotated, northeast-verging thrusts and folds. In contrast, the Purcell fan structure is conspicuously asymmetric, and the back-rotation that occurs in its narrower, steeper southwest flank appears to be a result of displacements on underlying younger, concave (listric) thrust faults within the same duplex structure, rather than detachment folding above some deeper structural discontinuity. The Dogtooth duplex/Purcell fan forms an easterly component of the larger Selkirk fan structure (Figs. 2-3, 2-4), which stretches from the Purcell thrust to the Columbia River fault zone at Revelstoke.

The Selkirk fan structure is part of a regional zone of structural divergence (bivergent thrusting and folding), about 1500 km long, extending from the Kootenay arc of northeastern Washington and southeastern British Columbia to the southeastern Yukon (Eisbacher et al., 1974; Price, 1986). This regional zone of bivergent compressional deformation follows the suture zone (interface) between Intermontane terrane and the miogeoclinal rocks that have been detached from the North American craton and accreted to Intermontane terrane. It formed as a result of the collision (compressional convergence) between the North American miogeocline and the Late Triassic-Early Jurassic Quesnellia magmatic arc and its underlying northeast-dipping subduction zone. This occurred during and following the collapse of the intervening Slide Mountain marginal basin.

Columbia River Fault

The Early to Middle Eocene Columbia River fault zone is a crustal scale, east-dipping, *extensional* detachment fault that follows the northwest-trending valley of the Columbia River near Revelstoke, B.C. (Figs. 2-3, 2-4; see also Fig. 2-16). It separates middle to lower crustal rocks of the Shuswap metamorphic core complex, which have been exhumed from a depth of about 25 km, from upper crustal low-grade metasedimentary rocks of the Selkirk Mountains, which have been within the upper 10 km of the crust since the Middle Jurassic. Displacement on the Columbia River fault has exhumed not.only the deepest levels of the allochthonous wedge of detached and displaced North American supracrustal rocks, but also the underlying parautochthonous North American basement rocks, which are known as the *Monashee Complex.* The Monashee Complex is exposed in a window eroded through the *Monashee décollement,* which is marked by thick zone of mylonitic rocks above parautochthonous basement rocks. It is the exhumed mid-crustal equivalent of the basal décollement of the Foreland thrust and fold belt under the Rocky Mountains.

Okanagan-Eagle River Fault Zone

The Okanagan-Eagle River fault zone (Fig. 1; see also Fig. 2-16) is an Early and Middle Eocene, west-northwest-verging, crustal-scale, *extensional* detachment fault that has exhumed the west side of the Shuswap metamorphic complex (Johnson and Brown, 1996). The rocks in the hanging wall include variably metamorphosed Lower Paleozoic and(?) upper Neoproterozoic supracrustal rocks of the pericratonic Kootenay terrane, intruded by Late Devonian-Early Mississippian granodioritic gneiss (Okulitch et al., 1975), and also late Paleozoic to Jurassic rocks of Intermontane terrane. As with the rocks in the western Selkirk Mountains, they have been deformed by tight, syn-metamorphic folds and penetrative foliation of Middle Jurassic to Early Cretaceous age (Johnson and Brown, 1996). The higher grade metamorphic rocks in the footwall are of uncertain stratigraphic correlation, but appear to comprise parts of the off-scraped North American supracrustal cover that have been metamorphosed and penetratively deformed at mid-crustal depths.

Reset odometer to 0 km.

0.0 km Junction of Highway 1 and Highway 95 at Golden.

15.2 km Beginning of a long roadcut in southwesterly overturned calcareous slates and interbedded

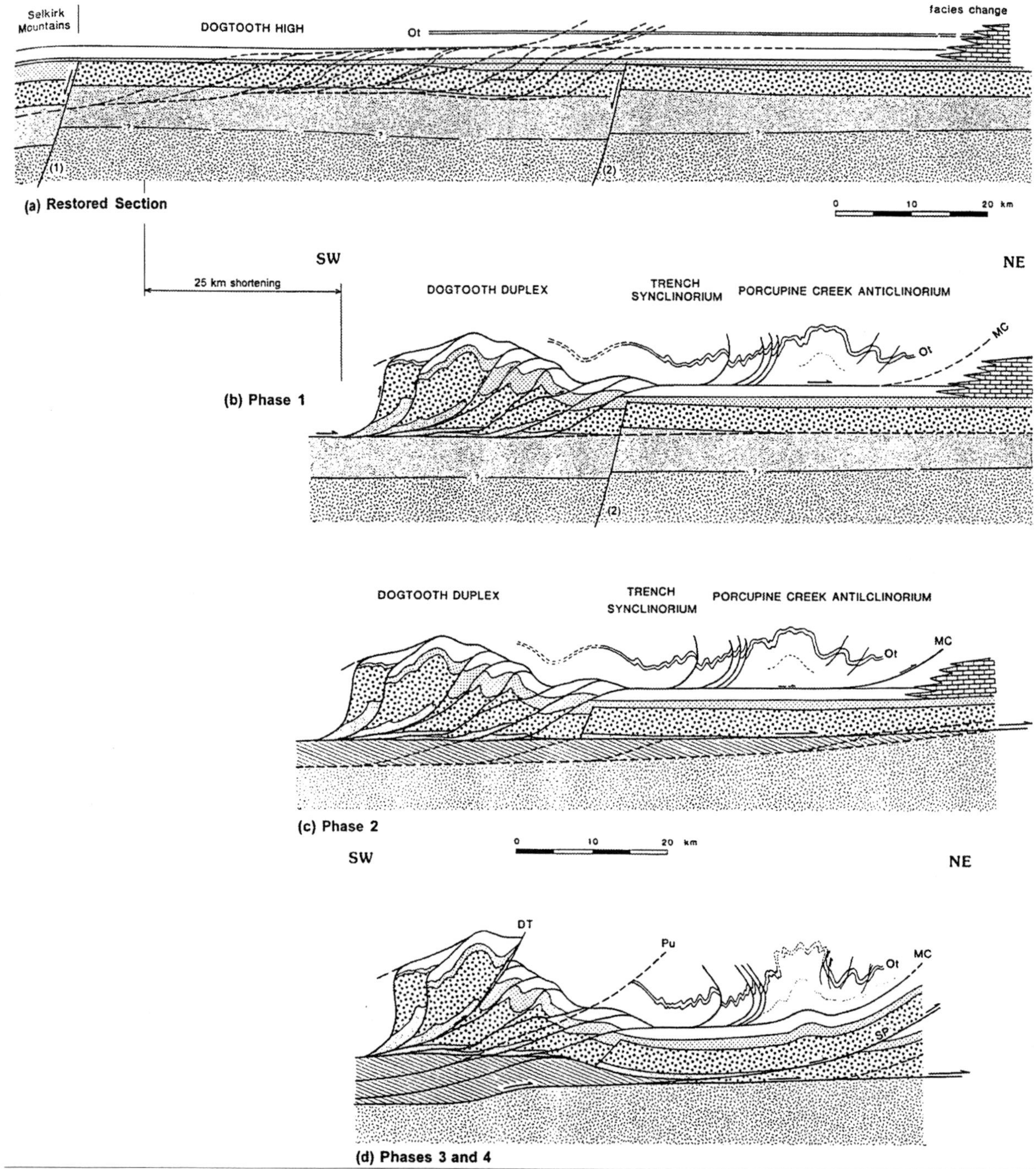

Figure 2-5. Schematic restored regional cross sections illustrating stages in the evolution of the Dogtooth duplex and the Porcupine Creek anticlinorium (Kubli and Simony, 1994). Legend as in Figure 2-2. Solid lines, faults; broken lines, incipient faults; (a) Restored section with syndepositional normal faults. (b) Phase 1: Development of the Dogtooth duplex above a lower detachment below the base of the Windermere Supergroup; simultaneous formation of the Porcupine Creek anticlinorium above an upper detachment within the Cambrian Chancellor Formation. (c) Phase 2: Northeastward transport of the Dogtooth duplex and the Porcupine Creek anticlinorium. (d) Phase 3: Continued northeastward transport accompanied by downward transfer of motion to a lower detachment level resulting in the formation of a "parautochthonous wedge" with internal duplex structure. Late-phase displacement on the Dogtooth-Quartz Creek thrust? Phase 4: 10–15 km of shortening accommodated on the Purcell thrust. Phase 5: Horizontal extension accommodated partly on new normal faults and partly by reactivation of pre-existing thrust faults to produce the structure shown in Figure 2-5.

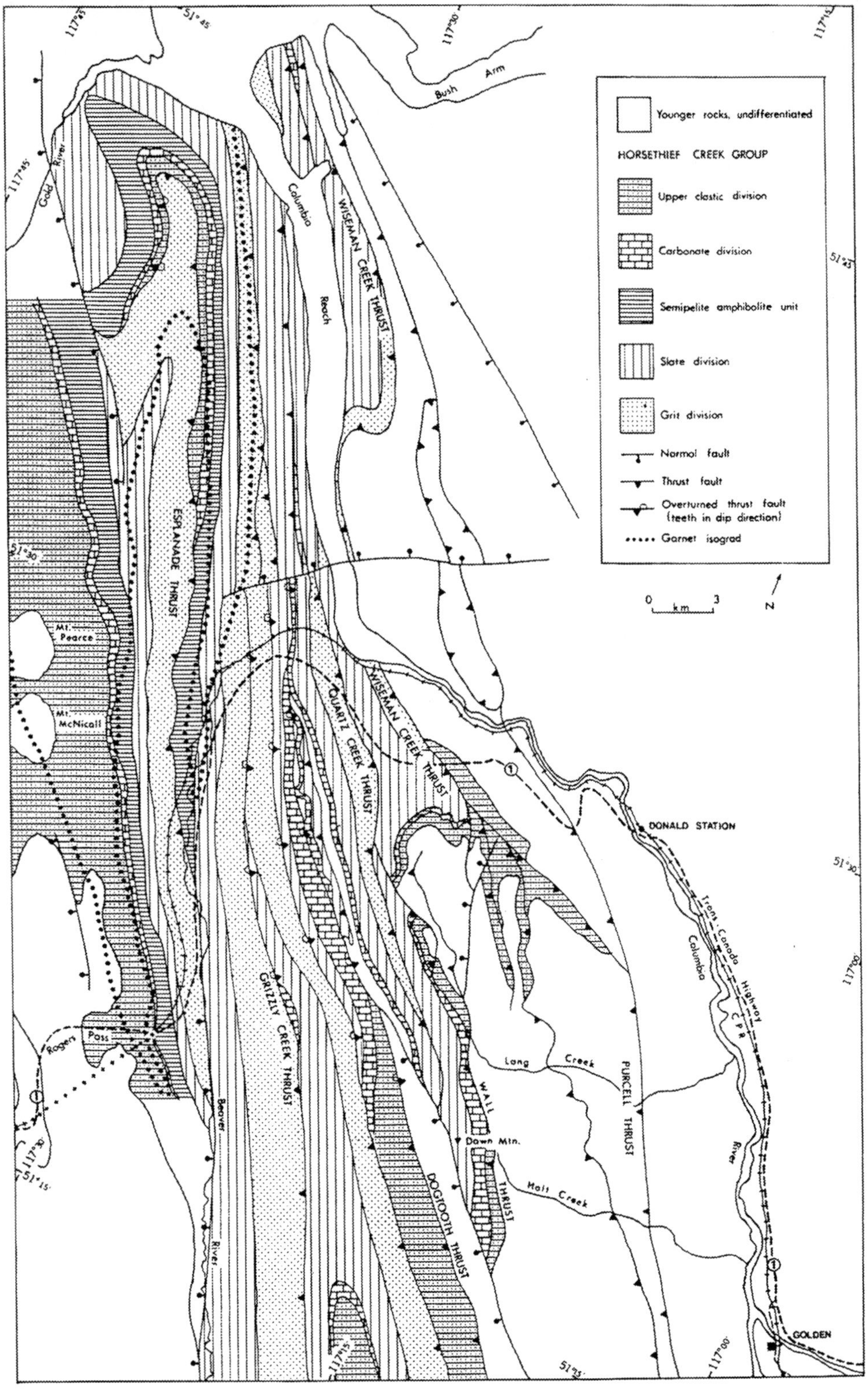

Figure 2-6. Geological map of the northern Dogtooth Mountains and adjacent eastern Selkirk Mountains (after Poulton and Simony, 1980).

limestones of the McKay Group (Upper Cambrian and Lower Ordovician). These rocks, which occur in the overturned southwest limb of the Porcupine Creek anticlinorium, have a northeast-dipping cleavage and southwest-verging folds and thrust faults.

15.7 km Bridge over Blaeberry River. The Blaeberry River has a gentle gradient to its headwaters at Howse Pass (1550 m). This broad, open pass separates the Blaeberry River valley from the valley of Howse River, which also has a low gradient, and is a major tributary of the North Saskatchewan River.

At the beginning of the nineteenth century, fur traders travelled into the Columbia River basin from Rocky Mountain House on the North Saskatchewan River via Howse Pass. However, the engineers who designed the trans-continental Canadian Pacific Railway selected the route from Calgary through Kicking Horse Pass (1590 m), in spite of the steep gradient westward from Kicking Horse Pass to Field, B.C.

26.6 km Bridge over Columbia River at Donald, B.C. (see Fig. 2-6).

27.6 km Roadcuts in calcareous slate and interbedded limestone of the McKay Group (Upper Cambrian and Lower Ordovician). Here the rocks are darker grey than they normally are in the Rocky Mountains, a difference which perhaps indicates a transition into the black phyllites of the Index Formation, which appears to occupy the same stratigraphic level farther west, within the Selkirk Mountains.

Stop 2-1 Deformed McKay Group rocks in footwall of Purcell thrust; panoramic view across the Rocky Mountain Trench into the western Rocky Mountains

30.3 km The following commentary is after Simony (1998).

The Roadcut

Here, on the southwest side of the Rocky Mountain Trench and close to the hanging wall of Purcell thrust fault (Fig. 2-7a), the McKay Group displays the same sequence of structures as occur on the northeast side of the Rocky Mountain trench in the west limb of the Porcupine Creek anticlinorium. Conspicuous southwest-verging folds and thrust faults deform older northeast-verging thrust faults, and are themselves truncated by the overriding, northeast-verging Purcell thrust fault. In the slates and calcareous slates, cleavage is pervasive and bedding is transposed and difficult to find. The limestone beds show thickening at fold hinges and thinning on the limbs, boudinage, and much veining, but primary sedimentary features (ooliths, intraclasts, trilobite fragments) are well preserved and little strained. The thrust and fold structures in this roadcut provide an outcrop-scale illustration of the larger scale structures in the southwest flank of the Porcupine Creek anticlinorium. As outlined in the sketch of the outcrop (Fig. 2-7b), an early northeast-verging thrust repeats a succession of buff, thin-bedded limestone, a limestone marker, and an overlying olive-grey slate. Northeast-verging folds, which deformed the thrust, were back-rotated. As the folds tightened, two thrusts emerged out of the synclines, and these in turn were folded. This structural panel is bounded on the east (left) by a detachment surface, and on the west by a fault of unknown throw.

The surface trace of the Purcell thrust is in a depression at the base of the mountains to the west. It can be followed to the railway tracks 300 m below the highway, and over the southern end of a ridge. From the topographic relief, the near-surface dip of the Purcell thrust is estimated to be 35° southwest. The footwall of the Purcell thrust follows a unit of soft black slates of the Canyon Creek Formation 25 km to the northwest and to the southwest of the Trans-Canada Highway. As the fault is approached from the footwall, the northeast-dipping cleavage passes through the vertical and dips steeply southwest, and is sheared in the fault zone. Slaty cleavage in Lower Cambrian slates of the Donald Formation in the hanging wall of the Purcell thrust dips southwest at a moderate angle. The Purcell thrust cuts deformed rocks in its hanging wall and in its footwall. It is out-of-sequence, and therefore unique net slip solutions are difficult to obtain. A reasonable, conservative estimate here is 10 km.

The Panorama

There is a good panoramic view from this locality, especially from the top of the roadcut. Northeastward, in the Rockies, is the Porcupine Creek anticlinorium. Northward there is a view along the Rocky Mountain Trench. Northwestward, a wooded ridge is the continuation of the Dogtooth duplex. Westward (in the distance) is the Esplanade Range of the eastern Selkirk Mountains, which consists of folded and thrust faulted Neoproterozoic strata containing biotite, garnet and staurolite. Southwestward is the Dogtooth Mountains, where Lower Cambrian and Neoproterozoic strata occur in the hanging wall of the Purcell thrust.

The Rocky Mountain Trench

North of Donald, the Purcell thrust runs into the Rocky Mountain Trench, where it is truncated by a

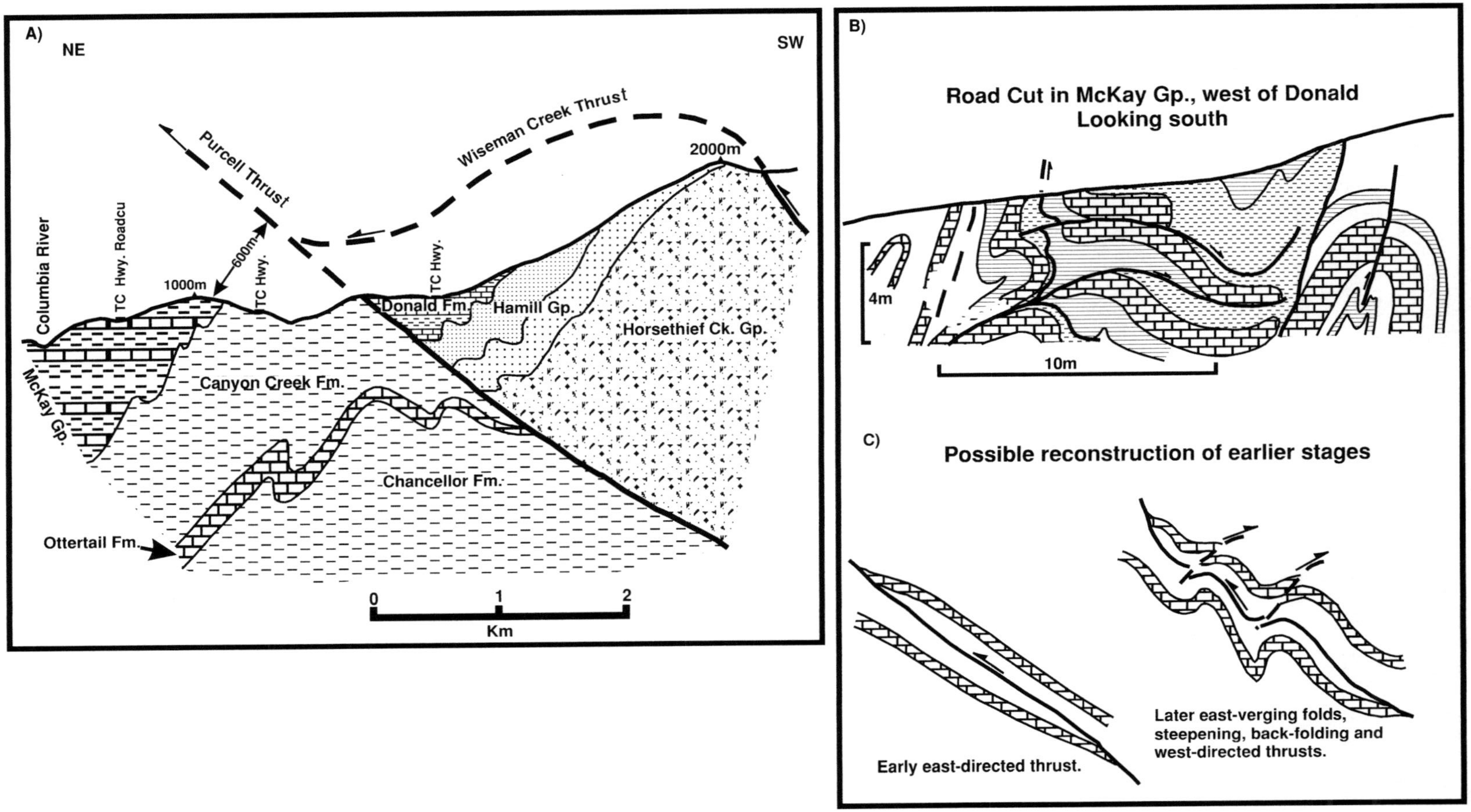

Figure 2-7: Southwest-verging thrust and fold structures in the McKay Group in a roadcut on the Trans-Canada Highway close to the Purcell thrust (modified from Simony, 1998).
A. Structure section showing location of this roadcut in relation to the Purcell thrust and to the roadcut in the Canyon Creek Formation between this roadcut and the Purcell thrust.
B. Sketch of the structures in the central part of the roadcut.
C. Possible reconstruction of earlier stages in the development of these structures.

southwest-dipping normal fault that emerges form the western Rocky Mountains to form the northeast wall of the trench. Lower Cambrian facies belts related to an Early Cambrian north-trending paleogeographic feature called the "Dogtooth High" (Kubli and Simony, 1992), and metamorphic isograds from the Esplanade Range of the eastern Selkirk Mountains, both cross the trench obliquely. Both are truncated by the fault in the northeast wall of the trench, but without significant strike separation.

31.4 km Non-calcareous slates of the Canyon Creek Formation. The Canyon Creek Formation contains Upper Cambrian trilobites, and is correlative with the calcareous slates and interbedded limestones in the Rocky Mountains to the east.

32.9 km The sharp curve in the highway is the approximate location of the trace of the Purcell thrust fault (Figs. 2-3 and 2-6).

34.8 km Dip slope of Hamill quartzite (Lower Cambrian) to the south.

37.1 km Landslide debris comprising blocks of Hamill quartzite.

38.3 km Roadcuts in Hamill Group (Lower Cambrian) quartzite with interbedded siltstone and slate. The contact with the Horsethief Creek Group (Neoproterozoic) is at the west end of the roadcut. There are porphyroblasts of chloritoid in the slates. Cleavage dips to the southwest. Structures in the hanging wall of the Purcell fault are northeast-verging, and contrast with the southwest-verging structures in the McKay Group and Canyon Creek Formation in the footwall of the Purcell fault.

40.0 km Roadcut in upper Horsethief Creek Group (Neoproterozoic).

40.9 km Approximate trace of the Wiseman Creek thrust fault.

42.8 km Roadcuts in slates of the Neoproterozoic Horsethief Creek Group.

45.2 km Roadcuts in interbedded slates and feldspathic grits of the Neoproterozoic Horsethief Creek Group.

46.8 km Roadcuts in crenulated slates of the Neoproterozoic Horsethief Creek Group.

48.8 km Junction with logging road to Beavermouth.

49.5 km Long roadcut in slates and feldspathic grits of the lower part (grit unit) of the Neoproterozoic Horsethief Creek Group.

50.3 km Roadcuts on both sides of the highway. Steep west-facing beds mark the boundary between the grits and overlying slate of the Neoproterozoic Horsethief Creek Group.

50.9 km Roadcuts in the slate division of the Neoproterozoic Horsethief Creek Group.

52.2 km The Heather Mountain thrust is (not labelled in Fig. 2-6) exposed at the sharp curve where the Trans-Canada Highway turns south up the valley of the Beaver River. It lies within the western part of the Dogtooth fan structure. The structures are steeply dipping as a result of back rotation during the formation of the fan structure, and at this locality, the overturned Heather Mountain thrust dips steeply northeast. It juxtaposes a steeply dipping, west-facing structural panel comprising strata of the upper slate unit of the Horsethief Creek Group, on the west, against a steeply dipping, west-facing thick, massive grey dolomite unit that overlies the upper slate unit in the western Dogtooth Mountains. The Heather Mountain thrust is similar to the back-rotated, overturned thrust faults that occur in the west flank of the Porcupine Creek fan structure, in the western Rocky Mountains, and in the west flank of the Selkirk fan structure in the western Selkirk Mountains.

53.3 km Roadcut in Neoproterozoic Horsethief Creek Group slates. Cleavage dips steeply to northeast here in the western part of the Dogtooth fan structure.

Stop 2-2 Purcell Trench viewpoint and roadcut in crenulated phyllites of Neoproterozoic Horsethief Creek Group

57.0 km The roadcut (after Simony, 1998). A very strong phyllitic cleavage is crenulated, and sandstone beds are deformed into meso-scale folds. Cleavage, which is also discernible in "grit" beds, is marked by an alignment of white mica, chlorite and strained quartz clasts. A local, north-plunging metamorphic culmination (garnet zone) occurs below the highway, in the Beaver River valley and eastern slopes of the Selkirk Mountains (Fig. 2-6).

Purcell Trench

The valley of the north-flowing Beaver River forms the north end of the Purcell Trench. The Purcell Trench separates the Selkirk and Purcell Mountains. It is a narrow, south-trending glaciated valley, now occupied by low-gradient streams that alternatively flow to the north and the south, and are separated by very low divides. The Purcell Trench is generally par-

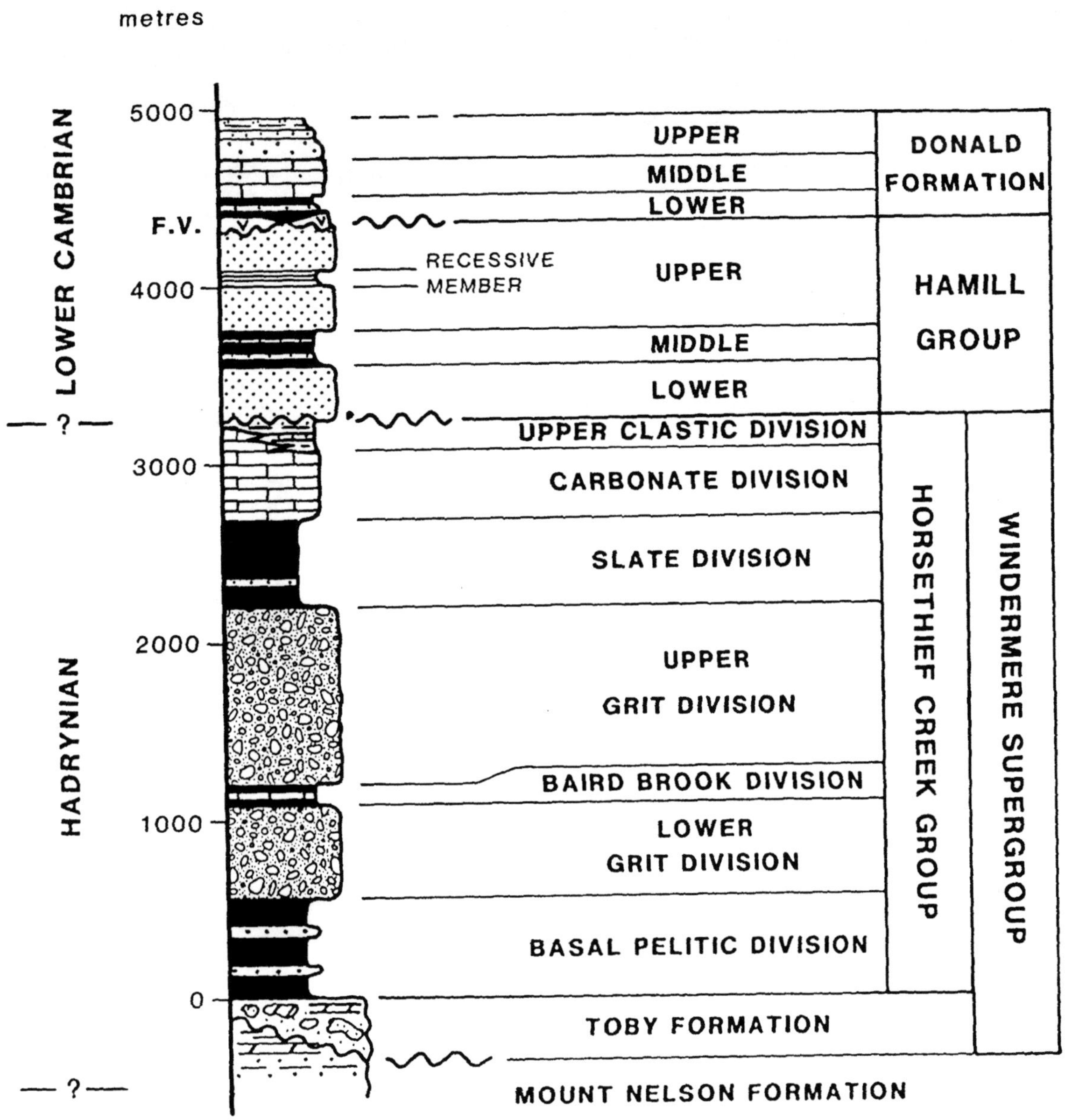

Figure 2-8. Composite stratigraphic section of Neoproterozoic and Lower Cambrian strata of the northern Purcell Mountains. F.V. = Fish Lake volcanics (Kubli and Simony, 1994).

allel with the regional structure, and locally coincident with faults. However, in southeastern B.C., it cuts across major structural features, without producing significant strike separation. Accordingly, is not the locus of a major strike-slip fault.

Structures of the Beaver River Valley

The Beaver River fault, a north-trending, west-side-down normal fault which extends southward beyond the headwaters of the Beaver River, dies out northward in the Beaver River valley at about this latitude. It offsets older, northeast-verging thrust faults that extend under the Selkirk fan structure, and that have juxtaposed northeast-verging structures in the east flank of the Selkirk fan structure over steeply dipping to overturned structures of the west flank of the Dogtooth fan structure (Figs. 2-2, 2-4, 2-6).

The zone between the northeast-verging structures of the eastern Selkirk Mountains and the backfolded, steeply northeast-dipping structures of the southwestern Dogtooth Mountains coincides with conspicuous changes in the stratigraphy of the Lower Cambrian and uppermost Neoproterozoic strata. These changes are indicative of Early Cambrian-latest

Neoproterozoic, syndepositional, down-to-the-southwest, normal faulting (Kubli and Simony, 1992; Colpron et al., 1998). In the northeastern Selkirk Mountains, the Hamill Quartzite is 2000 m thick, and the unconformably underlying siliciclastic unit, which is homotaxial with the upper clastic division of the Horsethief Creek Group in the Dogtooth Mountains (Fig. 2-8), is more than 1000 m thick. However, in the southwestern Dogtooth Mountains, the location of the "Dogtooth High" (Kubli and Simony, 1992), the Hamill Quartzite is very thin and locally absent, and the mafic Fish Lake volcanics of the lower Donald Formation overstep the sub-Hamill unconformity, and lie unconformably on the 300 m thick upper clastic unit of the Horsethief Creek Group (Poulton and Simony, 1980).

65.6 km Long roadcut in the Horsethief Creek Group.

66.7 km Bridge over Beaver River.

69.2 km Bottom of steep grade with roadcuts through Horsethief Creek grits and slates with steep southwest-dipping bedding and cleavage. These rocks are underlain by the northeast-verging Esplanade thrust fault (Fig. 2-3).

72.3 km Roadcuts in upper part of Horsethief Creek Group on both sides of the highway.

Stop 2-3 Schist of the upper part of the Horsethief Creek Group

73.7 km Examine exposures at the entrance to the first, easternmost, avalanche shed. Garnet-biotite-muscovite schists of the upper part of the Horsethief Creek Group are exposed in the roadcut near the avalanche shed. They lie within a southwest-plunging metamorphic culmination (Figs. 2-6 and 2-9) that expands northwestward to encompass the high-grade metamorphic rocks of the northern Selkirk Mountains. The large porphyroblasts of biotite are randomly oriented, and clearly not aligned with the schistosity. The isogradic surfaces, defined by the first appearances of biotite and garnet, are steeply dipping. They are sub-parallel with the dominant foliation but cut across the stratigraphic layering which is relatively flat.

77.5 km Good view of Mount MacDonald to the south. Bedding in resistant quartzites of the Lower Cambrian Hamill Group outlines the trough of a broad syncline. The base of the cliffs marks the top of the Horsethief Creek Group.

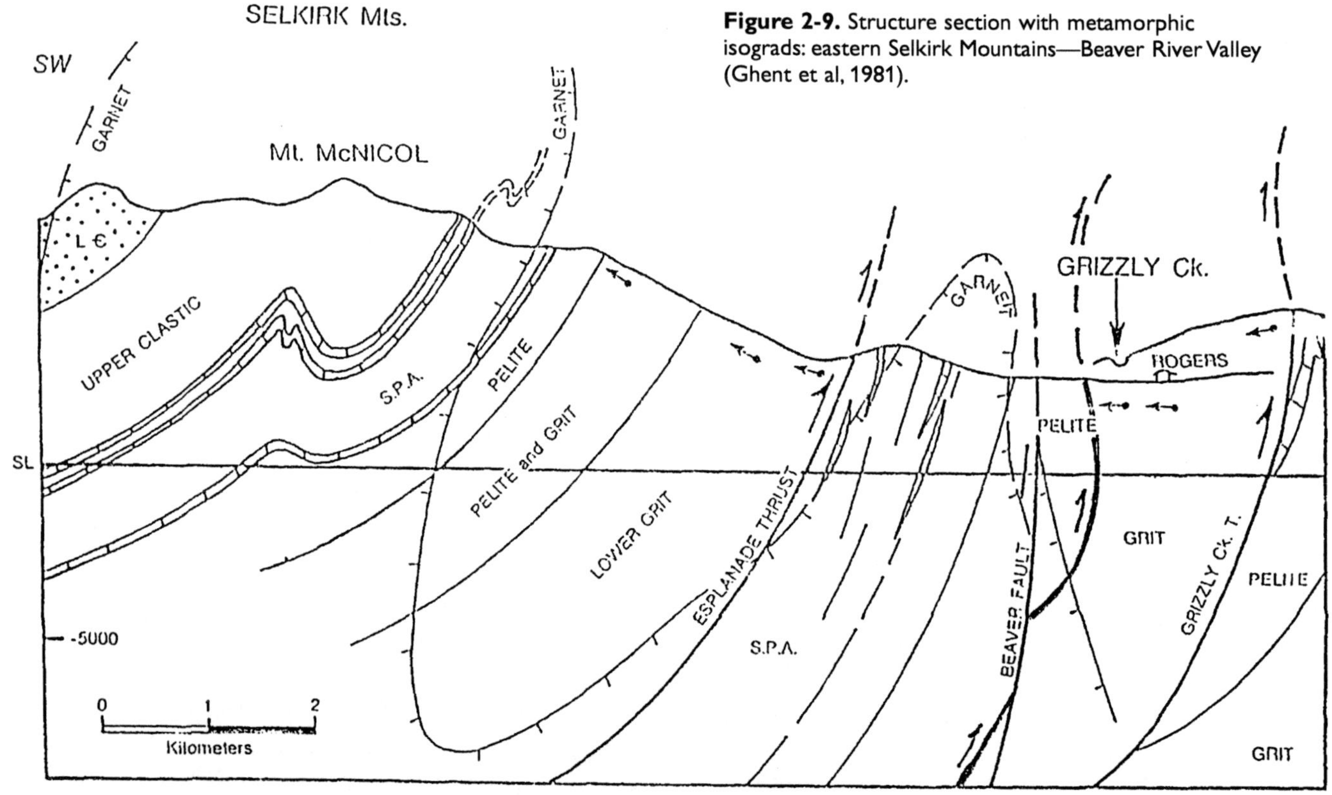

Figure 2-9. Structure section with metamorphic isograds: eastern Selkirk Mountains—Beaver River Valley (Ghent et al, 1981).

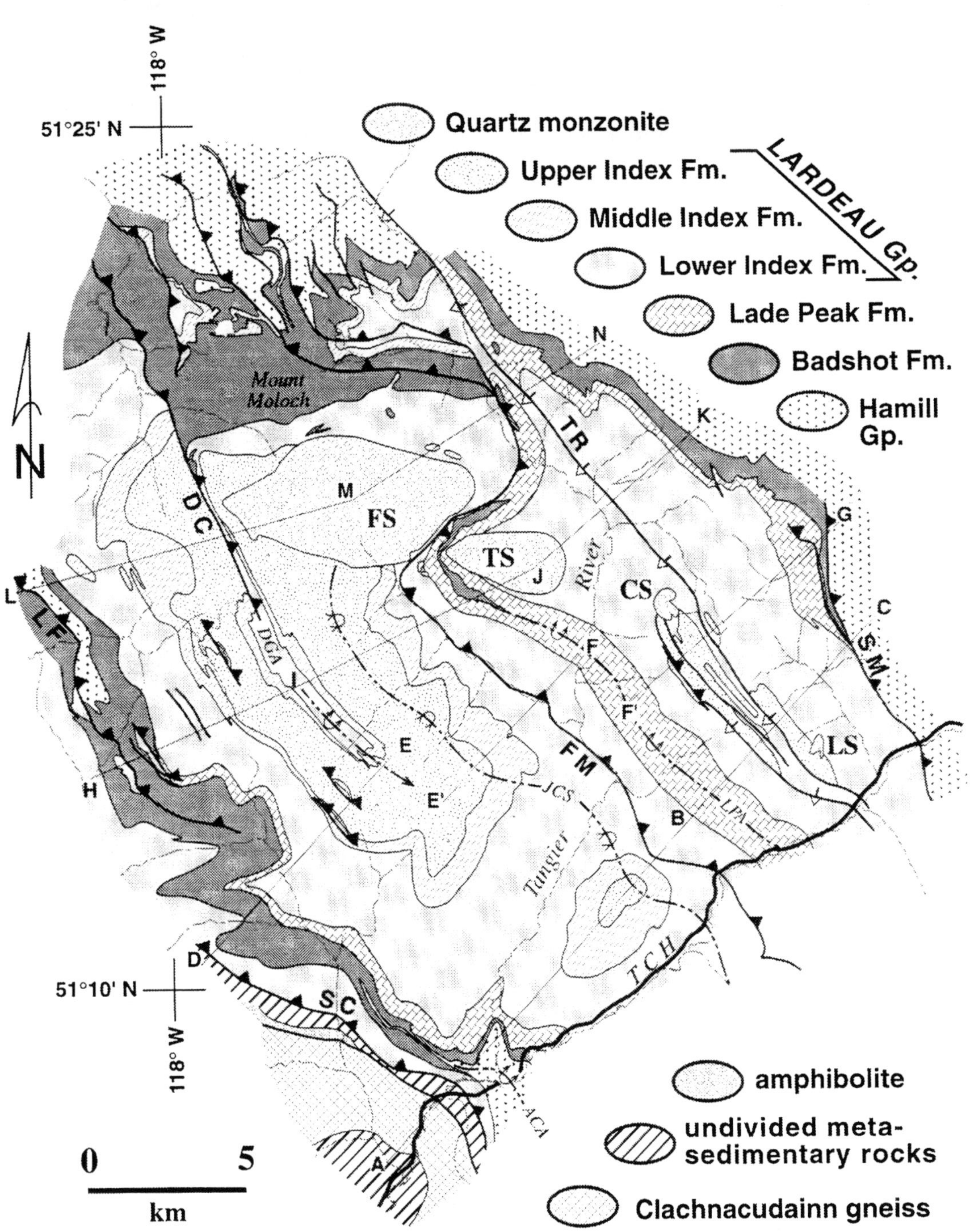

Figure 2-10. Geological map of the Illecillewaet synclinorium (Colpron et al., 1996). Plutons: FS = Fang stock, TS = Tangier stock, CS = Corbin stock, LS = Lanark stock. Folds: ACA = Albert Canyon anticline, DGA = Durand Glacier anticline, JCS = Jumping Creek syncline, LPA = Lanark Peak anticline. Faults: SC = Standfast Creek, LF = LaForme, DC = Downie Creek, FM = Fortitude Mountain, TR = Tangier River, SM = Slick Mountain. Black teeth on faults indicate northeast-dipping, southwest-verging thrust faults; white teeth on faults (with tick mark on lower plate) indicate overturned, northeast-dipping, but formerly northeast-verging, thrust faults. TCH = Trans-Canada Highway.

The Connaught Tunnel takes the Canadian Pacific Railway line beneath Mount MacDonald. Vestiges of abandoned older railway bridges and trestles that have survived the ravages of the avalanches can be seen locally through this part of the Rogers Pass.

81.5 km Summit of Rogers Pass at an elevation of 1330 m. The large, open, upright folds outlined by the cliff-forming quartzites of the Hamill Group lie just west of the axis of the Selkirk fan structure (Figs. 2-3, 2-4).

Overnight at Best Western Glacier Park Lodge.

83.6 km Roadcuts in Horsethief Creek Group with steep cleavage.

84.0 km Junction with road to Illecillewaet Campground.

85.6 km Approximate trace of the northeast-dipping Cheops thrust along which the Horsethief Creek Group in the core of the fan structure has over-ridden the Hamill Group in the southwest flank (Figs. 2-3, 2-4).

86.8 km Loop Creek, junction with road to Loop Creek Campground.

Stop 2-4 View of Ventigo syncline

87.1 km Stop at avalanche-control gun position, west of Loop Creek. View to the northwest into the core of the Ventigo syncline, a southwest-overfold of Hamill quartzite with Lower Cambrian Badshot marble (white) and Index graphitic phyllite (black) preserved in the core. The overturned east limb of the fold is partly cut out by a southwest-verging thrust fault.

Cross-bedded Hamill quartzite exposed in the roadcut on the south side of the highway has been deformed by penetrative flow and recrystallization; but this is much less intense than in equivalent strata to the southwest in the opposite limb of the Illecillewaet synclinorium (Figs. 2-3, 2-4).

87.6 km Bridge over Illecillewaet River.

93.3 km Hamill quartzite in landslide debris, approximately in the core of a southwest-verging anticline. To the north (see Figs. 2-3, 2-4), the Argentine fault, an overturned, northeast-dipping, but formerly southwest-dipping and northeast-verging thrust fault, dies out in the core of this anticline.

95.5 km Approximate trace of the northeast-dipping, southwest-verging Slick Mountain fault (Figs. 2-3, 2-4, 2-10, 2-11), along which Hamill quartzites have been thrust over black phyllites of the (Early Cambrian or younger) Lardeau Group.

96.0 km Roadcut in black phyllites of the Index Formation (Lardeau Group, Early Cambrian and(?) younger). Bedding and schistosity are steeply northeast-dipping. The carbonaceous (graphitic) pelitic rocks of the Lardeau Group evidently represent a deeper-water, anoxic, more condensed facies-equivalent of the thick sequence of light coloured lower Paleozoic carbonate rocks and shales that occurs in the Rocky Mountains and the Dogtooth Mountains.

97.4 km Bostock Creek.

98.6 km Long roadcut in black phyllites of the Lardeau Group.

100.2 km Bridge over Illecillewaet River.

101.2 km Bridge over Illecillewaet River; approximate trace of axial surface of Lanark Peak anticline (LPA in Fig. 2-10).

103.1 km Entrance to Twin Avalanche shed; approximate trace of Fortitude Mountain fault.

103.3 km Exit from Twin Avalanche shed. Long roadcut in calcareous black phyllite of the Lardeau Group in overturned southwest limb of Jumping Creek syncline (JCS in Fig. 2-10).

103.9 km Entrance to MacDonald avalanche shed.

Stop 2-5 Lardeau phyllite

104.8–105.8 km Long roadcut in black graphitic phyllite and phyllitic siltstone of the Lardeau Group (Index Formation, Early Cambrian and(?) younger). These rocks are in the overturned southwest limb of the Jumping Creek syncline, which is overlain by the northeast-dipping Fortitude Mountain fault (Figs. 2-3, 2-4, 2-10), a southwest-verging thrust fault. The Index Formation is underlain by black, graphitic, phyllitic limestone and calcareous phyllite of the Lade Peak Formation, which abruptly but gradationally overlies light coloured, fossiliferous (archeocyathid-bearing) Lower Cambrian shallow-water limestones and dolomites of the Badshot Formation. The carbonate-platform deposits of the Badshot Formation and the underlying shallow-water meta-quartz arenites of the upper part of the Hamill Group are important structural markers over a very large area in the Selkirk and Purcell Mountains. They overlap rift structures, outlined by facies and thickness variations in the underlying middle and lower parts of the Hamill Group and syn-rift volcanic rocks in the middle part of the Hamill Group. They have been interpreted as "the overlap assemblage" marking the "rift-drift transition" and the initiation of the thermally driven "passive" subsidence of the Cordilleran miogeocline (Devlin and Bond, 1988; Warren, 1997; Colpron et al., 1998). The base of the Lardeau Group marks the onset of the drowning and deep subsidence of the archeocyathid carbonate platform, and of the anoxic sedimentation that characterizes the outboard, deeper water facies of the Cordilleran miogeocline.

108.0–109.0 km Long roadcut in black phyllite and calcareous phyllite of the Lardeau Group (Index Formation, Early Cambrian and? younger?). The metamorphic grade increases from chlorite to biotite grade.

111.0 km Beginning of steep grade and long roadcut in lower part of the black phyllites of the Lardeau Group (Index Formation, Early Cambrian and(?) younger).

113.2 km Badshot Limestone (Lower Cambrian) is exposed to the north of the highway.

113.6 km Hamill quartzite is exposed in the core of the Albert Canyon anticline, a tightly appressed

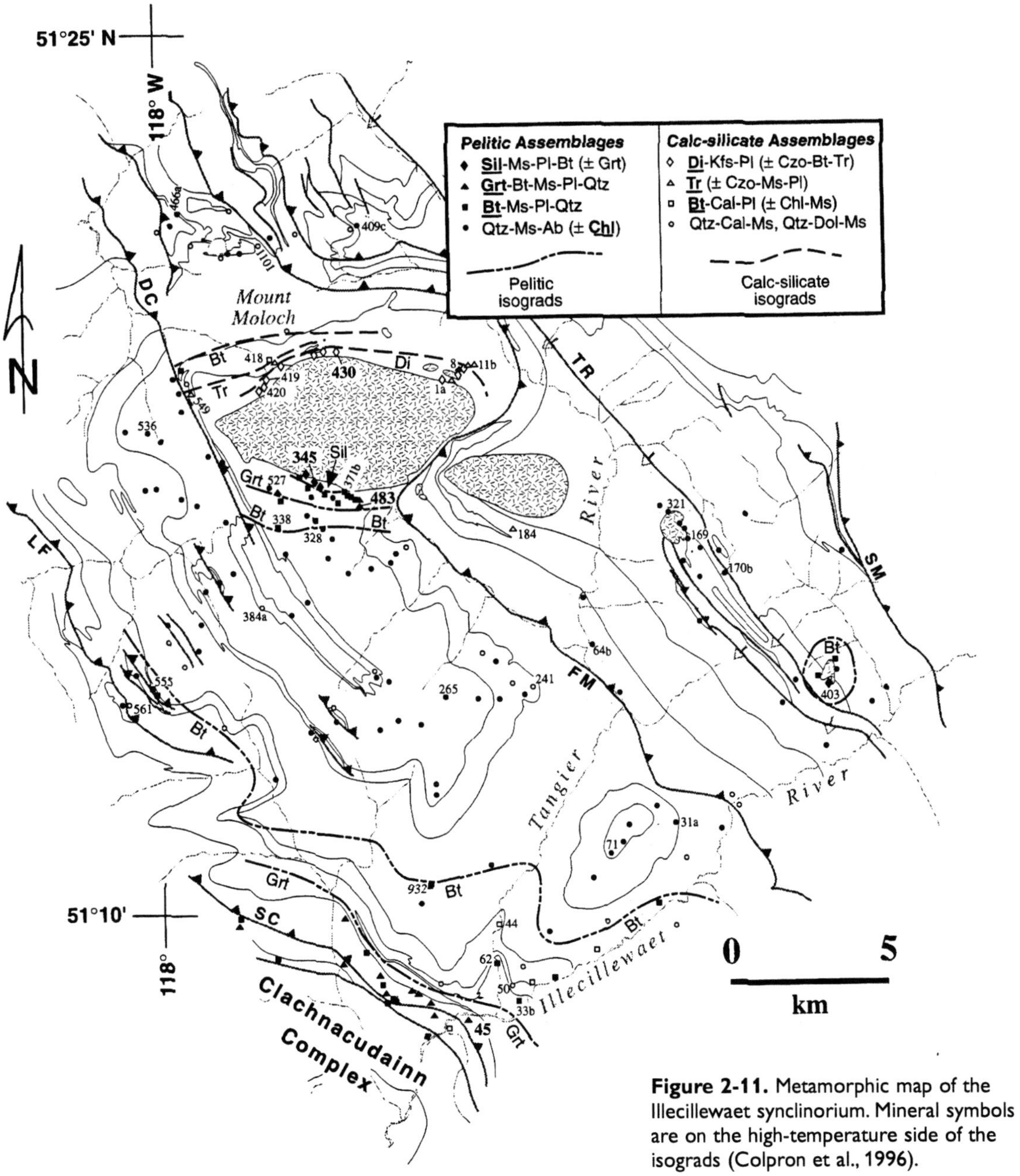

Figure 2-11. Metamorphic map of the Illecillewaet synclinorium. Mineral symbols are on the high-temperature side of the isograds (Colpron et al., 1996).

southwest-verging structure that is underlain by the southwest-verging Standfast Creek fault (Figs. 2-4, 2-10, 2-12). Metamorphic grade increases from biotite to garnet grade across this fold (Fig. 2-11).

114.5 km Bridge over the Illecillewaet River, close to the mouth of the Tangier River.

115.1 km Junction with road to Canyon Hot Springs.

115.3 km Lardeau Group biotite-muscovite schist in the overturned southwest-limb of the Albert Canyon anticline.

115.9 km Quartz-muscovite-biotite schist, locally with garnet and/or pseudomorphs of staurolite, in part phyllonitic. These rocks are close to the Standfast Creek fault, and probably belong to the Lardeau Group.

116.3 km Bridge over the Illecillewaet River.

Stop 2-6 Clachnacudainn complex and Standfast Creek fault

117.3 km The Clachnacudainn complex is a domal structural and metamorphic culmination, in which highly deformed, kyanite-grade and sillimanite-

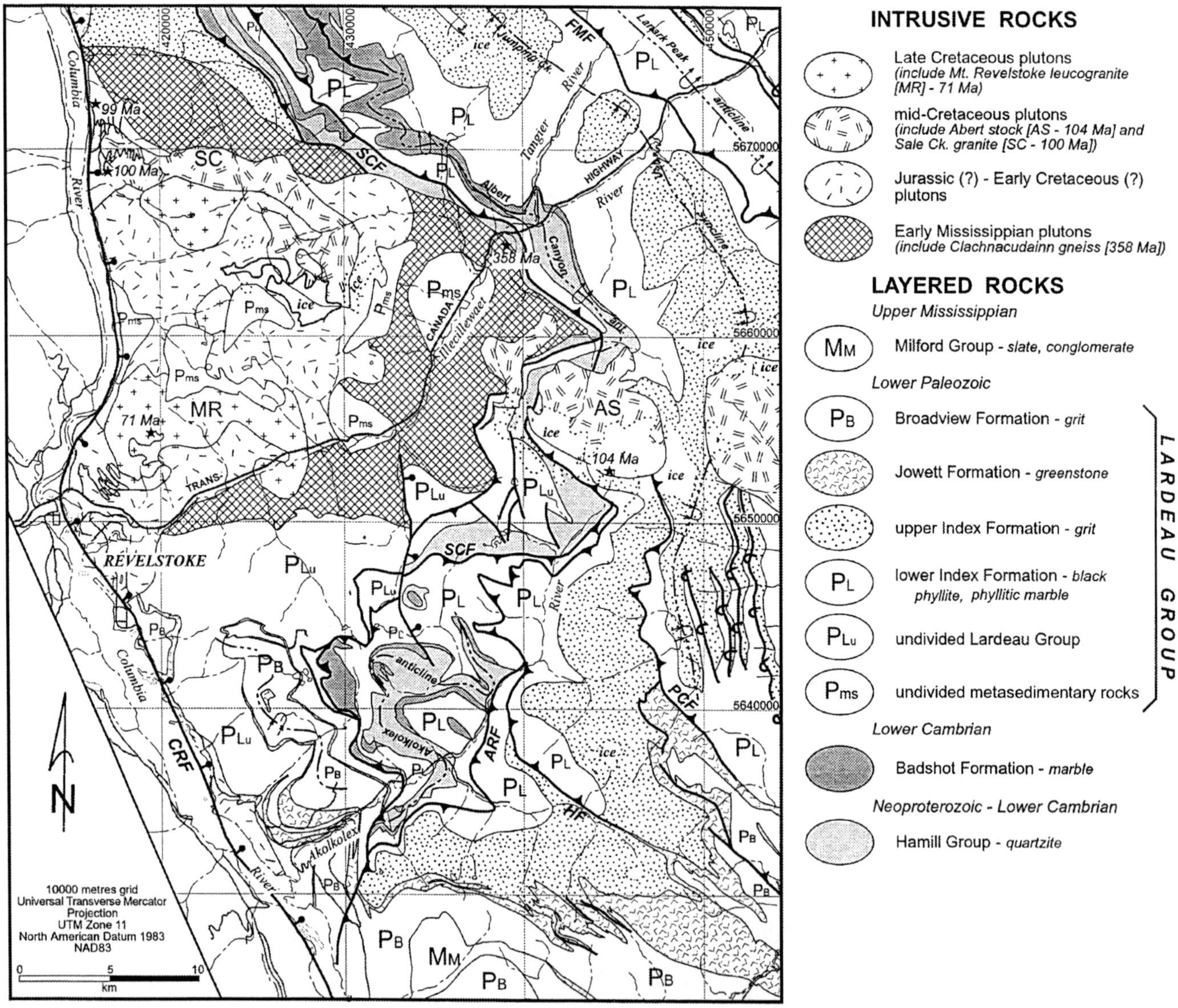

Figure 2-12. Geological map of the Clachnacudainn complex (Colpron et al., 1999). AS = Albert stock, MR = Mount Revelstoke leucogranite, SC = Sale Creek granite. Faults: ARF = Akolkolex River fault, CRF = Columbia River fault, FMF = Fortitude Mountain fault, = HF, Holyk fault, PCF = Pulley Creek fault, SCF = Standfast Creek fault.

grade metasedimentary rocks occur in association with a semi-concordant sheet of deformed granodiorite, and with discordant, cross-cutting Middle Jurassic (?), mid-Cretaceous, and Late Cretaceous granitic plutons (Figs. 2-3, 2-4, 2-12). Because of the higher metamorphic grade and the uncertainties about the correlation of the metasedimentary rocks, the Clachnacudainn complex has been described in the past as a local salient of the Shuswap metamorphic complex (Wheeler, 1965; Ross, 1968). However, it is separated from the core of the Shuswap metamorphic complex by the Columbia River fault (Read and Brown, 1981; Lane 1984), a major northeast-dipping Eocene normal fault that follows the valley of the Columbia River north and south of Revelstoke (Figs. 2-3, 2-12).

The rocks in the footwall of the Columbia River fault (Fig. 2-16) which were exhumed from much deeper crustal levels by Eocene crustal extension, include the parautochthonous Paleoproterozoic basement of the Cordilleran foreland thrust and fold belt (Monashee Complex basement) and unconformably overlying parautochthonous supracrustal cover rocks (Monashee cover assemblage). These rocks are exposed in tectonic window through a thick mylonite zone (the Monashee décollement), which is the exhumed basal décollement of the Cordilleran foreland thrust and fold belt (Brown et al., 1992). The metasedimentary rocks and Devono-Mississippian granodiorite of the Clachnacudainn complex may be the lateral equivalents of rocks assigned to Kootenay terrane (Eagle Bay assemblage, Tsalkom and Sicamous Formations, and Silver Creek assemblage) that occur on the west side of the Shuswap metamorphic complex (Figs. 7; 2-16).

Devono-Mississippian Granodiorite Gneiss

Gneiss exposed along the north side of the highway at this locality is part of a broadly folded sheet about 0.5 km thick that is underlain and overlain by metasedimentary rocks of uncertain stratigraphic correlation (Figs. 2-3, 2-4, 2-12). The granodiorite has been penetratively deformed along with the enclosing metasedimentary rocks and with some of the numerous pegmatitic veins that cut both the gneiss and the metasediments (Gilman, 1972). The sheet of granodiorite outlines a broad dome (Lauretta dome) in which a window about 5 km wide exposes the underlying metasedimentary rocks along the highway to the west to the southwest of this locality. A U-Pb zircon date of 356 ± 6 Ma (Parrish, 1992) from a nearby quarry (see "356 Ma" in Fig. 2-12) shows that the granodiorite is of about the same age as the Devono-Mississippian plutons that intrude the Kootenay terrane on the west side of the Shuswap metamorphic complex. At this locality, which is close to the northeast-dipping, southwest-verging Standfast Creek fault, the granodiorite is unusual in that it is mylonitic, highly fractured, cut by veins of pseudotachylite, and extensively chloritized.

Standfast Creek Fault

The northeast-dipping Standfast Creek fault has been interpreted as major southwest-verging thrust (Sears, 1979; Price et al., 1981, 1985; Price, 1986; Colpron et al., 1998), as a southwest-verging thrust on which there has been later northeast-verging normal displacement (Parrish et al., 1988), and as a very minor southwest-verging thrust (Crowley and Brown, 1994). South of the Trans-Canada Highway, the fault branches to form several separate faults (Fig. 2-12). The upper branch (the Pulley Creek fault) and the main branch (the Standfast Creek fault) are both cut by the Albert stock, which has been dated (U-Pb zircon) at 104 Ma (Crowley and Brown, 1994). The lower branch (unnamed), which was interpreted as an Eocene normal fault by Parrish et al. (1988) is now known to be cut by a satellite of the Albert stock, and therefore is not an Eocene normal fault. The Akolkolex River fault, which branches from the Standfast Creek fault just south of the Albert stock, separates two structural levels with contrasting styles of deformation (Sears, 1979): (1) an upper level with moderately to steeply dipping southwest-verging folds and thrust faults involving chlorite grade rocks of the middle and upper parts of the Lardeau Group; and (2) a lower level with recumbent isoclinal folds involving the biotite and garnet grade rocks of the Hamill Group, Badshot Limestone, and lower part of the Lardeau Group (e.g. Akolkolex anticline in Figure 2-12). The recumbent folds have been mapped as northeast-verging (Wheeler, 1965; Thompson, 1978; Sears, 1979); and they have been correlated with the older, northeast-verging thrusts and folds that were subsequently back-folded and overturned in the southwest flank of the Selkirk fan (e.g. Tangier River fault and Argentine fault in Figs. 2-3, 2-4). However, Crowley and Brown (1994) have concluded that the recumbent folds are southwest-verging.

Price (1986) has argued that the Standfast Creek fault is the upper detachment of a northeastward tapering, crustal-scale tectonic wedge that is responsible for the southwest-verging thrusting and folding of the Selkirk fan structure. A modified version of this interpretation has been incorporated in the structure section of Figure 2-4. Successive stages in the evolution of the Selkirk fan structure, based on this interpretation, are illustrated in Figure 2-13.

Evolution of the Selkirk Fan Structure

The evolution of the Selkirk fan structure comprised a sequence of three kinematically contrasting episodes of compressional deformation (Fig. 2-13).

(1) The first episode (Fig. 2-13a and b), which involved *northeast-verging thrusting and folding and the collapse of the Slide Mountain basin,* culminated in the obduction of the deeper-water, ocean-floor, supracrustal rocks of the Slide Mountain basin over the outboard part of the Cordilleran miogeocline. In southern British Columbia the timing of the obduction of Slide Mountain terrane, as constrained by biostratigraphic data and U-Pb zircon dating (Murphy et al., 1985), is mid-Toarcian (late Early Jurassic, ca. 187–185 Ma), and >181 Ma respectively.

The crustal-scale kinematics and the geodynamics of the obduction process are enigmatic, but the results appear to be straightforward. Metamorphic geobarometry and geochronometry from various

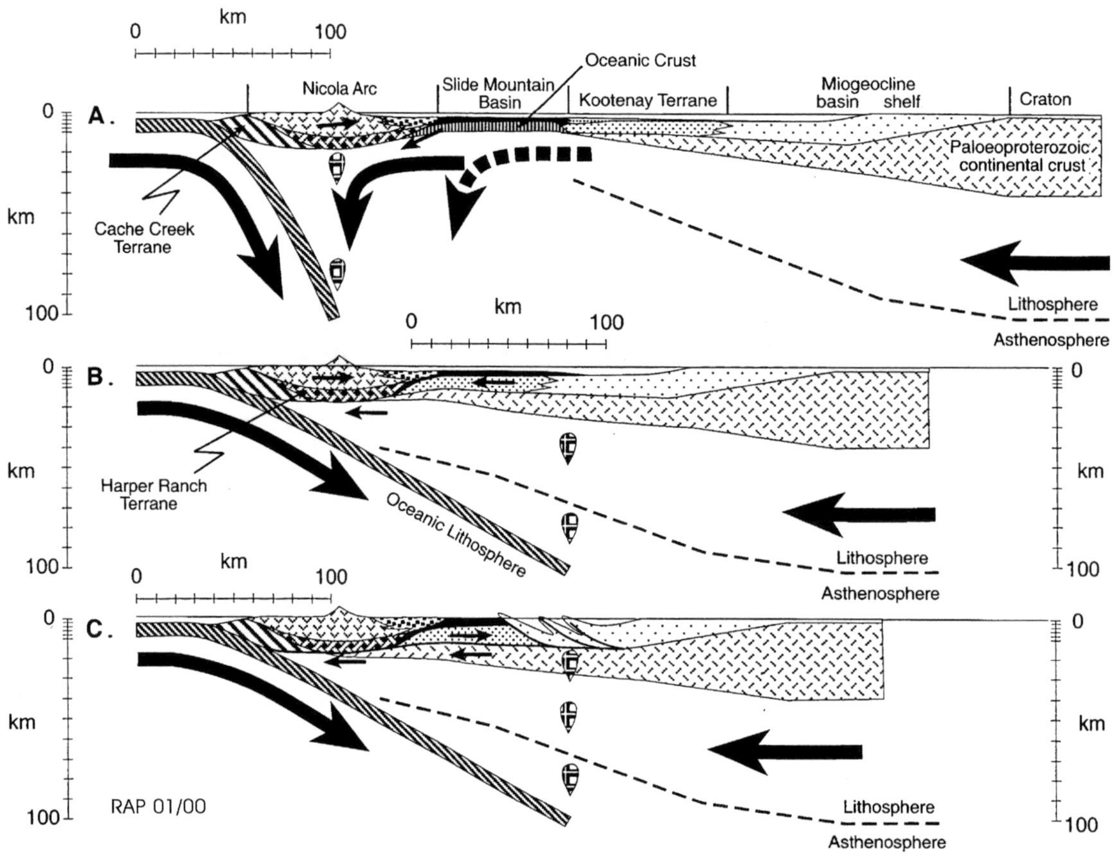

Figure 2-13. Cartoon of three stages in the Early and Middle Jurassic orogenic evolution of the western margin of North America in the southern Canadian Cordillera:
A. Early Jurassic convergence between North America and the subduction zone under the Nicola (Quesnellia) arc. Retrograde (roll-back) subduction of oceanic lithosphere of Slide Mountain basin and its entrainment in the main subduction zone resulted in the obduction of a tectonic flake of weak Slide Mountain supracrustal rocks (Slide Mountain terrane) over Kootenay terrane and the outboard part of the Cordilleran miogeocline.
B. late Early Jurassic tectonic wedging of distal North American lithosphere initiated the delamination of the Nicola arc from its lower crustal and upper mantle roots, which were entrained in the subduction zone; concurrently, the subduction zone flattened and the arc magmatism migrated northeastward into Kootenay terrane and the miogeoclinal rocks of the North American continental margin.
C. early Middle Jurassic detachment of Kootenay terrane from underlying North American lithosphere initiated tectonic wedging of Kootenay terrane between the miogeocline and the North American basement, and the delamination of the Quesnellia magmatic arc from its lower crustal and upper mantle roots; it also initiated the back thrusting and back folding of the outboard part of the miogeocline; as the Quesnellia tectonic flake continued to grow and to over-ride North American lithosphere, arc magmatism from a flattened subduction zone penetrated the evolving back thrusts and back folds as they were forming and being uplifted by advancing tectonic wedge.

localities along this zone in southeastern British Columbia (Archibald et al., 1983; Colpron et al., 1996, 1998) indicate that Lower Paleozoic shallow-water deposits of the outboard part of the Cordilleran miogeocline were transported to depths of about 20 km during the northeast-verging obduction of Slide Mountain terrane. The lower crustal and mantle lithosphere of the Slide Mountain basin were delaminated from the overlying supracrustal rocks of Slide Mountain terrane, and have disappeared into the deeper mantle, presumably by becoming entrained in the downward flow of the northeast-dipping subduction zone. The total amount of horizontal overlap between Slide Mountain terrane and the under-riding North American rocks is poorly constrained; but in the Cariboo Mountains (~54°N), where the Slide Mountain terrane has not been completely stripped away by subsequent erosion, the remaining overlap is about 70 km (Wheeler and McFeely, 1991). Northeast-verging thrust and fold structures formed in the outboard part of the Cordilleran miogeocline as it was displaced under Slide Mountain terrane.

(2) The obduction of Slide Mountain terrane was followed by a second kinematically distinct episode of deformation (Figs. 2-13c, 2-14C and 2-14D). At upper crustal levels it was expressed as a *reversal in tectonic vergence* that involved the development of southwest-verging thrusts, folds, and fold nappes in the outboard part of the Cordilleran miogeocline, and tectonic thickening, uplift, and exhumation of these rocks as they were being deformed (Colpron et al., 1998). The older northeast-verging structures were deformed by the younger southwest-verging structures; but in central and northern B.C. some of the southwest-verging thrusting was confined, by the basal detachment of Slide Mountain terrane, to the underlying rocks (Struik, 1988; Harms, 1986). These southwest-verging thrust displacements evidently occurred while Slide Mountain terrane was still overthrusting the miogeocline.

Two disparate hypotheses have been advanced to account for the southwest-verging structures (Fig. 2-15). One hypothesis (Fig. 2-15A) contends that the southwest-verging thrusts, folds and fold nappes are the near-surface expression of northeastward subduction of oceanic lithosphere beneath the Cordilleran miogeocline (Brown et al., 1986; Brown et al., 1993). Because these southwest-verging structures occur behind the Quesnellia magmatic arc (Fig. 2-13c), this hypothesis implies the existence of another subduction zone in addition to the one that dipped under the Quesnellia arc. The other hypothesis (Fig. 2-14B) maintains that the southwest-verging structures formed above an advancing tectonic wedge that was made up of Quesnellia and the pericratonic rocks of Kootenay terrane, and that was bounded below by a northeast-verging thrust zone that evolved into the basal detachment of the foreland thrust and fold belt, and above by a zone of southwest-verging back thrusting and folding (Price, 1986; Colpron et al., 1998). According to this hypothesis, the southwest-verging deformation was restricted to the upper side of the wedge, and it occurred as the supracrustal rocks of the outboard part miogeocline were being tectonically delaminated from their underlying basement, and uplifted and exhumed by the advance of the wedge. Thermobarometric and thermochronometric data from lower Paleozoic metasedimentary rocks and Middle Jurassic (168 Ma) granitic plutons in the Illecillewaet synclinorium, a broad regional southwest-verging structure in the southwest flank of the Selkirk fan north of the Trans-Canada Highway (Figs. 2-3, 2-4, 2-11) indicate that southwest-verging folding and thrusting were accompanied by rapid exhumation of these rocks from about 20–25 km to <10 km (Colpron et al., 1996). The available data indicate that all significant compressional deformation and most of the exhumation of the supracrustal rocks along the western flank of the Selkirk fan structure had occurred by late Middle Jurassic time (Colpron et al., 1998).

(3) The third kinematically distinct pattern of compressional deformation in the development of the Selkirk fan structure involved *northeast-verging thrusting and folding* along the northeast side of the fan structure (Fig. 2-14B and A). It did not begin until the tectonic wedging, and the accompanying back thrusting and back folding had generated sufficient crustal thickening and topographic (gravitational) potential to drive the initiation and propagation of the critical-taper thrust-and-fold wedge that became the Canadian Rockies. The crustal thickening occurred while the miogeocline was still situated outboard of the basement ramp marking the edge of the continental craton; and it was accomplished by horizontal compression and tectonic thickening of the miogeoclinal supracrustal above basal décollement, but not of the underlying crystalline basement, which lay beneath the basal décollement.

In the vicinity of the Trans-Canada Highway transect, the northeastward propagation of the tectonic wedge was temporarily arrested at a barrier (Fig. 2-14D and C) formed by the Early Cambrian southwest-dipping normal fault that separated the thinner section of Lower Cambrian and Windermere strata

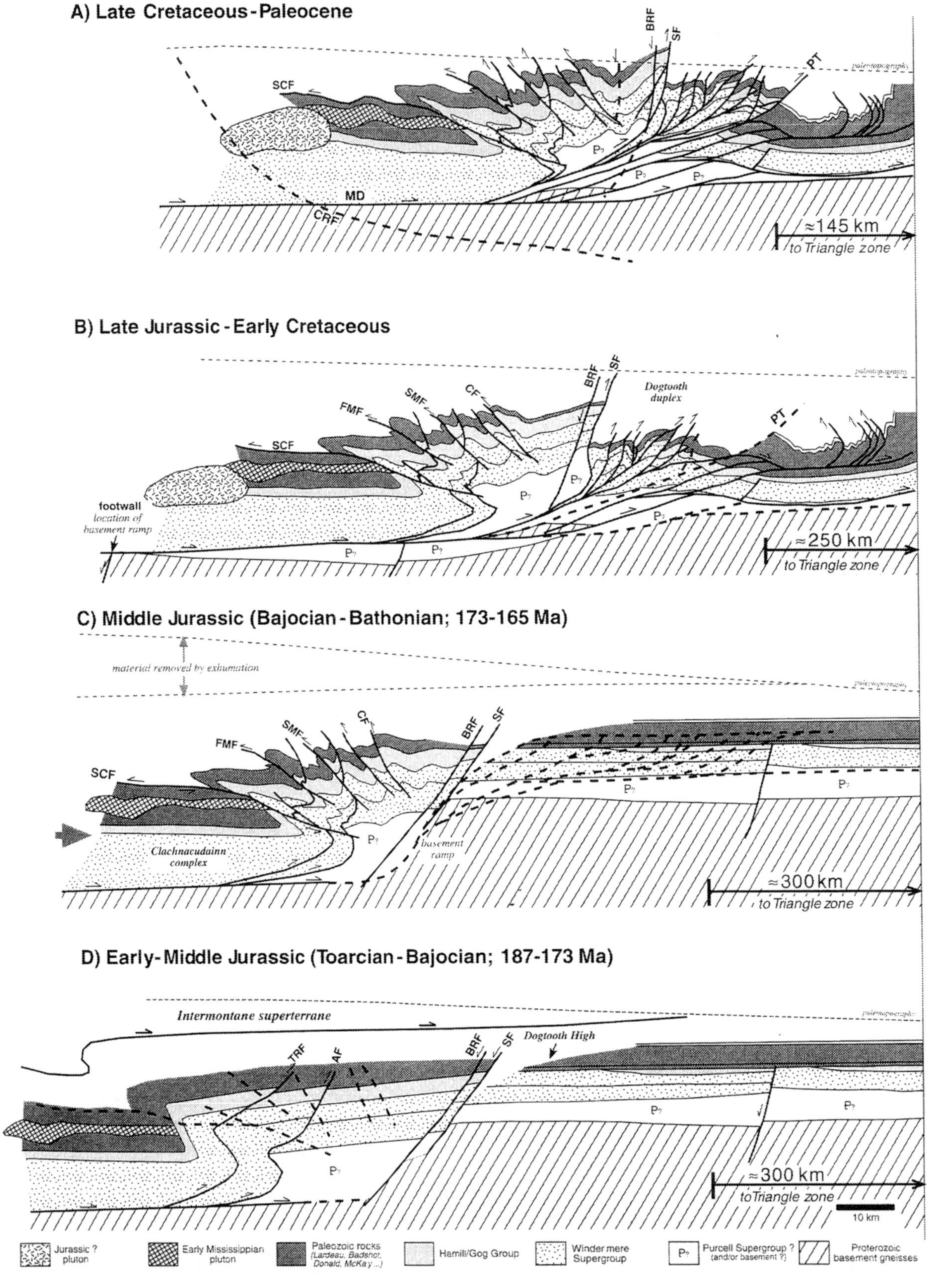

Figure 2-14. Schematic palinspastic restoration and sequential evolution of the Selkirk fan structure. Each step (A to D) represents the configuration of the Selkirk fan at the time indicated (Colpron et al., 1998).

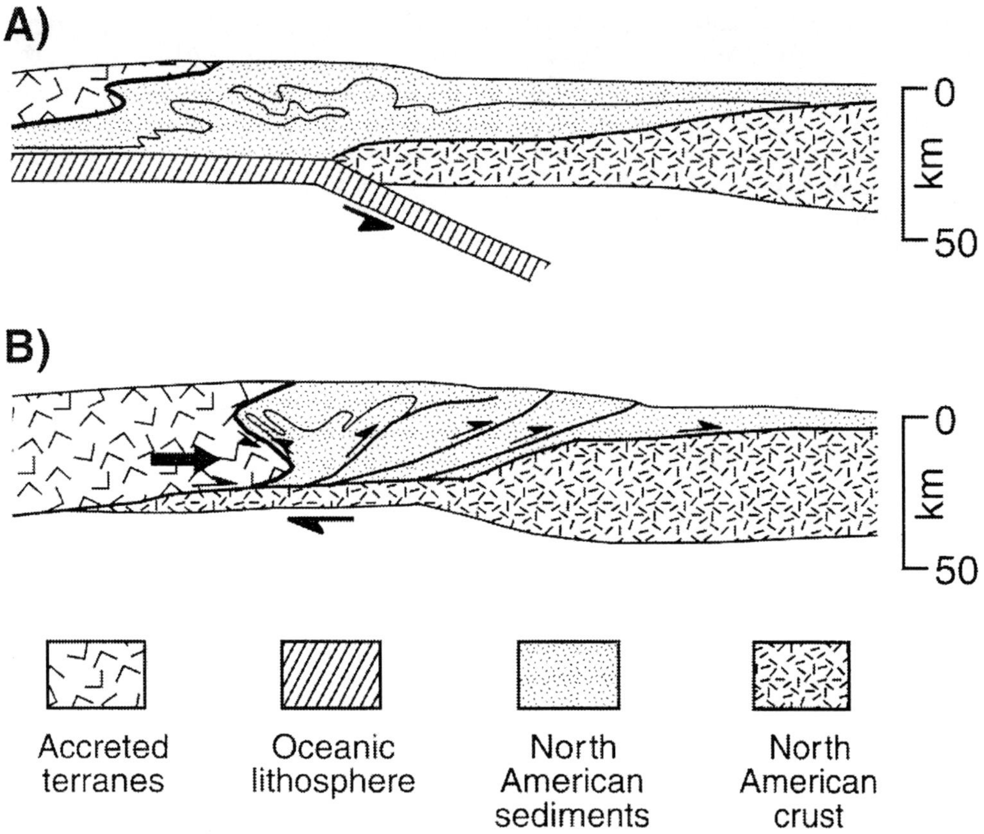

Figure 2-15. Two contrasting hypotheses to account for the southwest-verging structures in the southwest flank of the Selkirk fan (after Colpron et al., 1998): A. Northeastward subduction of oceanic lithosphere under the North American plate (Brown et al., 1993);
B. Northeastward tectonic wedging within the North American plate (Price, 1986).

in the "Dogtooth High" from the much thicker section that occurs in the Selkirk Mountain near Rogers Pass (Colpron, 1996; Colpron et al., 1998). However, with ongoing convergence between Intermontane superterrane and North America, there was continuing tectonic thickening of the supracrustal rocks outboard of the barrier. Eventually there was sufficient increase in gravitational potential to overcome the barrier, and to initiate the development of the foreland thrust and fold belt, first with northeast-verging thrusting and folding in the eastern Selkirk Mountains, and then with synchronous development of the northeast-verging Dogtooth thrust duplex and of the Porcupine Creek anticlinorium. Subsequently, the supracrustal rocks of the remainder of the Cordilleran miogeocline were scraped off the underriding North American basement and accreted to the over-riding Intermontane superterrane and the deformed miogeoclinal rocks that had already become attached to it and consolidated with it.

The crustal thickening associated with this phase of horizontal compression of the miogeoclinal supracrustal rocks caused the initial flexural subsidence of the foreland basin, and it produced the initial influx of Cordillera-derived sediment that was trapped in the foreland basin in the Late Jurassic (~155 Ma). The off-scraping and the accretion to Intermontane superterrane of the Neoproterozoic-Lower Cambrian succession of the Dogtooth Mountains, and the Mesoproterozoic rocks of the northern Purcell Mountains can be dated as pre-mid-Cretaceous. In the Purcell Mountains, south of Golden and Lake Louise, several transverse, northeast-striking faults that branch from the Purcell thrust (the Mt. Forster, Hall Lake and St. Mary faults in Fig. 1) are cut by granitic plutons that are of mid-Cretaceous age.

There is a substantial northwest-southeast overlap on the transverse northeast-trending segments of these pre-mid Cretaceous, foreland-verging thrust faults, and this implies a significant northwest-southeast component of displacement was associated with the Late Jurassic to mid-Cretaceous convergence between Intermontane superterrane and North America. On the basis of these and other considerations, the late Jurassic-Early Cretaceous thrusting and folding in the eastern Selkirk Mountains, Purcell Mountains, and western Rocky Mountains has been interpreted as an episode of *left-lateral transpression* (Price, 1981, 1994; Price and Sears, 2000). This left-lateral transpression contrasts with the *right-lateral transpression* associated with Late Cretaceous and Paleocene convergence within the Cordillera, caused by the interaction of the Kula oceanic plate and the North American craton (Price, 1994, Price and Sears, 2000).

118.3 km Roadcut in granodioritic gneiss that is laced with veins of pegmatite.

119.5 km Bridge over Woolsey Creek, approximate contact of the sheet of granodioritic gneiss and the underlying metasedimentary rocks in the Woolsey Creek window (Fig. 2-12).

123.0 km Metasedimentary rocks in the Woolsey Creek window, laced with veins of pegmatite.

125.2 km Granodioritic gneiss on southwest side of the Woolsey Creek window.

127.5 km Granodioritic gneiss.

129.9 km Metasedimentary rocks engulfed in granite. View to the south of marble bands beneath the Standfast Creek fault at the head of Twin Creek.

132.0 km National Park gate.

150.3 km Town of Revelstoke, junction with Highway 23 (north to Revelstoke and Mica Creek dams).

The east-dipping Columbia River fault follows the valley of the Columbia River, which separates the Selkirk Mountains from the Monashee Mountains to the west.

The Columbia River flows southward from here into the State of Washington. A series of dams were built on the Columbia River and its major tributary the Kootenay River to control the supply of water for use in the northwestern United States, and to generate electricity in British Columbia. The reservoir behind the Revelstoke dam, which is located just north of Revelstoke, stretches northward to the Mica Creek dam, which is at the north end of the Selkirk Mountains. Lake Kinbasket, the reservoir behind the Mica Creek dam, occupies a 200-km long segment of the Rocky Mountain trench, extending from near the Trans Canada Highway at Donald to Valemont. This was formerly drained by the northwest-flowing the Columbia River and the southeast-flowing Canoe River respectively.

150.6 km Junction with road to Revelstoke business district.

View to the southwest of the quartzites and paragneisses of the Monashee Complex "cover sequence" on Mount Begbie. The Monashee Complex has been exhumed from a depth of about 20 km below the Selkirk Mountains as a result of Early and Middle Eocene displacement on the Columbia River normal fault.

150.8 km Bridge over the south-flowing Columbia River.

151.5 km Junction with Highway 23 to Galena Bay ferry on Upper Arrow Lake) and to Highway 6 at Nakusp.

152.6 km Roadcuts in Monashee Complex cover gneisses: banded biotite-quartz-feldspar gneiss with lenses and veins of pegmatite.

162.9 km Roadcuts in Monashee Complex cover gneisses: layered quartz-feldspar-biotite-muscovite paragneiss with bands of hornblende gneiss and amphibolite, all laced with pegmatite.

Stop 2-7 Victor Lake picnic site: Monashee Complex "cover gneisses"

165.1 km This is locality ViC on Fig. 2-16; stop in parking area. The Monashee Complex (Fig. 2-17) consists of two parts. (1) A "*basement assemblage*" ("core gneisses") comprises Paleoproterozoic paragneiss, granitoid orthogneiss, and leucogranite. (2) An unconformably overlying "*cover assemblage*" ("cover gneisses") consists of quartzite, marble, pelitic and semipelitic schist, and amphibolite, and contains deformed 740 Ma syenite, 540 Ma amphibolite, 360 Ma carbonatite, and undeformed Late Paleocene (55–57 Ma) pegmatite. The "basement assemblage" contains granitic rocks that crystallized at 2077 ± 4 Ma and 1862 ± 1 Ma, and tectonite fabrics that developed before 1848 ± 3 Ma. It has been correlated (Crowley, 1999) with Paleoproterozoic basement rocks under the Western Canada Sedimentary Basin in central Alberta (Fig. 6).

The lower part of the "cover assemblage" contains 1.99 Ga detrital zircons and is intruded by 1852 ± 4 Ma pegmatite, 1762 ± 6 Ma leucogranite, and 724 ± 5 Ma syenitic gneiss; the upper part contains detrital zircons dated at ~1.21 Ga (Crowley, 1997). Combining these ages with the probable Paleozoic depositional ages for the middle and upper parts of the "cover assemblage" (Fig. 2-15) constrains the thickness of Mesoproterozoic and/or Neoproterozoic in this succession to <0.2 km, which is considerably less than that of coeval rocks in the hanging wall of the Monashee décollement. This suggests that the depo-

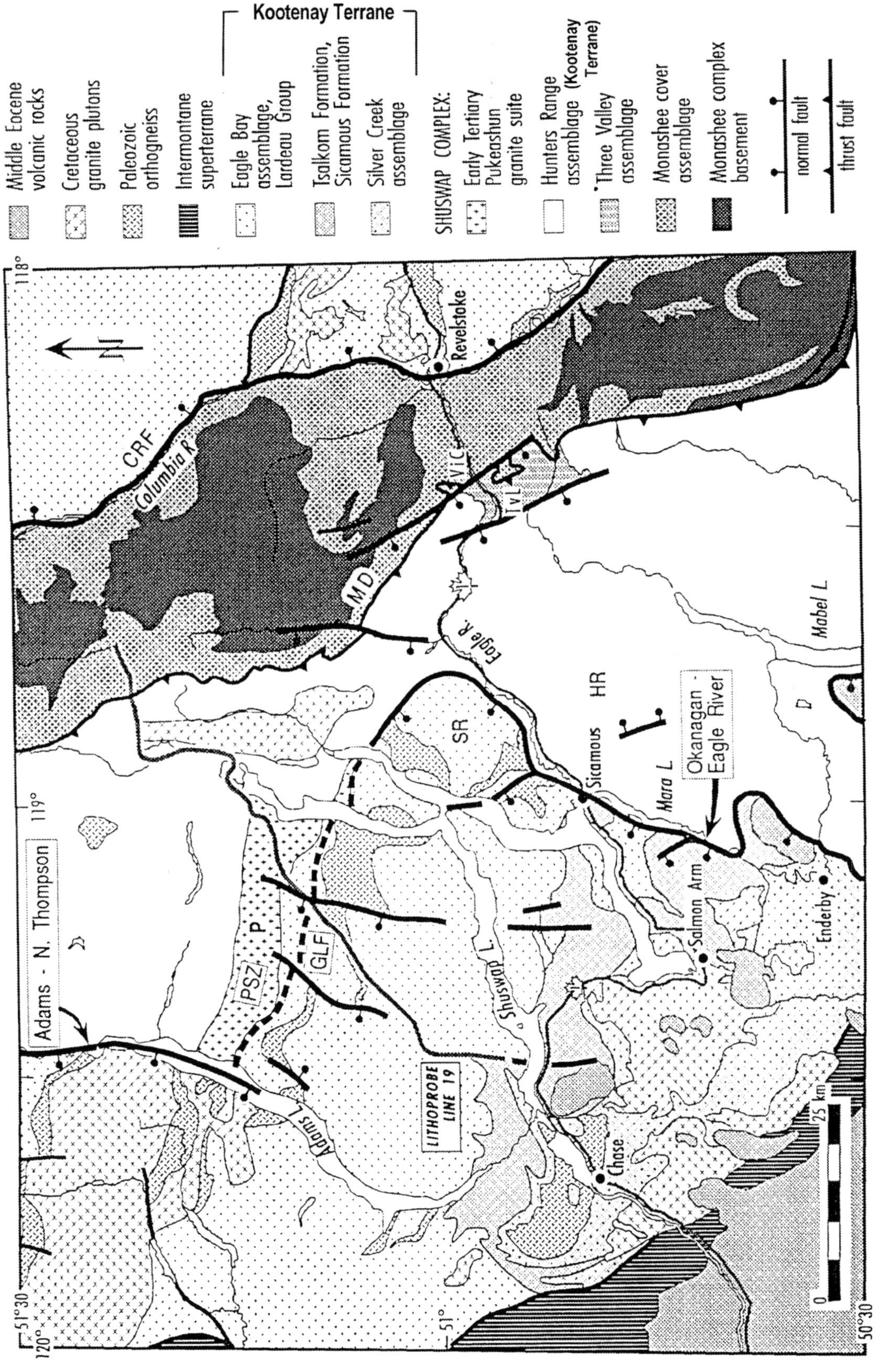

Figure 2-16. Geological map of part of the Omineca belt along the Trans-Canada Highway between Revelstoke and Chase, B.C. (from Brown, Crowley, and Johnson, 1993; and Johnson and Brown, 1996). The Pukeashun shear zone (PSZ) lies within the early Tertiary Pukeashun granite. CRF = Columbia River fault, GLF = Grizzly Lake transfer fault (inferred), HR = Hunters Range, MD = Monashee Décollement, P = Pukeashun Mountain, SR = Shuswap Range, TvL = Three Valley Lake, ViC = Victor Lake.

sition above and below the décollement occurred in different parts of the miogeocline (Crowley, 1997).

The Monashee décollement separates the Monashee Complex from overlying highly sheared metasedimentary rocks and amphibolites. The Trans-Canada Highway transect of the Monashee Complex is situated in a structural depression between areas of extensive exposure of Paleoproterozoic basement rocks that lie to the north, in the Frenchman Cap culmination, and to the south, in the Thor-Odin culmination (Fig. 2-16) therefore, only the cover assemblage is exposed close to the highway. The Monashee décollement is not exposed along the highway, because it has been down-dropped to the west along a relatively minor west-dipping normal fault (Victor Lake fault) that crosses the highway just west of this locality; however, the rock types, the metamorphic mineral assemblages, and the tectonic fabrics related to the Monashee décollement can be viewed in the roadcuts on the north side of the highway. The rocks above the Monashee décollement contain sillimanite and K-feldspar; but kyanite has been collected at this locality. Most kinematic indicators record a top-to-the-northeast sense of shear.

170.1 km Three Valley Lake Resort.

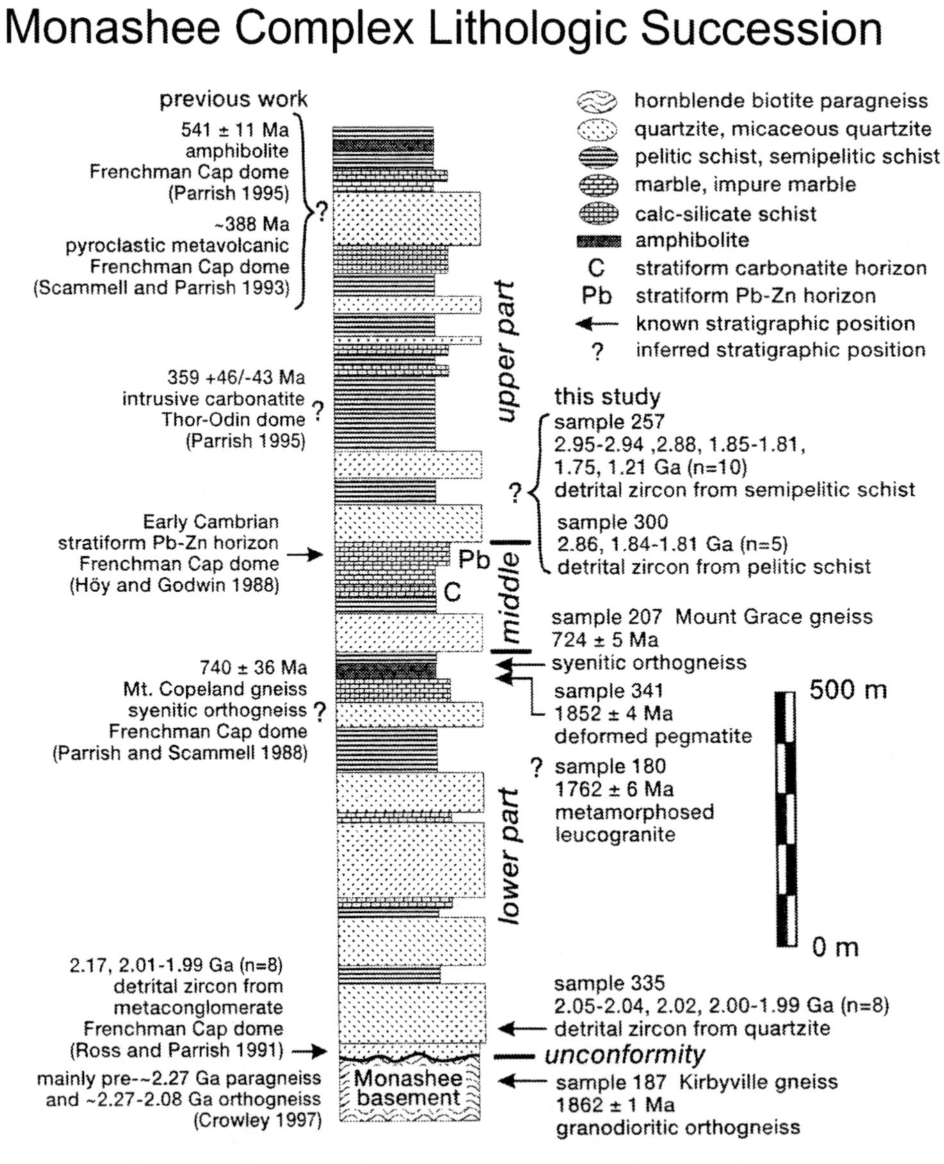

Figure 2-17. Monashee complex "cover assemblage" lithologic succession and geochronology (from Crowley, 1997).

Stop 2-8 Three Valley Lake; high shear strain zone and boudinage in the hanging wall of the Monashee décollement

170.5 km This is locality TvL on Fig. 2-16; park along lake shore about 0.5 km west of Three Valley Lake Resort.

The roadcuts along the south side of the highway and in the high cliffs on both sides of Three Valley Lake provide a magnificent display of boudinage involving amphibolite layers that are intercalated with pelitic gneisses and mica schists. The rocks contain abundant lenses, veins, and pods of pegmatite that are in various stages of transposition into the dominant foliation, which is generally relatively shallow-dipping and contains a strong northeast-trending stretching lineation. The transposed compositional layering and stretching in the boudinage attest to very high levels of northeast-verging shear.

Johnson and Brown (1996) have separated this distinctive assemblage of metamorphic rocks, which they called the "Three Valley assemblage," from other metamorphic rocks that occur in the hanging wall of the Monashee décollement in the vicinity of the Trans-Canada Highway, which they have called the "Hunters Range assemblage." Brown et al. (1993b) reported that U-Pb and Sm-Nd data produced by David Parkinson and Dick Armstrong suggest that the mafic rocks of the boudins are at least 1.5 Ga old.

Near-vertical, north-northeast trending mafic dykes with chilled margins fill brittle extension fractures that cut across the ductile fabrics. The dykes are parallel with the large-scale Late Paleocene to Middle Eocene extensional culminations and depressions, and perpendicular to the associated stretching lineations along the west side of the Shuswap metamorphic complex. They are oblique to the northeast-trending stretching lineations in the Monashee décollement zone and to the northwest trend of the structures in the Selkirk fan and in the more external parts of the foreland thrust and fold belt. The dykes evidently were emplaced during Eocene crustal extension, but after the exhumation and refrigeration of the amphibolites and paragneisses.

Optional stop Craigellachie: The Last Spike (historic site and washrooms)

195.9 km The Canadian Pacific Railway track was built from the east and from the west. Craigellachie is the site of the "Last Spike" where, on November 5, 1885, the rails were joined, linking the provinces of Canada together. The original ceremonial spike was golden, but now (1999) is gilded iron! The cairn marking the site contains rock specimens from all of the Canadian Provinces, together with one from Craigellachie in Scotland.

Craigellachie also is located near the gradational and somewhat arbitrary physiographic boundary separating the Monashee Mountains, the westernmost division of Columbia Mountains at this latitude, from the physiographically more subdued Shuswap Highlands to the west (Fig. 2). Much farther west, near Pritchard (323 km) the latter grade into the Interior Plateaus.

196.5 km Cross Gorge Creek, roadcuts in pegmatitic gneiss just west of bridge.

197.4 km Cross Canadian Pacific Railway overpass.

201.0 km Cross bridge over Eagle River.

217.1 km Eagle River to north of Highway 1.

Stop 2-9 High-grade metamorphic rocks close to the footwall of the Okanagan-Eagle River extensional detachment fault

218.3 km Park at east end of long roadcut in pegmatitic gneiss on south side of the Trans-Canada Highway. These high-grade metamorphic rocks of the Hunters Range assemblage are close to the footwall of the Okanagan-Eagle River extensional detachment fault. They

> ... include paragneiss, migmatitic garnet-sillimanite-biotite schist, some biotite-hornblende schist and calc-silicate gneisses, and lots of pegmatite. The pegmatite is strained and is concordant with the layering in the other rocks. The migmatites are mylonitic and display a number of excellent indicators of west-directed shear. These include C-S fabrics, asymmetrical extensional shear bands, and feldspar porphyroclast systems. The peak metamorphic mineral assemblage grt-sil-bt-pl-Kfs-qtz is synkinematic with the west-directed shearing. Ductile shearing in the footwall of the Okanagan Valley fault system to the south is known to have occurred in the Middle Eocene, mylonitic synextensional leucogranites to the northwest are Early Eocene or possibly older. Monazites from the pegmatite at this outcrop have given Late Cretaceous U-Pb dates... [Brown et al., 1993b]

Two mylonitic fabrics with opposing sense of shear occur in the panel of the high-grade metamorphic rocks between the Monashee décollement and the Okanagan-Eagle River extensional detachment fault (Johnson and Brown, 1996). In the Three Valley assemblage and the lower part of the Hunters Range assemblage, northeast-verging shear is associated with an annealed mylonitic foliation with a lineation that trends 065°; in the upper part of the Hunters Range assemblage, close to the Okanagan-Eagle River fault, mylonitic paragneisses that consistently indicate west-verging shear are associated with a strong stretching lineation defined by sillimanite and quartz-feldspar aggregates that trend 275°.

Horizontal Extension and Crustal Stretching in the South-Central Canadian Cordillera

The east-west crustal stretching that exhumed the Shuswap metamorphic core complex was oriented obliquely to, and superimposed discordantly on the crustal thickening and the fold and thrust structures that had been produced by the preceding northeast-southwest horizontal compression. It began in the Late Paleocene, at about 58 Ma (Carr, 1992). It was linked northwestward to right-lateral displacement on the Tintina-Northern Rocky Mountain Trench fault system (Fig. 2-18), and southwestward to right-lateral displacement on the Yalakom-Ross Lake and Fraser River-Straight Creek fault systems (Price, 1979; 1994).

The crustal stretching was conspicuously inhomogeneous. It comprised crustal-scale boudinage involving the formation of north-northeast-trending linear culminations that expose exhumed mid-crustal rocks and are separated by linear structural depressions that contain less deformed supracrustal rocks and unconformably overlying syntectonic Eocene sedimentary and volcanic strata (Figs. 1a and 2-18). Lithoprobe seismic reflection imaging (Figs. 2-19, 2-20) has documented a remarkable feature of this aspect of the crustal stretching. The inhomogeneous, east-southeast to west-northwest crustal stretching has generated about 20 km of structural relief in the upper crust between the tops of the structural culminations and the bottoms of the structural depressions. However, at the base of the crust, the Moho is relatively planar and changes elevation by <5 km between the west side of the Monashee Complex and the Fraser River fault. In spite of the large lateral variations in the amount of horizontal extension in the upper crust, the total thickness of the crust remains relatively constant over the interval from the Monashee Complex to the Fraser River fault, because there is a reciprocal relationship between the stretching of the middle and lower crust, on one hand, and the stretching of the upper part of the crust on the other (Fig. 2-21). Where the upper part of the crust has been strongly attenuated, the middle and lower crust are relatively unstretched; conversely, where the upper part of the crust is relatively unstretched, the lower and middle crust is strongly attenuated.

The Monashee Complex, which is exposed in the core of the largest culmination, comprises a layer of Paleoproterozoic continental crust that is about 40 km thick. Palinspastic reconstructions of the shortening due to thrusting and folding in the Rocky Mountains (Figs. 2-22, 2-23) show that the hinge zone between miogeocline and the platform lies approximately along the west side of the Monashee Complex. This hinge zone separated the Paleoproterozoic continental crust under the miogeocline, that was stretched by rifting during the Neoproterozoic and the Early Cambrian, from Paleoproterozoic continental crust under the platform that was not stretched. Accordingly, the present thickness of about 40 km of continental crust under the Monashee Complex, which matches the thickness under the autochthonous cover in the vicinity of Calgary, implies that there has been relatively little internal horizontal compression or horizontal extension of the crust under the Monashee Complex. The Monashee Complex was exhumed by simple isostatic buoyant uplift as the detached supracrustal cover was displaced into an adjacent extensional structural depression under the Selkirk Mountains.

Early Jurassic to Paleocene horizontal compression and Late Paleocene to Middle Eocene horizontal extension have profoundly modified the structure and thickness of the continental crust within the Cordillera. However, the late Neoproterozoic-Early Cambrian basin-margin structures that controlled thickness and facies changes in the Cordilleran miogeocline may still be discernible in the large-scale structure of the crust in the southern Cordillera. Figure 2-23 is a palinspastic restoration of the part of the foreland thrust and fold belt that lies within the Rocky Mountains. The map shows the present positions of the surface traces of the main thrust faults in the Rocky Mountain and the southern Purcell Mountains. It also shows the locations of two present-day crustal-scale monoclinal flexures across which there is a change in the level of exposure involving a stratigraphic separation of about 10 km (c.f. Fig. 1). The *Kootenay arc* marks a change in structural level from the lower part of the Mesoproterozoic Purcell Supergroup, on the east, to the upper Paleozoic and lower Mesozoic rocks of Slide Mountain and Quesnel terranes, on the west. The *Vernon monocline* marks a change in structural level from highly metamorphosed lower Paleozoic and (?) Neoproterozoic rocks on the north to less metamor-

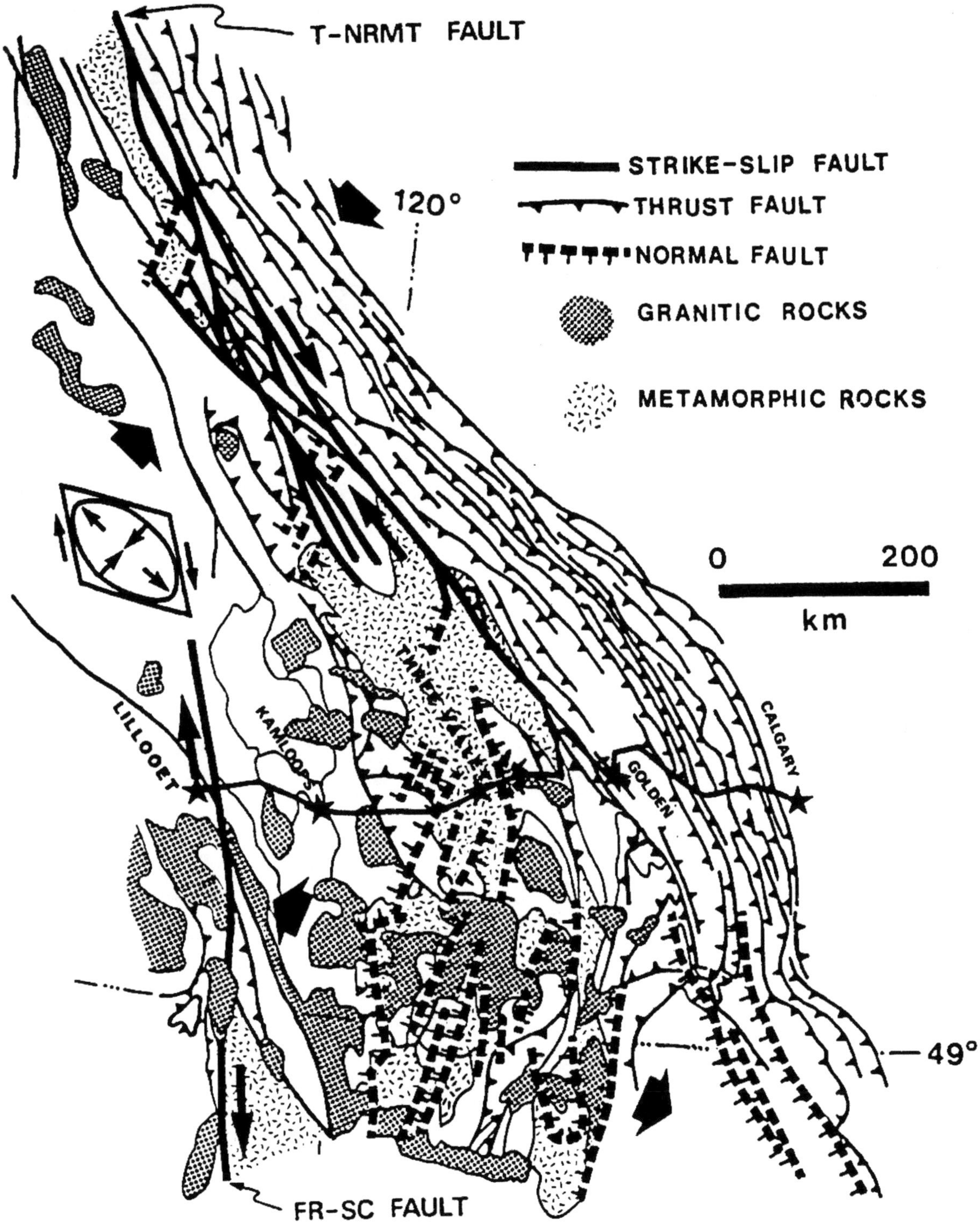

Figure 2-18. Transtensional deformation of the south-central Canadian Cordillera. Ductile crustal stretching linked to en echelon dextral strike-slip faults, and to the Early and Middle Eocene counter-clockwise rotation of Intermontane and Insular terranes relative to North America (From Price et al., 1985).

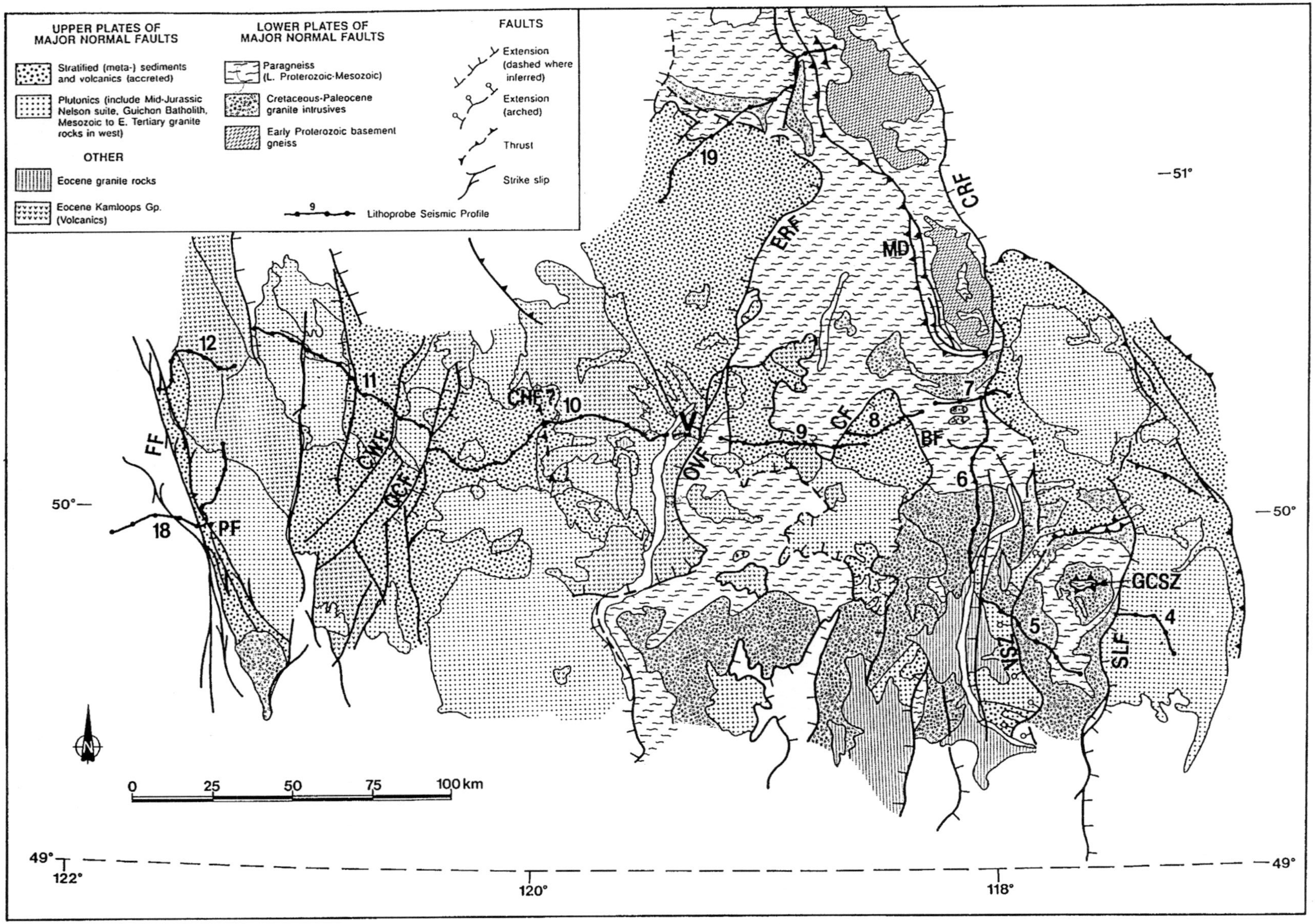

Figure 2-19. Index map of Lithoprobe seismic reflection profiles: south-central Canadian Cordillera (after Cook et al., 1992). Faults: AF = Ainsworth fault, BF = Beaven fault, CF = Cherryville fault, CHF? = Chapperon fault (?), CRF = Columbia River fault, CWF = Coldwater fault, ERF = Eagle River fault, FF = Fraser River fault, GCSZ = Gwillim Creek shear zone, MD = Monashee décollement, OVF = Okanagan Valley fault, PF = Pasayten fault, QCF = Quilchena Creek fault, SLF = Slocan Lake fault, VSZ = Valkyr shear zone.

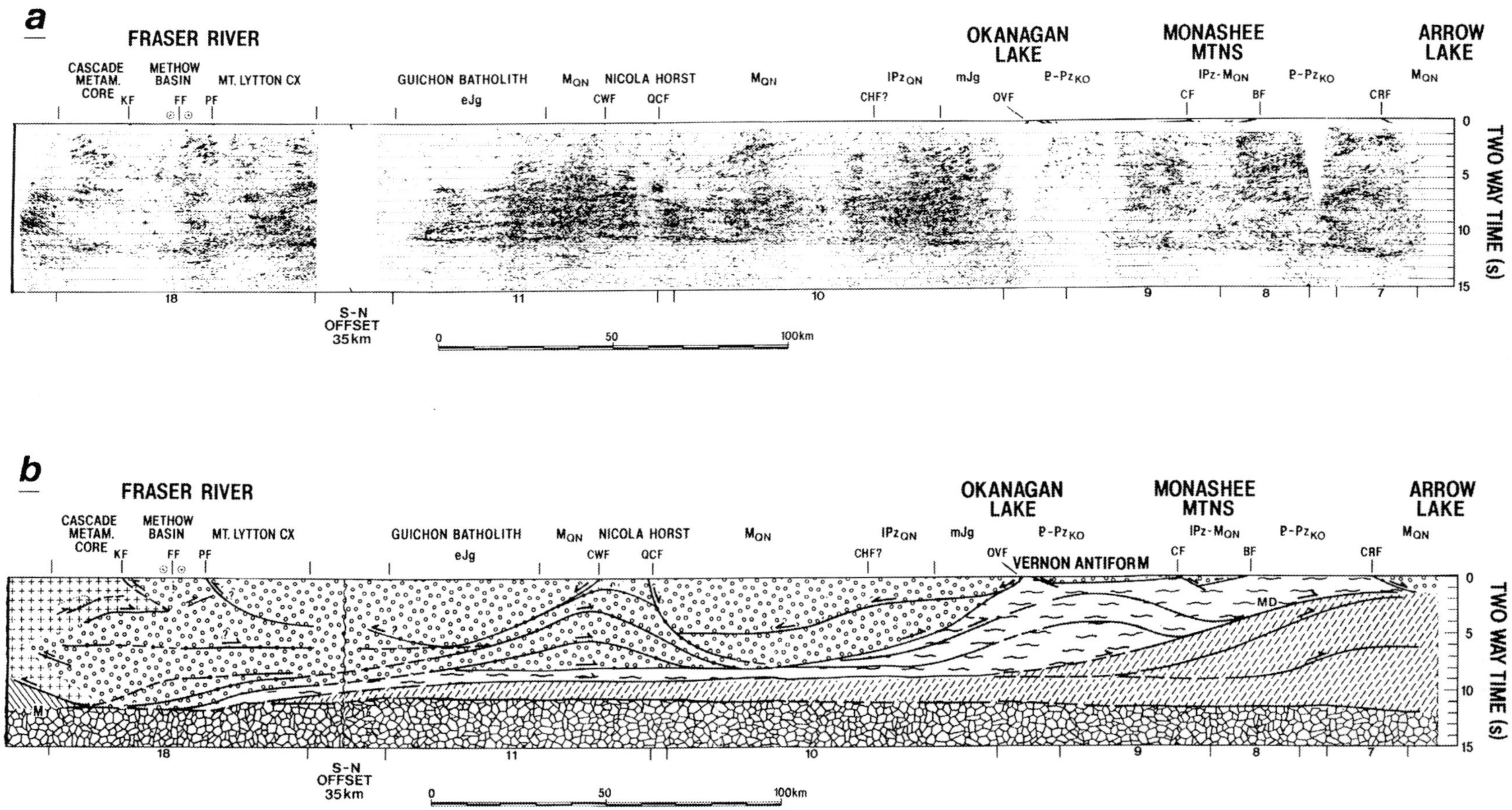

Figure 2-20. Lithoprobe seismic reflection images and interpretive crustal section of the interior of the southern Canadian Cordillera between the Columbia River fault and the Fraser River fault (after Cook et al., 1992, with modifications). Locations of seismic reflection lines are given in Figure 2-19.

a. Migrated seismic reflection cross section. Data are plotted to 15.0 s, have been coherency filtered to attenuate background random noise, and are displayed with no vertical exaggeration for a velocity of 6.0 km/s.

b. Geological interpretation of the seismic reflection cross section. Patterns: diagonal dash = North American basement; wavy dash = allochthonous North American rocks; circles = accreted terranes; plus signs = rocks of the Cascade orogen; polygons = upper mantle.

phosed rocks of Slide Mountain terrane on the south. The Vernon monocline is a conspicuous feature on the regional geological maps of the southern Canadian Cordillera because it is the zone in which the boundary between the North American rocks and Intermontane terranes swings northwestward across the Shuswap metamorphic complex from the Kootenay arc to where is crosses the Trans-Canada Highway near Pritchard, B.C. (Fig. 1).

The palinspastic restoration which compensates for the effects of the horizontal translation and compression of the supracrustal rocks, is focused on two features of the lower Paleozoic miogeocline. The location of the boundary between the displaced rocks of the miogeocline and those of the cratonic platform is approximately coincident with the Bourgeau thrust fault from near 53°N latitude southward to the Hosmer nappe and the Moyie fault. From there, it is offset ~250 km to the southwest into northeastern Washington, beyond which it swings southward. The separation between it and the present-day location of the carbonate bank margin that marks the northeastern border of the deeply subsided lower Paleozoic shale basin decreases southeastward toward the Moyie fault. The total amount of shortening across the southern Rocky Mountains also decreases southeastward (Price and Sears, 2000). The palinspastically restored location of the boundary between the cra-

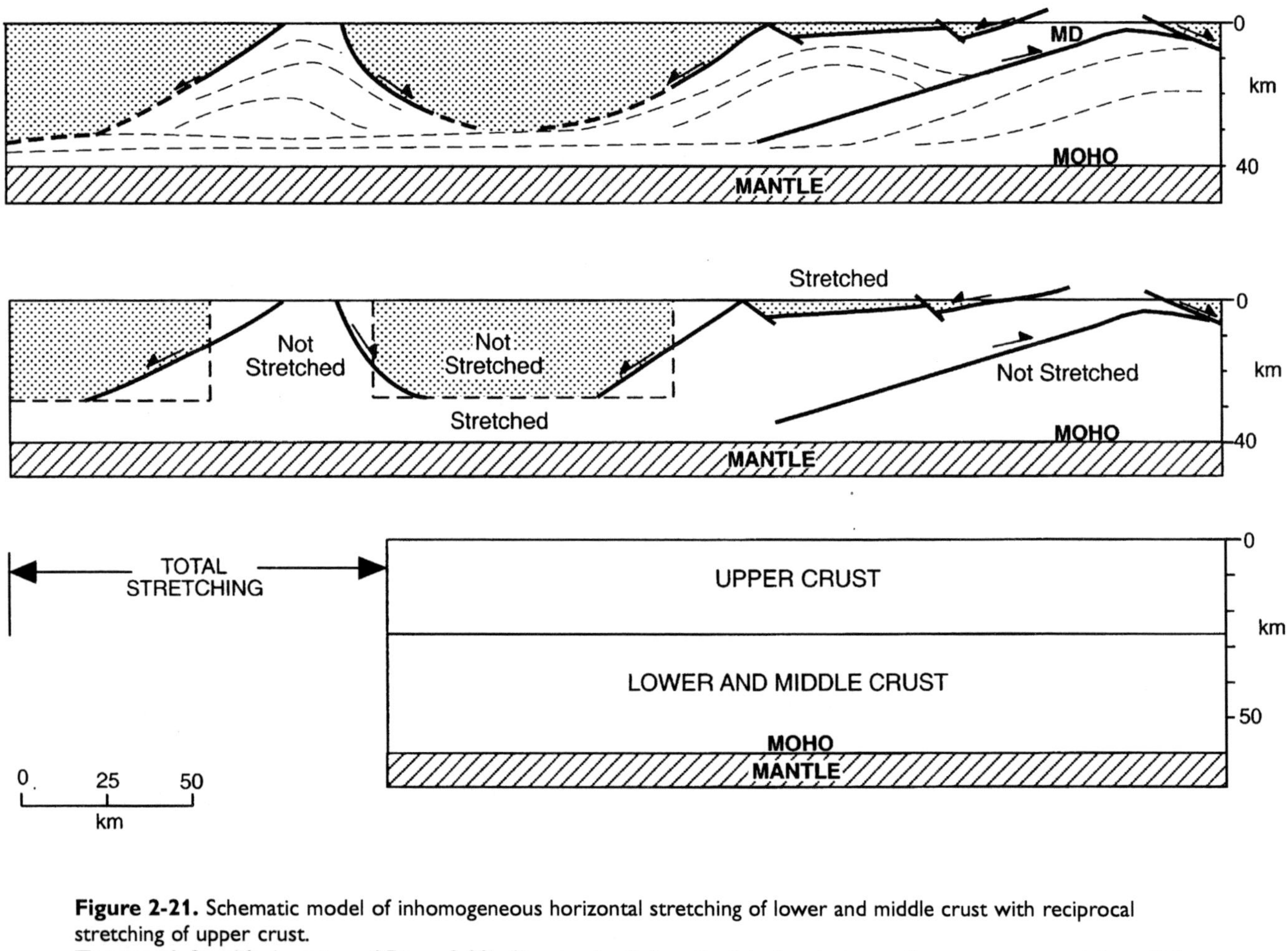

Figure 2-21. Schematic model of inhomogeneous horizontal stretching of lower and middle crust with reciprocal stretching of upper crust.

Top panel. Simplified version of Figure 2-20b. Structural relief on the interface between the upper crust and the middle and lower crust is ~25 km, but structural relief at the Moho is <5 km.

Middle panel. Model of reciprocal stretching; the upper crust is severely attenuated above crustal boudins in which the middle and lower crust is "not stretched;" and the middle and lower crust is severely attenuated below "neck zones" within which the upper crust is "not stretched."

Lower panel. Crustal thickness prior to extension (an area-balanced "palinspastic restoration" of the middle panel). Total amount of crustal thinning is relatively constant across the section in spite of large lateral variations in amount of thinning of both the upper crust and the lower and middle crust.

tonic platform and the miogeocline (Fig. 2-23) coincides approximately with the steep crustal ramp on the southwest side of the Monashee Complex Fig. 2-20b). The palinspastically restored location of the carbonate bank margin that marks the eastern border of the deeply subsided lower Paleozoic shale basin coincides approximately with the Vernon monocline (Fig. 2-23). Lithoprobe seismic reflection line 6 (Fig. 2-24), which trends north-south across the Vernon monocline (see Fig. 2-19 for location) shows that the Vernon monocline coincides with a crustal ramp along which the Monashee décollement descends southward from where it is exposed in the high peaks of the Monashee Mountains in the Thor-Odin culmination to depth of about 17 km below sea level. This thrust ramp appears to have developed along a pre-existing lower Paleozoic basin-margin ramp that was produced by the rifting that gave birth to the Cordilleran miogeocline.

221.6 km Town of Sicamous.

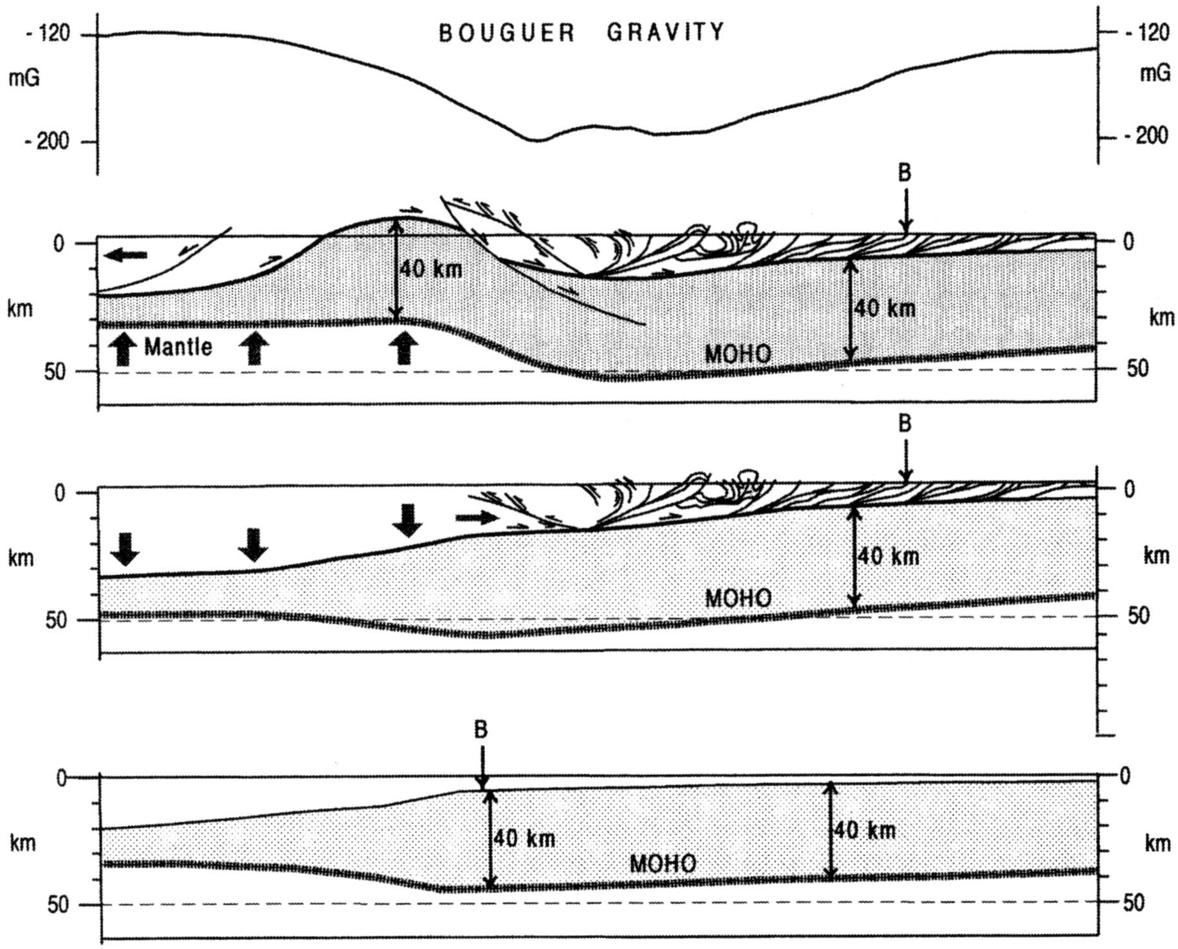

Figure 2-22. Cartoon of evolving vertical displacement of the continental crust in the Monashee complex.
Bottom panel. Crustal section in Early Jurassic, before the onset of compressional collapse of the Slide Mountain marginal basin and the formation of the Selkirk fan structure and the foreland thrust and fold belt; B: palinspastically restored position of the rocks, which mark the transition from the cratonic platform to the Cordilleran miogeocline, and are now exposed in the toe of the Bourgeau thrust sheet.
Middle panel. Crustal section in Late Paleocene, after the formation of the northeast-tapering foreland-thrust-and-fold-belt accretionary wedge and the tectonic burial of the Monashee complex, but prior to onset of Late Paleocene to Middle Eocene transtension and crustal stretching.
Top panel. Present crustal section showing extensional exhumation of the Monashee complex in relation to Bouguer gravity anomaly curve.

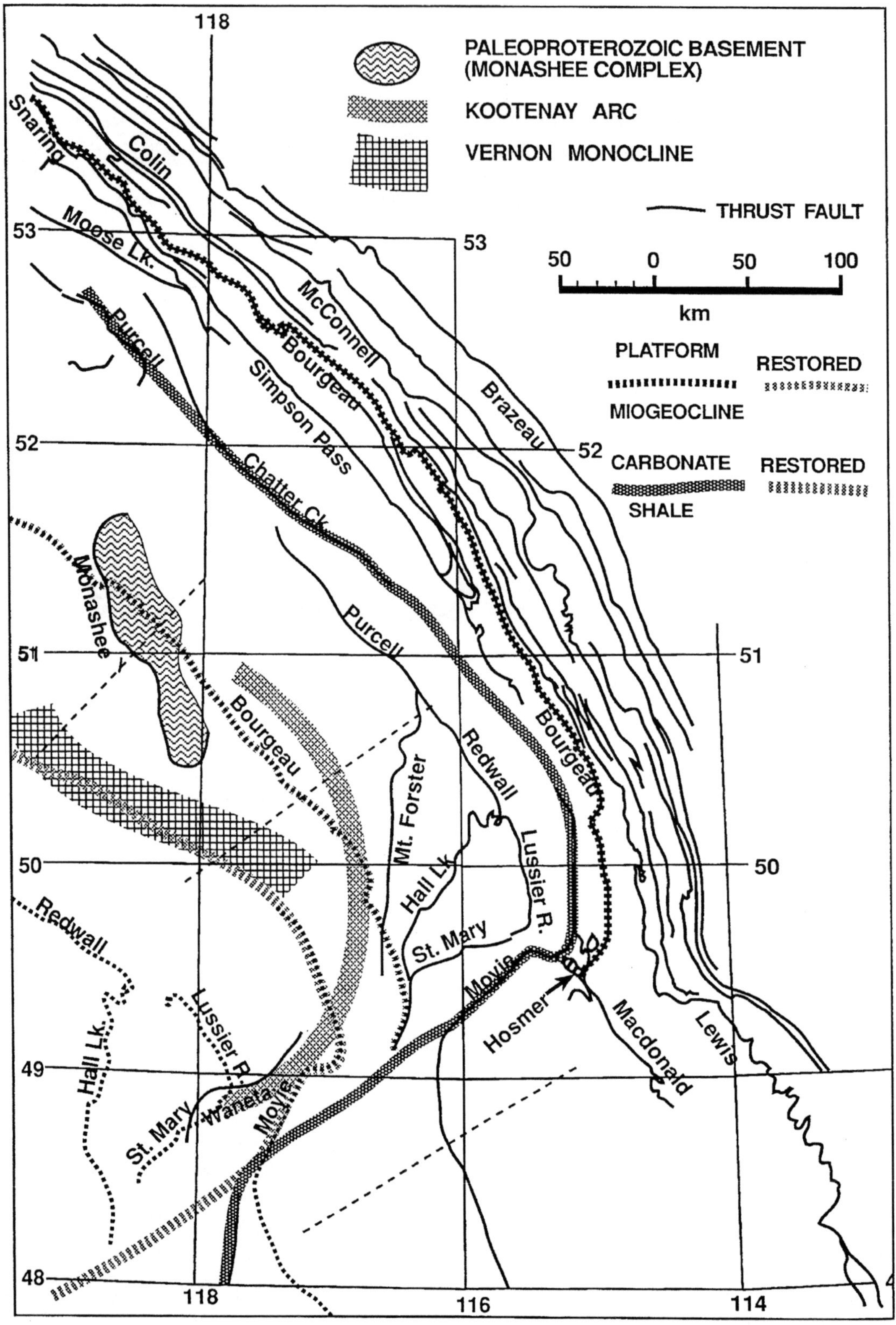

Figure 2-23. Palinspastic reconstruction of the hinge zone of the Cordilleran miogeocline and of the outboard edge of the Lower Paleozoic miogeoclinal carbonate platform. The reconstruction is based on palinspastically restored balanced structure sections by Price and Mountjoy (1970), Mountjoy (1980), Price (1981), Price and Fermor (1985), and Monger et al. (1985). See text for explanation of patterns.

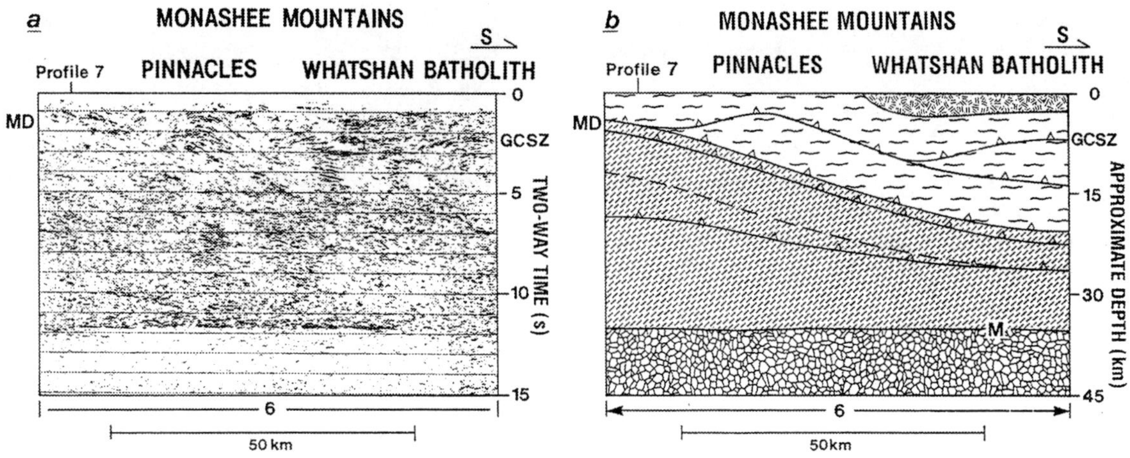

Figure 2-24. Lithoprobe seismic reflection image and interpretive crustal section of the Vernon monocline (seismic profile 6 in Figure 2-19) (after Cook et al., 1992):
a. Migrated and coherency-filtered seismic reflection data from north-south profile 6. Data are plotted to 15.0 s, and are displayed with no vertical exaggeration for a velocity of 6.0 km/s.
b. Geological interpretation. Patterns: diagonal dash = North American basement; wavy dash = allochthonous North American rocks; circles = accreted terranes; polygons = upper mantle.

SICAMOUS TO PRITCHARD

INTRODUCTION

The western part of the Omineca Belt traversed on Highway 1 between Sicamous (221.6 km) and Pritchard (322.9 km) consists of penetratively deformed clastic, carbonate and volcanic rocks of mainly greenschist and lower amphibolite grade with Proterozoic and Paleozoic protoliths, and Devonian and Middle Jurassic through early Tertiary granitic plutons. Sillimanite, which is widespread in the pegmatite-rich rocks along Highway 1 between the eastern border of the Clachnacudainn complex (~116 km) and Sicamous (221.6 km) is rare. The rocks underwent possible Paleozoic and definite Jurassic through earliest Tertiary episodes of contractional deformation, and Eocene extension.

While there is little doubt that the rocks in the region formed in distal parts of the ancient North American continental margin, their complex, intense deformation (e.g. Fyson, 1970) and the general lack of age control means that the stratigraphic order of most of the rock units is uncertain. Probable and possible correlations of some rock units in the region may be made with those of: (1) the high-grade metamorphic rocks along Highway 1 between Victor Lake and Sicamous (which can be traced north northwestward into the lower grade rocks); (2) the Clachnacudainn complex east of Revelstoke (Devonian granitic rocks in common); (3) the stratigraphic succession of the Kootenay arc of southeastern British Columbia, about 200 km southeast of here (although that succession lacks known Devonian magmatic rocks; Fig. 2-25); (4) the distal continental margin strata crossed on Highway 1 between the Purcell Thrust and the Clachnacudainn complex (Neoproterozoic Horsethief Creek Group; the early Paleozoic Hamill and Badshot formations and Lardeau Group; again, that section lacks Devonian magmatic rocks). Largely because of their uncertain stratigraphy and correlations, rocks in the western Omineca Belt together with those in (1) and (2) above, are presently "parked" in the pericratonic Kootenay terrane (Fig. 1, Fig. 7, Table 2 Wheeler and McFeely, 1991; Colpron and Price, 1995).

Lithologies and Rock Units

Most rock unit names in the region originated with Jones (1959), were modified by Schiarizza and Preto (1984, 1987) and Okulitch (1989), and currently are being mapped and further modified (Thompson and Daughtry, 1998; R.I. Thompson, pers. comm. 1999).

The *Silver Creek Formation* comprises quartzite, quartzofeldspathic and mica schist, with local sillimanite, and minor amphibolite and marble. It is exposed in the core of a major structure called the Chase antiform, and probably is the oldest unit in the area. Lithologically it could be derived from the Neoproterozoic Windermere succession (e.g. Horsethief Creek Group crossed on Highway 1). The Silver Creek Formation is intruded by the *Little Shuswap orthogneiss* exposed along Highway 1 near Chase

(~300 km), which although shown as Ordovician-Silurian on Fig. 1, is more probably Devonian (A.V. Okulitch, pers. comm. 1999). The Silver Creek Formation is overlain by biotite and hornblende schist, phyllite, marble and deformed conglomerate containing granitic cobbles and boulders dated at 555 Ma (R.I. Thompson, pers. comm. 1999). The inference drawn from this relationship is that the coarse conglomerate is derived from a nearby but unseen intrusion in the Silver Creek Formation, which therefore must be Precambrian.

The *Tsalkom Formation* (Stop 2-11) comprises mafic metavolcanic and local ultramafic rock and is in contact with the Silver Creek Formation. Such lithologies also occur in the early Paleozoic Jowett Formation and the Permian Kaslo Formation in the Kootenay Arc (Fig. 2-25) and in the Carboniferous and Permian Paleozoic Fennell Formation to the northwest.

The *Sicamous Formation* (Stop 2-10) consists of calcareous and graphitic pelite and large masses of recrystallized limestone. Lithological correlations with both the Cambro-Ordovician Index Formation of the Lardeau Group and with the Triassic Slocan Formation in the Kootenay Arc (Fig. 2-25) have been suggested. The former is currently favoured, not least because the Sicamous has been extensively sampled for conodonts and none have been found. The presence of the limestone masses suggests relatively shallow water, which implies underlying thick crust (R.I. Thompson, pers. comm. 1999). Along the regional strike, farther north in the Cariboo Mountains (latitude ~53°N), Struik (1987) suggested that latest Precambrian extension created an irregular, "lumpy" basement topography, which controlled the early Paleozoic deposition in that area. Such may be the case here.

The *Eagle Bay Formation* contains sericite and

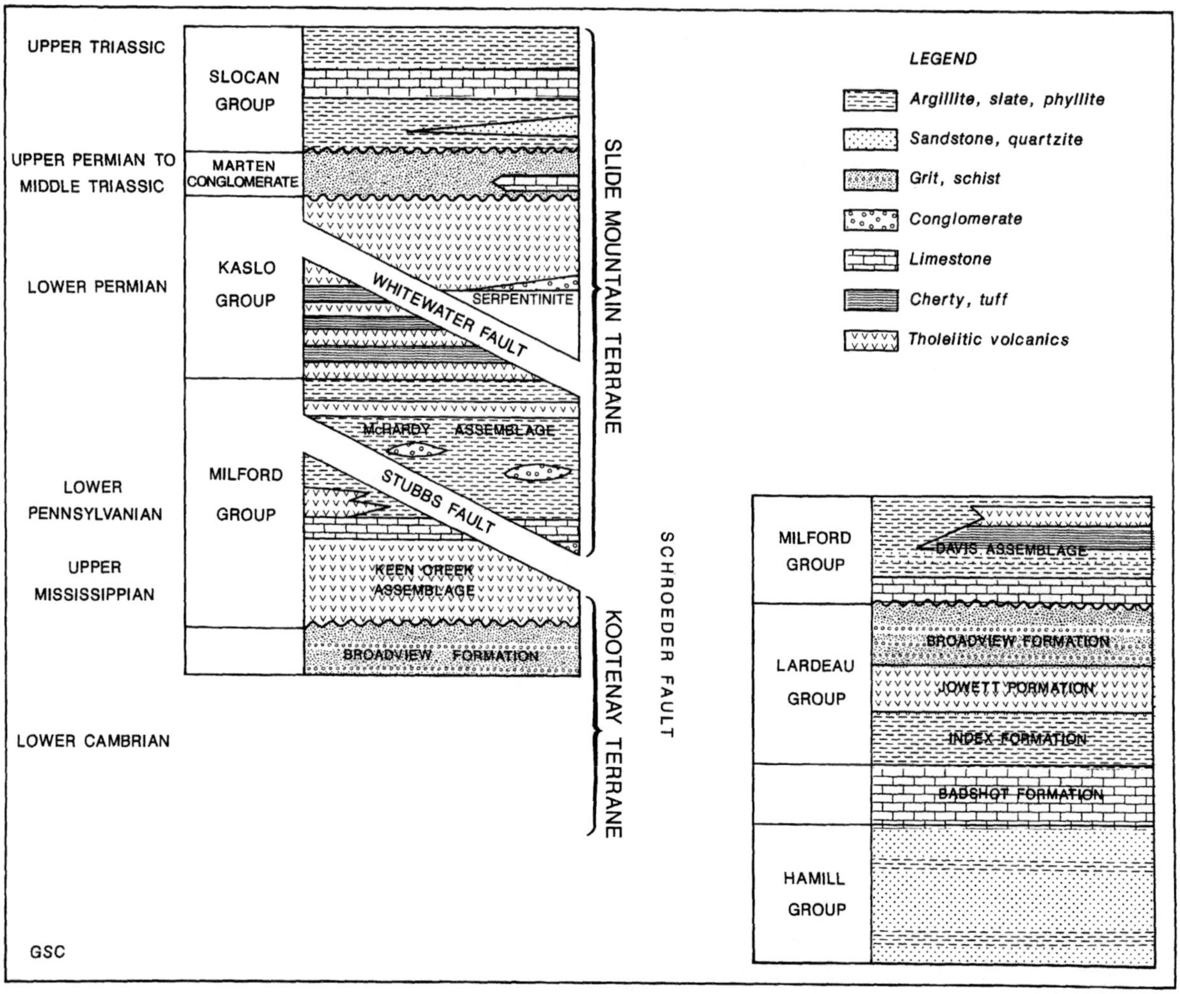

Figure 2-25. Stratigraphy of the Kootenay arc region (modified from Klepacki and Wheeler, 1985).

chlorite schist derived from felsic and mafic volcanic rocks, phyllite, and local marble that at one locality (Tshinakin limestone) contains archeocyathids of Early Cambrian age (also found in the Badshot Formation in the Kootenay Arc; Fig. 2-25). The felsic volcanics are at least in part of Middle Devonian age (387 ± 6 Ma). The Late Devonian (365–355 Ma) foliated granitic rocks that are widely distributed in this area may be partly intrusive equivalents; if so, more dating is needed to close the 20-million-year gap between intrusive and extrusive ages. Carboniferous pelite and carbonate, although included in the Eagle Bay Formation by Schiarizza and Preto (1987), in fact may unconformably overlie it and be correlated with the Milford Group in the Kootenay Arc (Fig. 2-25; R.I. Thompson, pers. comm. 1999).

Most recently, Thompson and Daughtry (1998), in a reflection of the current state of stratigraphic knowledge in this area, include Tsalkom, Sicamous and (most of) Eagle Bay formations in an undifferentiated Paleozoic "amphibolitic schist-marble-quartzite unit" that unconformably overlies the Neoproterozoic Silver Creek Formation.

Linkages to East and West

The rocks of the western Omineca Belt potentially contain the link between the craton margin deposits and those of the accreted terranes. Some possible correlations between these rocks and the distal miogeoclinal deposits were noted earlier. A further tie between the Late Devonian magmatic rocks in this area and cratonal strata is provided by Late Devonian (Famennian, ~370 Ma) tuff horizons within the Exshaw Formation near the front of the Rocky Mountains (near Stop 1-3; Richards et al., 1994), and by the ~368 Ma alkalic rocks of the Ice River Complex (see Stop 1-11). However, because Devonian magmatic rocks occur in continental margin and pericratonal settings from California to Arctic Alaska, this says little about the original (latitudinal) position of Kootenay terrane along the ancient continental margin.

The Fennell Formation is located about 70 km north of where Highway 1 crosses the boundary between Omineca and Intermontane belts (322.9 km; DTrs of Fig. 1). It features abundant mafic volcanics with MORB chemistry, chert and argillite, and ranges in known biostratigraphic age from Early Mississippian through Early Permian. Locally, dated Devonian quartz feldspar porphyry intrudes mafic volcanic rocks assigned the Fennell (Schiarizza and Preto, 1987), which suggests a linkage between the Fennell and the Devonian magmatic rocks of the Kootenay terrane. The Tsalkom Formation *within* Kootenay terrane has similar lithologies to the Fennell, but as noted above, there are alternative correlations.

The Fennell Formation is one of a series of aligned but discrete rock assemblages with similar lithologies and age ranges that have been included in the *Slide Mountain terrane,* the easternmost accreted terrane (Fig. 7; Table 2). In places, sedimentological and stratigraphic evidence show that some rocks presently included in Slide Mountain terrane were deposited on or near the old North American continental margin (e.g. Monger, 1977; Klepacki, 1985; Fig. 2-25). However, paleomagnetic data from other places (latitudes 53°N and 59°N) within Pennsylvanian-Permian rocks of Slide Mountain terrane indicate that those parts of the terrane have been displaced by ~20° of latitude northward with respect to the craton margin, probably in the Jurassic (Richards et al., 1993). The seeming latitudinal discrepancy, combined with possible links with Kootenay terrane, thus raises questions about the pre-Jurassic paleogeographic positions of both Slide Mountain and Kootenay terranes, and at the present level of knowledge justifies inclusion of the rocks in Kootenay terrane as "suspect."

ROAD LOG

223.0 km Highway 1 crosses the bridge over a narrow arm of Shuswap Lake. The valley follows the trace of a fault, mapped originally by Jones (1959), and later regarded a west-dipping normal fault (Eagle River Fault; Fig. 2-19; Parrish et al., 1988), across which sillimanite-grade rocks to the east (seen at 218.3 km) are juxtaposed with greenschist and lower amphibolite grade rocks to the west.

As noted in the discussion following Stop 2-9, the Eagle River Fault is one of a series of aligned, mainly north northeast-trending faults, some of which are known to be of Eocene age, that may link up with the major, west-dipping, normal, Okanagan Valley Fault, which crosses the International Boundary, ~220 km south southeast of here (Fig. 2-19; Parrish et al., 1988). These faults generally mark the western boundary of the region in southeastern British Columbia and adjoining parts of the United States that underwent extreme, basin-and-range style extension between ~55–45 Ma. To the west, coeval transtension is more common. In addition, farther south near the International Boundary, the Okanagan Valley Fault is used as the western boundary of the Omineca Belt.

Stop 2-10 Sicamous Formation

223.2 km Immediately after bridge, turn south (left) across traffic on Highway 1, on to side road.

The outcrops of dark calcareous phyllite with quartz segregations are mapped as Sicamous Formation, although elsewhere the formation is mainly carbonate (view from 242.1 km). The high greenschist-lowest amphibolite metamorphic grade of these rocks is in marked contrast with the pegmatite-rich, sillimanite-bearing rocks east of Sicamous, which are visible across the Shuswap River from the southern end of the outcrop area.

223.2–227.9 km Rejoin Highway 1 and drive west. On south (left) side of bus are long roadcuts in Sicamous Formation, on the right is Salmon Arm of Shuswap Lake. According to K.L. Daughtry and R.I. Thompson (pers. comm. 2000) a gradational contact between Sicamous and the Silver Creek formations can be seen near the top of the hill.

Drainage Pattern

Highway 1 follows a generally westward-draining system of interconnected rivers and lakes almost as far as the village of Cache Creek. Shuswap Lake has four arms (Fig. 1). West of Sicamous, Highway 1 follows the south shore of Salmon Arm of Shuswap Lake for ~40 km to a location west of the town of Salmon Arm. It then crosses the neck of land separating Salmon Arm from the western arm of Shuswap Lake, to rejoin the lake near Sorrento (285.8 km). At its west end, Shuswap Lake narrows, is joined by the south-flowing Adams River and the water from both bodies flows southwestward into Little Shuswap Lake. That lake drains southwestward into the South Thompson River, which joins the south-flowing North Thompson River at the town of Kamloops. The combined rivers then drain westward through Kamloops Lake, into the Thompson River. The Thompson River temporarily diverges from Highway 1 about 12 km east of the village of Cache Creek, and flows southwesterly to join the Fraser River at the town of Lytton.

~227.5–236.1 km Intermittent roadcuts in pelitic and sericite and muscovite schist, considered by Thompson and Daughtry (1998) to belong to the Neoproterozoic Silver Creek Formation.

238.4 km Highway 1 goes downhill to the level of Shuswap Lake.

238.9–239.3 km Roadcuts are in undated (Jurassic?) granitic rock on south (left) side of highway.

240.8 km Sicamous Formation(?) pelite in roadcut.

242.1 km Settlement of Canoe, dominated by the lumber mill. To the north, across Salmon Arm of Shuswap Lake is Bastion Mountain composed of carbonate of the Sicamous Formation.

252.8 km Business section of the town of Salmon Arm.

256.3 km Cross bridge over north-flowing Salmon River.

258.7–264.9 km Roadcuts for the next 7 km are in granitic rock of the Salmon Arm pluton, which is ~15 × 15 km in areal extent and underlies Tappen Mountain to the east and south (left) of Highway 1. Discordant K-Ar dates of 76 Ma (bi) and 108 Ma (hb) have been obtained from the Salmon Arm pluton, but a nearby pluton of similar composition provides a Late Jurassic U-Pb date (R.I. Thompson, pers. comm. 1999). The granitic rocks are overlain non-conformably by Eocene volcanic rocks of the Kamloops Group.

258.9 km View of Bastion Mountain across Shuswap Lake to northeast, where a structural thickness of >1000 m of marble is exposed.

267.3 km Bridge over Canadian Pacific Railway track; at 268.2 km settlement of Tappen.

277.3 km Hill straight ahead is formed of dark, calcareous pelitic rocks of the Sicamous Formation, which are exposed discontinuously in roadcuts along Highway 1 for the next ~8 km.

285.8 km Settlement of Sorrento, located on the south side of the western arm of Shuswap Lake.

289.4–292.8 km Roadcuts in green mafic volcanics of the Tsalkom Formation.

Stop 2-11 Tsalkom Formation

290.7 km Pull off road at second parking area on right (north). **Take care when crossing highway!** Roadcuts are in greenschist facies mafic volcanics of the Tsalkom Formation. Generally few primary fabrics are preserved in the rocks, although rare relict pyroxenes, remnants of pillows and breccias, and intercalations of pelite and marble are locally present. The age of the formation is unknown. Lithologically, it is similar to the Cambro-Ordovician Jowett Formation of the Lardeau Group in the Kootenay Arc, to the Permian Kaslo Formation in the Kootenay Arc, or to the late Paleozoic Fennell Formation ~60 km northeast of here. The greenstone is cut by buff intrusions of probable early Tertiary age that are feeders to stratigraphically overlying Eocene volcanics of the Kamloops Group, exposed higher up on Squilax Mountain south of Highway 1.

293.2 km Outflow of Shuswap Lake.

294.8 km Roadcuts in buff-weathering felsic rock, which are probably subvolcanic to the Eocene Kamloops Group, which underlies upper parts of Squilax Mountain southeast of here.

295.1 km Turnoff and overpass at Squilax.

297.2 km Little Shuswap Lake on north (right) side of Highway 1.

297.2–304.1 km Roadcuts in white, felsic, Little Shuswap orthogneiss. The gneiss is mainly of igneous origin, but may include paragneiss. The original U-Pb dates suggested an Ordovician age, but recent U-Pb dating has shown that the systematics are complicated and the most probable age is Devonian (A.V. Okulitch, pers. comm. 1999). The body occurs in the axis of the Chase Antiform and intrudes Neoproterozoic quartz-feldspar biotite-muscovite schist mapped as Silver Creek Formation.

305.7 km Chase Plaza north (right) of Highway 1.

306.1–308.1 km Buff weathering roadcuts to left are mapped as Silver Creek schist.

312.9 km Highway 1 is along the valley of the South Thompson River. Note the diminished local relief in the westernmost part of the physiographic division called Shuswap Highland (Fig. 2).

315.6–317.0 km Granitic rock of the Cretaceous Nisconlith Pluton, which is overlain non-conformably by Eocene volcanics of Kamloops Group.

317.9 km The terraces developed low on the valley sides across South Thompson River are prominent on the lower sides of the valley west as far as Kamloops. The terraces are carved in white-buff varved silts that were deposited in a syn- and post-glacial lake behind an ice dam located farther to the west. Such silt terraces are found in many of the larger interior valleys in southern British Columbia.

322.9 km Pritchard; junction of Highway 1 with road to the south to the village of Falkland. This is 100 km west on Highway 1 from the bridge over the arm of Shuswap Lake at Sicamous, and becomes 0 km for Section 3, the Intermontane Belt.

Louis and Bolean Creek Faults: the Boundary Between Omineca and Intermontane Belts Near Highway 1

North and south of Highway 1 near Pritchard, but not obvious from the highway, are aligned north-northwest trending valleys. That to the north contains northward-flowing Louis Creek, and that to the south contains southward-flowing Bolean Creek. Both creeks approximately follow traces of a fault system. The northern fault segment is called the Louis Creek Fault. (Note: this fault is mis-labelled on the Tectonic Assemblage Map (Fig. 1), where the name is incorrectly applied to a fault 25 km farther west). The southern segment is called the Bolean Creek Fault. East of the fault system are mostly greenschist and higher grade rocks mainly derived from the Neoproterozoic and Paleozoic strata and included in the Omineca Belt, whereas west of the fault are the fossiliferous, mainly subgreenschist grade, latest Devonian and younger strata of the Intermontane Belt.

The fault history may be complex. Eocene volcanics of the Kamloops Group are voluminous west of the fault but relatively minor to the east of it, suggesting that there was west-side-down Tertiary movement on the fault, that may be related to the regional extension discussed in at Stop 2-9. The fault also may be an older west-side-down reverse fault that is congruent with west-vergent folds and faults mapped ~20 km north of the highway by Schiarizza and Preto (1984), and on the ridge north of the South Thompson River ~15 km east of Kamloops by Smith (1979). These west-vergent structures are later than the east-vergent thrust faults that imbricate the Fennell Formation, are associated with major contraction in the Omineca Belt (the west-vergent deformation in the evolution of the Selkirk Fan (Fig. 2-14; Clachnacudainn complex discussion ~117 km)) and probably are of Middle Jurassic age.

Reset odometer to 0 km.

3. INTERMONTANE BELT

PRITCHARD TO FRASER RIVER

INTRODUCTION

On physiographic maps (Fig. 2; Holland, 1964; Mathews, 1986) the valleys containing Louis and Bolean creeks mark the somewhat arbitrary boundary near Highway 1 between the Shuswap Highland to the east, and the even more subdued physiography of the Interior Plateaus to the west. In general, the southern Intermontane Belt is a region of relatively low relief that merges gradually with the flanking, mountainous, southern Omineca and Coast belts. It typically features rolling topography and plateaus ("Interior Plateaus") where landforms are partly controlled by a pre-Late Miocene erosion surface at an elevation of ±1000 m and by overlying flat-lying Late Miocene (~10 Ma) basaltic lava flows.

The Intermontane Belt is an orogen-parallel structural depression whose long axis extends for ~1500 km (from latitudes ~48°N to ~62°N), and which is up to ~200 km wide, although only ~125 km wide where crossed on Highway 1 (Fig. 4A; Table 1). It features marine and non-marine volcanic and sedimentary rocks and intrusive rocks that are coeval with at least some of the volcanics. The metamorphic grade typically is subgreenschist (prehnite-pumpellyite or zeolite facies), rarely higher, and blueschist is locally preserved (Fig. 4b). Structures within the Intermontane Belt are mostly brittle and record pre-, syn- and post-accretionary events.

Within the southern Intermontane Belt near Highway 1, the known ages of volcanic and sedimentary rocks range from latest Devonian (Famennian, ~365 Ma) to Quaternary, and granitic rocks range in age from Permian to Tertiary (Figs. 1; 3-1). The isotope geochemistry of pre-Middle Jurassic magmatic rocks typically is juvenile, with $Sr_i \leq 0.704$ and positive ε_{Nd} values. The younger magmatic rocks reflect varying amounts of old crustal components (Armstrong, 1988; Ghosh, 1995), and some of the late Paleozoic and Triassic sedimentary rocks contain mixed, juvenile and evolved isotopic components (Patchett and Gehrels, 1998; R. Creaser, pers. comm. 1999). Interpretations of seismic reflection and refraction data show that the crust of the southern Intermontane Belt is ~32–36 km thick, with attenuated old continental crust extending as a thin (~2–10 thick) layer in the lowest crust as far west as the Fraser Fault (Fig. 5; Clowes et al., 1995).

Rock Units and Terranes

Pre-Late Jurassic stratified rocks are mostly marine and disposed in Slide Mountain, Quesnel and Cache Creek terranes, whereas most younger strata are non-marine.

Slide Mountain terrane (Fig. 7; Table 2) forms a discontinuous series of outcrop areas that straddle the boundary between Intermontane and Omineca belts. These areas contain mafic volcanic rock, in places with MORB chemistry, chert, argillite and locally coarser clastics, minor carbonate and most are associated with alpine-type ultramafic rocks. The known age range of rocks in this terrane is from Early Mississippian through Permian. Where there is detailed mapping and good age control, the terrane is known to be internally complexly imbricated; any stratigraphic sections are within fault slices and there is no detailed correlation between slices. The imbricated stack locally includes slices of continentally-derived clastics and arc volcanics. Most of the rocks are inferred to have formed in a marginal basin setting. The Fennell Formation (Fig. 3-1), which is included in Slide Mountain terrane, outcrops ~70 km north of Highway 1 (Fig. 1).

Quesnel terrane (or *Quesnellia*, a name taken from the town of Quesnel located ~300 km to the north) is given integrity by the Upper and locally Middle Triassic Nicola Group, which consists of arc-related volcanic and sedimentary rocks, associated with coeval intrusions (Figs. 3-2A). Along Highway 1, the Nicola Group is exposed for a distance of ~120 km across the regional strike. The Nicola Group unconformably overlaps (at least) two different upper Paleozoic stratigraphic packages (Fig. 3-1). One, called the Harper Ranch Group (Fig. 3-2B; DTrH on Fig. 1), is of latest Devonian through Permian age and consists of arc-related volcaniclastic rocks, argillite, siltstone, and limestone. The other, exposed south of Highway 1 (CPA on Fig. 1) is Carboniferous and Permian and has the same range of lithologies as the Slide Mountain terrane.

In places, the pre-Middle Triassic rocks are more deformed and metamorphosed than the unconformably overlying Triassic strata (Read and Okulitch, 1977). Furthermore, Upper Triassic clastic strata that are physically continous with the eastern, sedimentary facies of the Nicola Group (part of Quesnel terrane) can be traced more-or-less continuously in a southeasterly direction along the Vernon

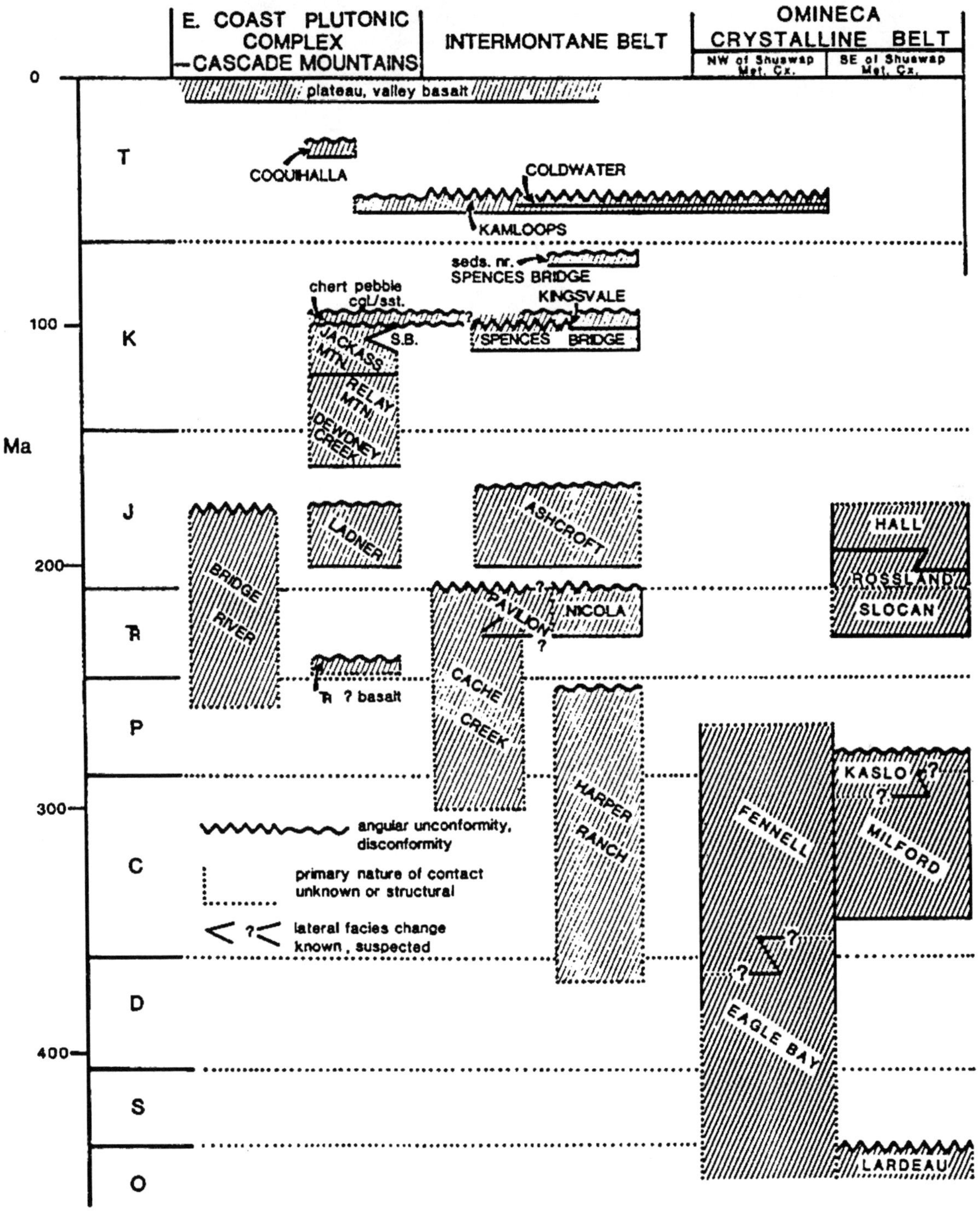

Figure 3-1. Correlation of rock units in southern Intermontane Belt with those in adjacent regions. This figure was made in ~1985; since then, rocks lithologically similar to the Fennell Formation (= Slide Mountain terrane) are recognized below the Nicola Group (as well as the Harper Ranch Group). The western Cache Creek (at that time called "Pavilion beds") is as young as Early, maybe Middle, Jurassic; and the Omineca Crystalline Belt and Coast Plutonic Complex have "metamorphosed" into Omineca and Coast belts (from Price et al., 1985).

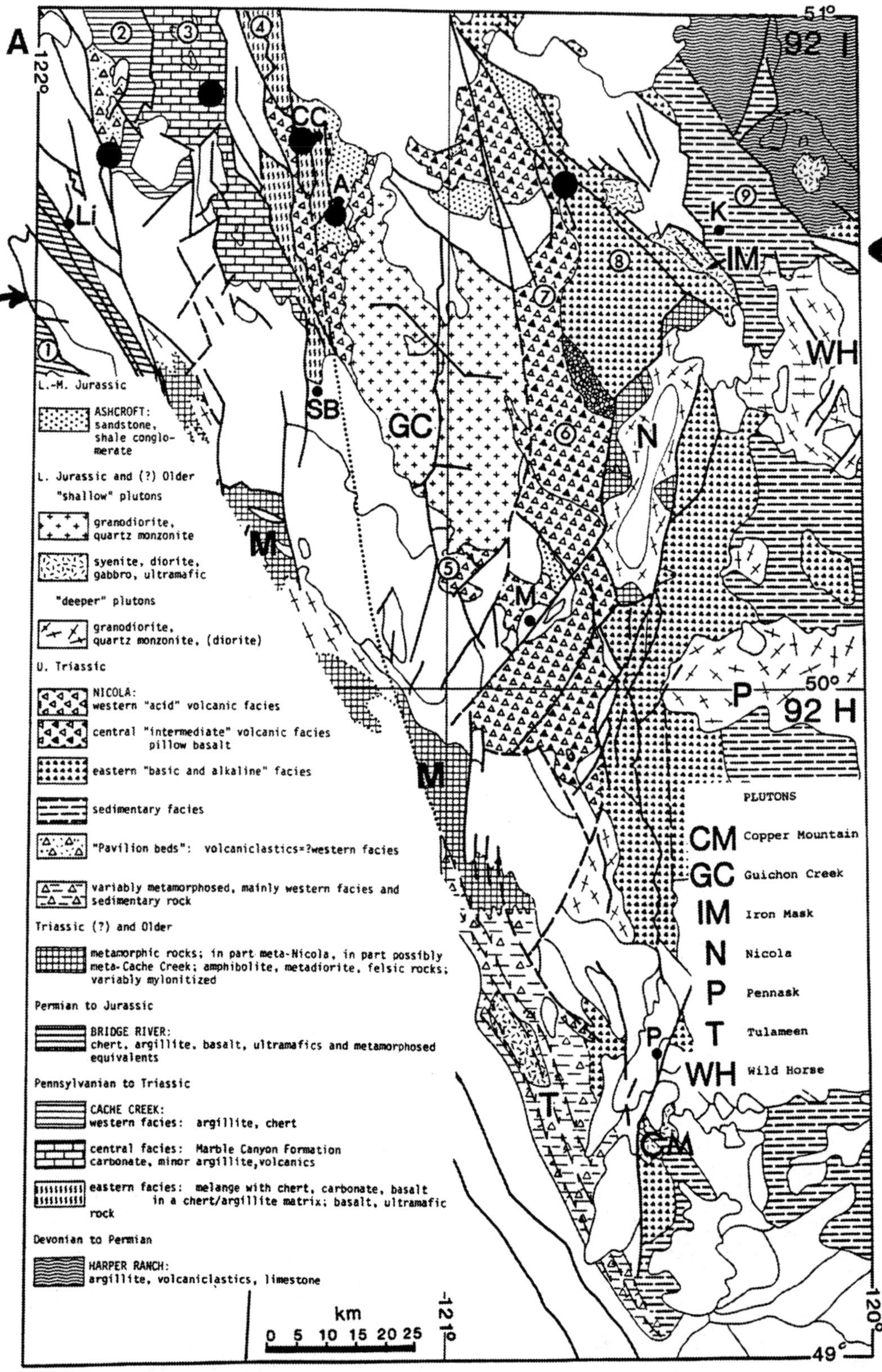

Figure 3-2. Maps of the southern Intermontane Belt; heavy black dots in northern part of map show locations of field trip stops 3-1 to 3-5 sequentially from east to west. Arrows in margins indicate ~position of route. Settlements: A = Ashcroft; CC = Cache Creek; K = Kamloops; Li = Lillooet; Ly = Lytton; K = Kamloops; M = Merritt; P = Princeton; SB = Spences Bridge; from Monger (1985).

Figure 3-2 A. Late Paleozoic and early Mesozoic rocks; Triassic and older metamorphic rocks are in the Mount Lytton Complex (M) and Nicola batholith (N); circled numbers refer to sections in Fig. 3-3. Note the position of the Late Triassic (212 Ma) calcalkaline Guichon Creek (GC) batholith relative to Late Triassic alkaline-subalkaline Iron Mask (IM), Tulameen (T) and Copper Mountain (CM) plutons and to the Early Jurassic (~195 Ma) calc-alkaline Wild Horse Creek (WH) and Pennask (P) batholiths, which are emplaced mostly in Triassic sedimentary rocks, just east of the eastern volcanic facies of the Nicola arc.

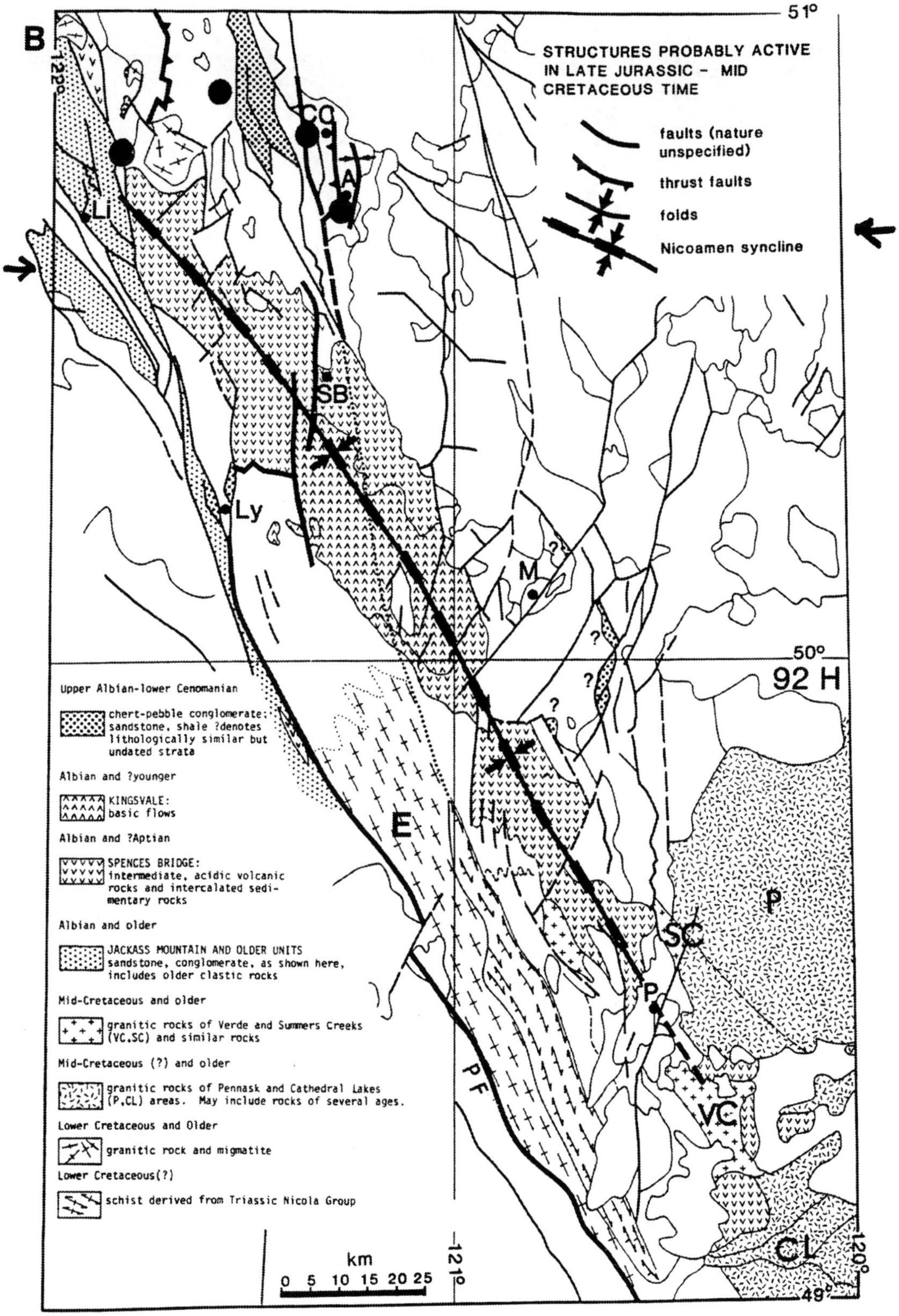

Figure 3-2 B. Distribution of Cretaceous rocks and Jurassic to Cretaceous structures; Early Cretaceous and older rocks belong to the Eagle Plutonic Complex (E); dotted line connects eastern boundary of the Eagle Complex with its southwest dipping gneissosity and adjacent concordant Nicola schist belt with the east-vergent structures near Ashcroft (for more, refer to Figure 3-7); PF = Pasayten Fault was active as a sinistral fault at 105 Ma and a reverse fault in early Tertiary. The pluton labelled P is now known as the Osprey Lake pluton and is of late Middle Jurassic age (~165 Ma); that labelled CL is late Early Cretaceous (105 Ma) and the same age as VC, SC.

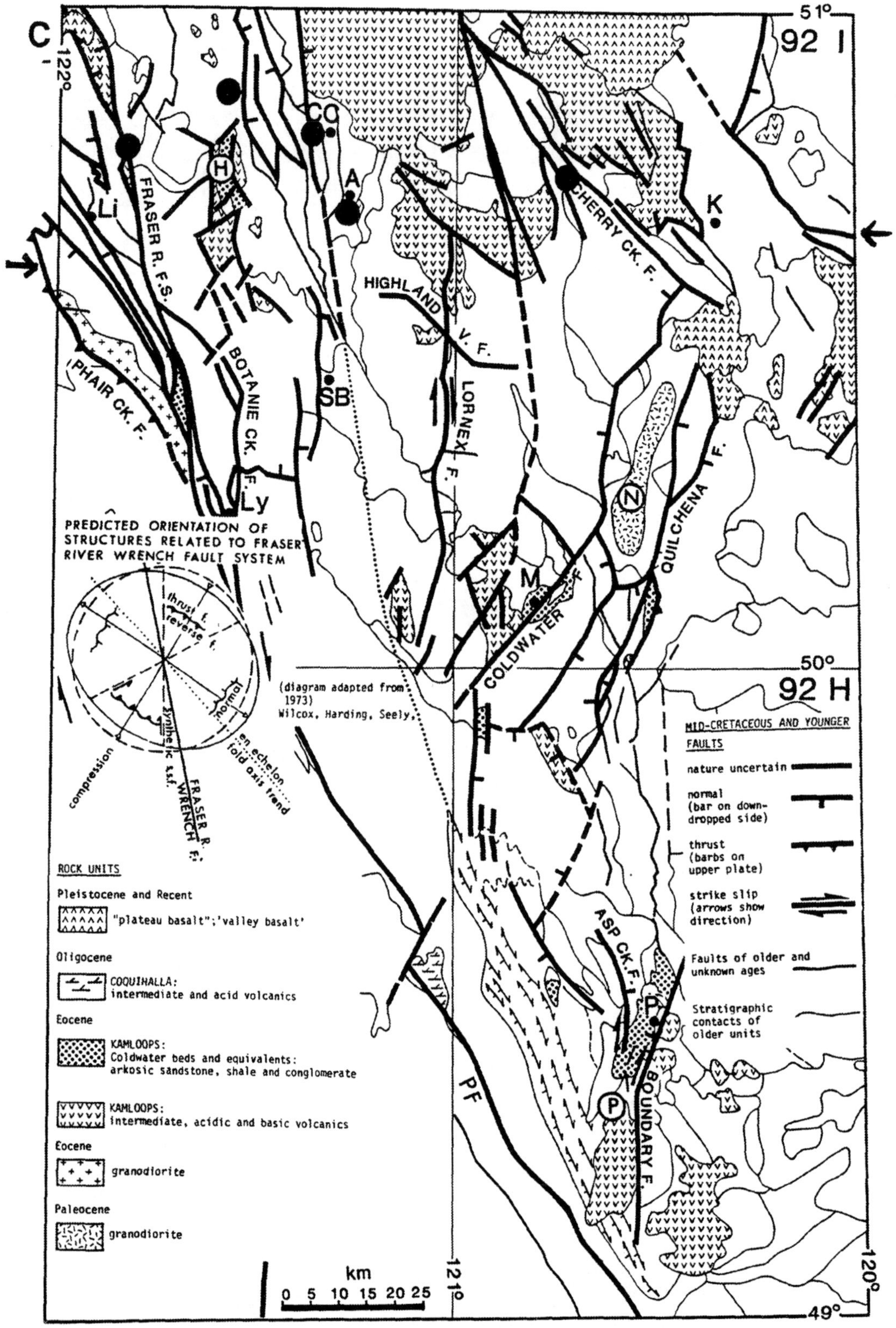

Figure 3-2 C. Distribution of Tertiary rocks and post-Early Cretaceous faults in the southern Intermontane Belt and adjacent parts of Coast Belt. Note that most structures can be related to dextral strike-slip faulting on the Fraser Fault. H = Hat Creek "graben"; N = Nicola horst; P = Princeton basin.

monocline (see Stop 2-9 discussion), and around the south end of the high-grade metamorphic rocks crossed on Highway 1 between Revelstoke and Sicamous, and into the Slocan Formation of the Kootenay Arc (Figs. 1, 2-23, 2-25; Thompson and Daughtry, 1998). These two relationships, one structural and one spatial, may mean that Okanagan, Harper Ranch subterranes and Kootenay terrane came together prior to the Middle Triassic in an event that at least temporally can be correlated with the Sonoma orogeny of Nevada.

Cache Creek terrane (or Cache Creek Complex) is located on the western side of the southern Intermontane Belt. It comprises disrupted radiolarian chert, argillite, mafic volcanic rocks, and local ultramafic rocks and distinctive large masses of carbonate, and ranges in age from Pennsylvanian to Early Jurassic (Fig. 3-3). It formed as the accretionary complex that likely accompanied the early Mesozoic Quesnellian magmatic arc (Figs. 3-2A, 3-4a).

The youngest known marine rocks, in the clastic Ashcroft Formation, are of late Middle Jurassic (Callovian, ~157 Ma) age. Most Middle Jurassic and later Mesozoic rocks in the Intermontane Belt evolved within a continental arc regime that featured contraction, tectonic uplift, erosion, and clastic deposition (Fig. 3-2B, Fig. 3-4b, c).

Early Tertiary continental arc rocks (Kamloops Group) evolved in a transtensional setting dominated by normal and strike-slip faults. These faults largely control the map-pattern of the southern Intermontane Belt (Figs. 3-2C, 3-4d).

Grounds for Paleogeographic Suspicion

All pre-Middle Jurassic rocks in the southern Intermontane Belt are *suspect* in the paleogeographic sense of Coney et al. (1980).

As discussed earlier in the Omineca Belt section (following 218.3 km) and shown in Fig. 2-23, stratigraphic and structural evidence permits *orogen-normal* restoration of the miogeoclinal rocks presently residing in the western Foreland Belt to an original location along Highway 1 near the (present) boundary of the Omineca and Intermontane belts. To this

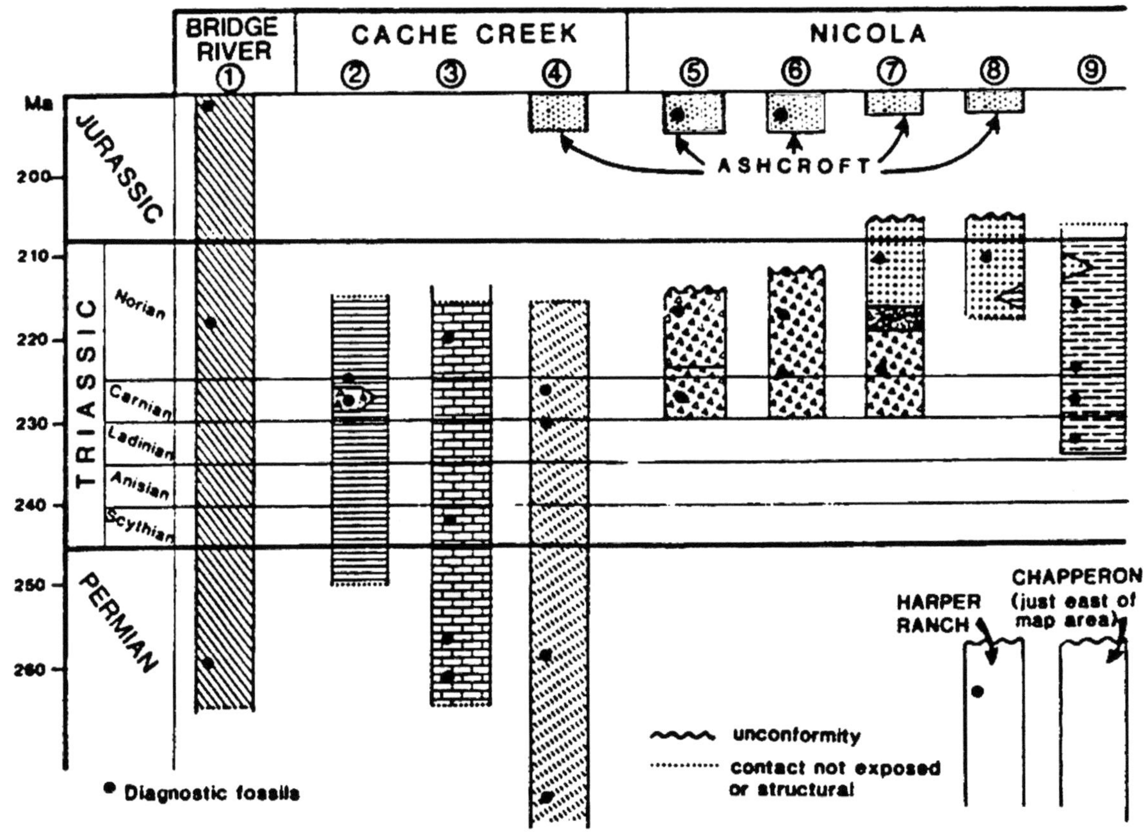

Figure 3-3. Age ranges of the early Mesozoic to late Paleozoic sections located by circled numbers in Figure 3-2A. The western Cache Creek is now known to be as young as Early, possibly Middle, Jurassic.

must be added undetermined amounts of shortening within the western Omineca Belt and within the internally imbricated Slide Mountain terrane. The longevity, from Early Mississippian to at least Middle Permian time (~90 million years), of the marginal basin represented by the Fennell Formation argues that it probably was wide. A second argument, favouring the existence of a wide basin between late Paleozoic and Triassic arcs in Quesnel terrane and the craton margin, is that no tuff is known in Triassic (or Permo-Carboniferous) continental margin strata in the southern Rocky Mountains (Ross et al., 1997; G.M. Ross, pers. comm. 1999), although present in the latest Devonian Exshaw Formation (Richards, et al., 1994) and in Early Jurassic rocks of the Fernie Group (Hall et al., 1998).

Paleomagnetic interpretations suggest that there was ~2000 km south-to-north *orogen-parallel* displacement for Pennsylvanian-Permian strata of at least parts of Slide Mountain terrane (near latitudes 53°N and 59°N) with respect to the craton (Richards et al., 1993). Paleobiogeographic comparisons of Permian, Triassic and Early Jurassic (Sinemurian-Pliensbachian) faunas in Quesnel terrane and of a Middle Permian fauna in arc-related rocks of a fault slice in Slide Mountain terrane in northern British Columbia, with coeval faunas on the craton margin, suggest south-to-north terrane displacements of ≥1000 km (respectively, Ross and Ross, 1983; Tozer, 1982; Smith and Tipper, 1986; M. Aberhahn, pers. comm., 2000; Ross, 1969). These faunas have affinities with those to the south in the western conterminous United States, and probably lived in lower latitudes within or near the margin of eastern Panthalassa. By contrast, the Middle Permian faunas in Cache Creek terrane (Stop 3-4) are exotic, comparable to faunas in the Paleo-Tethys, and occur in blocks within the accretionary complex that probably accompanied early Mesozoic Quesnellian arc magmatism.

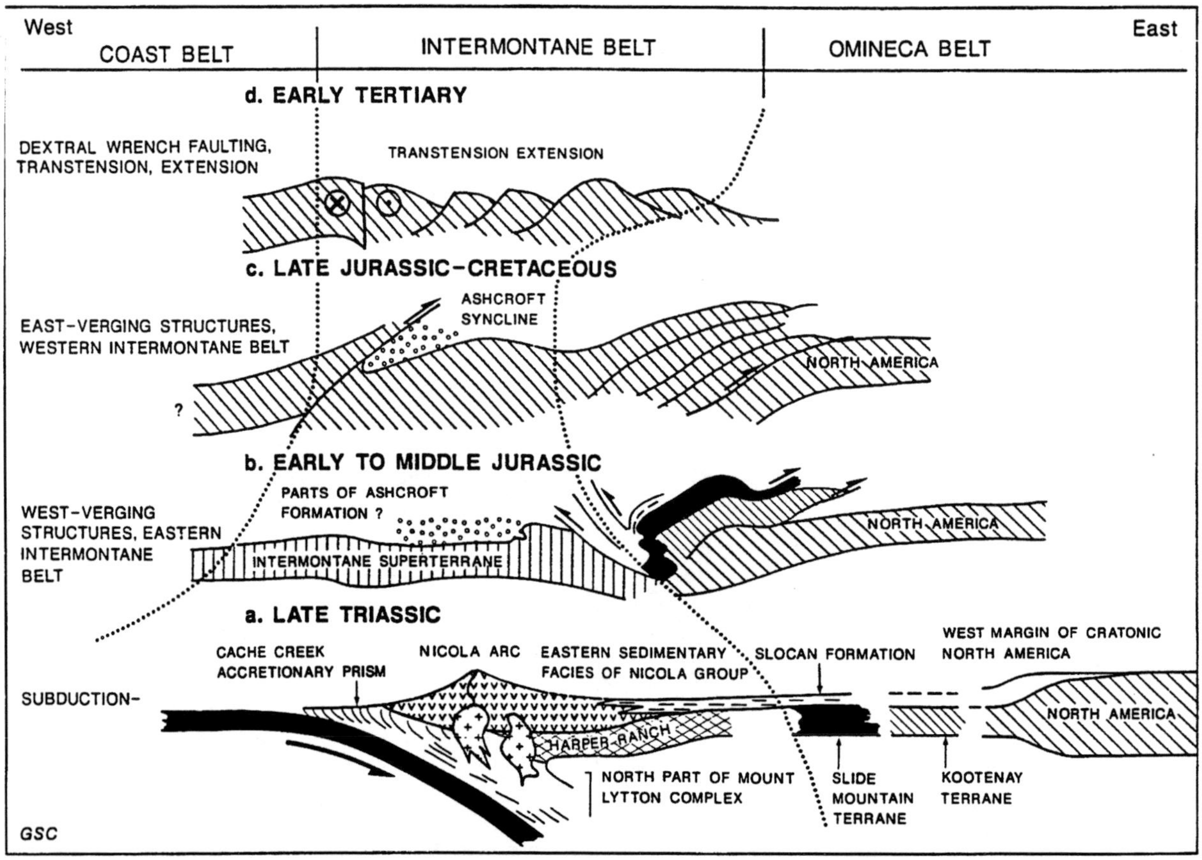

Figure 3-4. Cartoons illustrating the evolution of the Intermontane Belt and its relationship to the Omineca Belt. Possibly the rocks including and west of Slide Mountain terrane were positioned by south to north movements in late Early Jurassic time (~185 Ma) immediately before their accretion to the old continental margin. The west-directed structures at the boundary between Omineca and Intermontane belts are reported by Schiarizza and Preto (1984) and Smith (1979) and may include the Louis Creek-Bolean Creek fault system crossed on Highway 1, and may be congruent with Middle Jurassic west-directed structures in the Selkirk Fan (Fig. 2-14).

PRITCHARD TO CACHE CREEK

ROAD LOG

0.0 km Pritchard; junction of Highway 1 with road to Falkland.

7.2 km Roadcuts are in undated, dark green, locally epidotized, massive pyroxene-bearing mafic breccias. These are lithologically similar to rocks of the eastern, youngest, volcanic part of the Nicola Group. Similar rocks form the bluffs on the north side of the South Thompson River. The rocks are also similar lithologically to Early Jurassic volcanics along strike ~100 km to the north northeast, and to the Early Jurassic Rossland Group in the Kootenay Arc, ~200 km to the southeast. As dated Triassic strata along strike south of Highway 1 are mainly sedimentary, the question of the precise age of the volcanic rocks here along Highway 1 remains open.

From this locality, westwards for ~110 km, almost as far as the village of Cache Creek, Triassic strata of the Nicola Group are common along Highway 1. Although the Nicola Group changes laterally in lithology, from mainly argillites and siltstones south of this part of Highway 1 to mainly volcanic rocks 40 km farther west, it varies relatively little in its sub-greenschist metamorphic grade.

The bluff southwest of Highway 1 is volcanic rock of the Eocene Kamloops Group, which unconformably overlies the Nicola Group in many places in the southern Intermontane Belt (Figs. 3-1, 3-2C).

10.8 km Junction of Highway 1 with Highway 97 at Monte Creek. The eastern boundary of the maps in Fig. 3-2 is ~6 km west of here.

17.7 km The basal contact of undated volcanic rocks correlated lithologically with the Nicola Group is exposed near the top of the north side of South Thompson River valley, northeast of the Canada Lafarge cement quarry. There, massive augite porphyry grades down into a breccia unconformably overlying Early Pennsylvanian limestone of the Harper Ranch Group.

Harper Ranch Group

West of 17.7 km for ~8 km, rocks of the Harper Ranch Group are exposed on the north side of the valley above the glacial lake silt terraces in the valley bottom. The Harper Ranch Group includes latest Devonian (Famennian, ~360 Ma) chert-pebble conglomerate, sandstone and shale with brachiopods and plant fragments, and prominent Late Mississippian, Early Pennsylvanian(?), Middle Pennsylvanian, Early and Middle Permian limestone bodies hosted by more recessive and undated volcaniclastic rocks and graded siliceous argillite and siltstone. The rocks are faulted and folded, and although fragmentary sections can be measured, no continuous stratigraphic succession embracing all dated units is recognized.

Permian fusulinid faunas in the Harper Ranch Group are dominated by schwagerinids and can be correlated with the (North American) mid-continent-Andean fusulinid fauna. Although there are some schubertellid fusulinids and representatives of other phyla that are more typically Tethyan, the distinctive Middle Permian Tethyan verbeekinid fusulinids in the Cache Creek Complex (Stop 3-4) are absent from the Middle Permian fusulinid fauna in the Harper Ranch (Orchard and Danner, 1991a, b). The Permian faunas in the Harper Ranch are most like those of Stikine terrane in northwestern British Columbia (Fig. 7), of the arc-affinity slice within the Sylvester Group (=Slide Mountain terrane) of northernmost British Columbia, from some sequences in central Oregon, and from the Nosoni and McCloud limestones of northern California. The faunas are thought to have lived at low latitudes in eastern Panthalassa (Ross and Ross, 1983; Orchard and Danner, 1991a).

19.3 km Overpass to Canada Lafarge quarry and cement plant on the north side of Thompson River. The large quarry is in Permian limestone.

23.7 km Low on north side of river is an olistolith(?) of Late Mississippian limestone in a matrix of siliceous argillite and volcaniclastic rock.

24.7 km On ridge top on north side of the South Thompson valley is the contact between west-dipping Harper Ranch strata (to the east) and Middle Triassic and younger volcanic sandstone and argillite of the Nicola Group. The contact is a reverse fault; Nicola strata near the fault form the west limb of a large, overturned anticline (Smith, 1979).

Wild Horse Creek Batholith: Eastward Arc Migration Between Triassic and Jurassic Time?

South of Highway 1, the higher areas are underlain by granodiorite of the Wild Horse Creek Batholith, which is emplaced in argillite and siltstone of the eastern sedimentary facies of the Triassic Nicola Group (Fig. 3-2A). The granitic rock is U-Pb dated at 196 Ma (Parrish and Monger, 1992), and is one of a north-south aligned belt of Early Jurassic calc-alkaline plutons that extends southwards for ~150 km. The belt is coeval with alkaline/subalkaline volcanic rocks of the Early Jurassic Rossland Group and its comagmatic intrusions, located ~250 km southeast of here, in the Kootenay Arc.

The relationship of this belt of calc-alkaline plutons to the slightly older, mainly Upper Triassic, volcanic rocks of the Nicola Group and its co-magmatic, calc-alkaline and alkaline plutons (respectively Guichon Creek and Iron Mask batholiths, located ~60 km and ~15 km west of here; Fig. 3-2A) raises interesting questions. As discussed later, the Nicola Group and associated plutons belong to a west-facing arc. Arguably, the Wildhorse Creek and related plutons to the south are the western calc-alkaline roots of the Early Jurassic Rossland arc, whose volcanic edifice was removed by erosion.

The Rossland arc lies mainly east of the Late Triassic Nicola arc, and appears to be far wider (although the effects of Eocene extension in southern British Columbia must be taken into account). The subduction complex related to both arcs is probably the Cache Creek Complex, located ~100 km west of here. Parrish and Monger (1992) suggested that the apparent eastward migration and widening of the arc was a response to flattening of the subduction zone, which indicates that the upper, North American plate carrying the arc on its leading edge started to move towards and over the subduction zone between Triassic and Jurassic time (200 Ma). Note that this advance took place *prior* to the Toarcian (183–178 Ma) obduction/accretion process (Fig. 2-13; Murphy et al., 1995) which initiated Cordilleran mountain building, so that the initial hints of "collision" took place at the North American plate boundary, and not near the ancient continental margin.

36.9 km Junction of Highway 1 with Highway 5 to the north (to Jasper). Road ascends around the southeast side of the town of Kamloops, which is located on the site of a former (1800s) Hudson's Bay trading post located at the confluence of the south-flowing North Thompson River with the west-flowing South Thompson River.

39.1 km Highway 1 crosses a bridge over a gully. Contorted argillite, siltstone and volcanic sandstone visible at the head of the gully to the south (left) are the westernmost exposures along Highway 1 of the eastern sedimentary facies of the Nicola Group (Fig. 3-2A).

41.7 km Roadcut in volcanic rocks of the Nicola Group.

Nicola Group

The Nicola Group and associated, coeval plutonic rocks are relatively well studied, partly because they host important copper deposits. The group can be divided into four facies belts (Fig. 3-2A; Preto, 1979; Morrison, 1980; Monger, 1985; Mortimer, 1987). (1) The western volcanic facies belt comprises Ladinian, Carnian and early Norian felsic to intermediate volcaniclastics, carbonate, and pelite. (2) The central volcanic facies belt contains mainly early Norian, intermediate, fine-grained feldspar and feldspar augite porphyry volcaniclastics, including volcanic conglomerate. (3) The eastern volcanic facies belt is late Norian, mafic augite porphyry volcaniclastics and local pillowed basalt, interbedded with tuffaceous shale. (4) The easternmost sedimentary facies belt ranges in age from Ladinian to Norian. The western and central facies are intruded by the only slightly younger (210 Ma) calc-alkaline Guichon Creek Batholith, and the eastern facies by the alkaline Iron Mask Batholith (K-Ar dates 209–194 Ma).

42.3 km Junction of Highway 1 with Hillside.

42.5–44.6 km Roadcuts in massive purple-green Nicola volcanics.

47.3–48.6 km South (left) above Highway are exposures of the Iron Mask Batholith, which ranges in composition from syenite, through diorite and gabbro to ultramafic rocks. Arguably it is at least in part the intrusive equivalent of the augite porphyry flows and breccias of the eastern volcanic belt of the Nicola Group. An old mine dump is visible from Highway 1.

48.3 km Continue westward on Highway 1, which here diverges from Highway 5 (to Merritt).

49.1 km Roadcuts in rusty, altered marginal parts of Iron Mask Batholith.

52.0 km Mill buildings of Afton Mine on left.

54.1 km Top of Afton open pit mine visible on south (left) of Highway 1.

Afton Mine

The Afton copper mine is a currently (1999) inactive open pit mine which, according to Carr and Read (1976) contained ~31 million tons of copper ore at 1% grade with 0.25% cut-off, and recoverable gold and silver. Current world copper prices make production uneconomical. Production was mainly from fine-grained, supergene native copper in the northwestern extremity of the Iron Mask Batholith, which is emplaced in the eastern volcanic facies of the Nicola Group. The deposit is bounded on the north by a fault, north of which are down-dropped Eocene volcanics of the Kamloops Group, that post-date supergene enrichment.

55.0 km Brown weathering Eocene volcanics of the Kamloops Group are exposed north of Highway 1. Ewing (1980) mapped in this area and first proposed a model of Eocene tectonics involving regional extension and dextral wrench faulting. To the south of here, mid-crustal, mainly granitic rocks ranging in known age from Triassic to Paleocene are exposed in the north northeast-trending Nicola Horst, whose northern boundary is the northwest-trending, dextral Cherry Creek Fault (Figs.1, 3-2C). In this region, Eocene basin and range style extension, widespread east of the Okanagan Valley Fault, is replaced by Eocene transtension.

56.4 km Rusty outcrops of Upper Triassic Nicola volcanics on right of road.

57.2 km Valley of Cherry Creek on left; this valley follows the trace of the northwest-trending Cherry Creek Fault, which continues northwestward across Kamloops Lake.

58.7 km Roadcut on right exposes reddish conglomerate of uncertain, possibly mid-Cretaceous age, juxtaposed with Nicola volcanics.

59.6–63.9 km Broad valley containing Cherry Creek ahead; bluffs on north (right) side of Highway 1 are in Nicola volcanics; at 65.7 km cross Cherry Creek.

~66 km View across Kamloops Lake. On the north side of the lake, to the east (right) are massive brownish cliffs of Eocene Kamloops Group volcanics. Low down, and east of the valley (of Copper Creek) across the lake, the Kamloops Group overlies red and green volcanics with abundant pyroxene and altered olivine phenocrysts, which are included in the eastern volcanic facies of the Nicola. The Cherry Creek Fault bounds these rocks on the west. To the west (left) of the valley is a down-dropped block featuring continental chert pebble conglomerate and sandstone of probable Cretaceous age that stratigraphically overlies the Nicola Group. These chert grain clastics are tentatively correlated with mid-Cretaceous continental sandstones passed on the first part of Day 3 along Highway 99 near Marble Canyon.

72.2–76.5 km Highway 1 climbs and then descends, with numerous roadcuts in volcanics of the eastern belt of the Nicola Group, locally interbedded and interfaulted with sedimentary rocks.

Stop 3-1 Eastern volcanic facies of the Nicola Group

77.0 km Park in area on north (right) side of Highway 1. **Take care crossing the highway** to examine the large roadcut featuring sedimentary rocks in its eastern part and volcanic rocks in its western part. The rock association is typical of the eastern volcanic facies of the Nicola Group.

The black to grey carbonaceous, calcareous argillite and siltstone locally contains the small bivalve *Monotis salinaria* of late Norian age (uppermost Triassic, ~205 Ma) that according to Tozer (1982) is of Tethyan affinity. In fault contact with these rocks at the west end of the outcrop are volcaniclastics with graded beds and slump structures, containing lithic volcanic fragments and euhedral and broken pyroxene crystals. Such rocks, as well as coarser volcanic breccias, and relatively rare pillowed augite porphyry basalt flows, are characteristic of the eastern, alkaline-subalkaline volcanic facies of the Nicola Group.

77.4–79.3 km Rusty and locally reddish volcanics in Nicola Group.

79.9 km Settlement of Savona; road to Logan Lake to south.

82.2–82.7 km Roadcuts in coarse volcanic breccia of Nicola Group

84.2 km Bridge over Canadian Pacific Railway track; outcrops of Nicola volcanics above track.

84.8 km Outflow of Kamloops Lake into the Thompson River.

85.1 km Highway 1 crosses bridge over the Thompson River.

87.2 km The enormous deposits of gravel and sand in this area were deposited in a delta that formed during fluvio-glacial outflow from the Deadman River valley to north.

88.5 km Cross Deadman River. On the east bank of the valley are exposures of highly faulted Nicola volcanic and sedimentary rocks, including small carbonate blocks (olistoliths?) of Late Triassic (late Carnian-early Norian) limestone.

90.2–93.0 km Road climbs western valley wall; roadcuts are red and green, variably altered, fractured Nicola volcanic rocks of the central facies.

96.2 km To south (left) across Thompson River near Walhachin, are (lowermost) a quarry in altered Nicola volcanic rocks (material quarried for railway bed), above which is a series of more-or-less flat-lying benches formed of sandstone, local conglomerate and shale of the Ashcroft Formation. Ammonites from the upper part of the hillside are Aalenian (early Middle Jurassic, ~175 Ma). The contact between the

Ashcroft Formation and the Nicola Group is not exposed here, although it can be located to within a few tens of metres above the quarry. The Early and Middle Jurassic (Pliensbachian-Callovian, ~195–156 Ma) Ashcroft Formation is a marine, mainly clastic unit that unconformably lies on the Nicola Group and on the Guichon Creek Batholith.

Guichon Creek Batholith

Roughly 10 km south of this point on Highway 1 is the northern contact of the Guichon Creek Batholith. This is a large (areally ~30 × 50 km), mainly granodiorite to quartz monzonite body that has yielded numerous K-Ar dates of 200–210 Ma, and one U-Pb date of 210 Ma (Fig. 3-2A; Mortimer et al., 1990). The intrusion was emplaced at shallow depths (≤4 km) within the Nicola Group and is probably comagmatic with its youngest volcanic rocks, being about the same age as the youngest biostratigraphic age (~210 Ma) assigned to the enclosing western and central facies of the Nicola. Like the Nicola Group, its isotopic chemistry is juvenile, and it appears to be the calc-alkaline intrusive frontal part of the Nicola arc, just as the Iron Mask Batholith (passed at 48 km) is the intrusive equivalent of the alkaline/sub-alkaline eastern, back-of-the-arc facies of the Nicola.

The batholith is probably the most intensively mapped granitic body in the Canadian Cordillera as it contains large economic deposits located in Highland Valley in the central part of the batholith, which have been worked since the late 1950s. The deposits are classical porphyry coppers, with chalcopyrite, minor bornite and molybdenite finely disseminated through the innermost, most leucocratic, and youngest phase of the intrusion and its associated dyke swarms. Total ore production, and proven and probable reserves, amount to about 2 billion tons of 0.45% copper equivalent.

As noted earlier, the Guichon Creek Batholith is non-conformably overlain by the Ashcroft Formation. Historically, the long-recognized stratigraphic relationship between the batholith and Lower Jurassic strata of the Ashcroft Formation was used in the 200 Ma calibration of the age assigned to the Triassic-Jurassic boundary. Later, this boundary was considered to be older (~210 Ma). The latest determination, from elsewhere in British Columbia, suggests that the age of the boundary is 199.6 ± 0.7 Ma (Pàlfy, 1999).

98.2–101.4 km Roadcuts in central belt of the Nicola volcanics.

102.2–102.8 km Large roadcut in bluff on north (right) side of Highway 1 is in greenish-purple andesitic breccia with fine-grained feldspar porphyry clasts, that is characteristic of the central facies of the Nicola Group.

108.4 km View of the valley of the Thompson River to south.

109.6 km Red-brown, columnar jointed, basaltic andesite flows and phreatic breccias of the Eocene Kamloops Group, which are the remains of a continental arc. Here, the volcanics unconformably overlie the Nicola Group.

110.9 km Low hills on south (left) side of Highway 1 are underlain by diorite. This is the northernmost, marginal phase of the Late Triassic Guichon Creek Batholith, and here intrudes Nicola volcanics.

113.3 km Red-brown cliffs of the southern Arrowstone Hills on north side of Highway 1 of Eocene Kamloops volcanics.

117.0 km The low, gullied, Rattlesnake Hills on the south side of Highway 1 are mainly underlain by the Early and Middle Jurassic (here, Pliensbachian and Bajocian, ~190–170 Ma) conglomerate, sandstone, calcareous siltstone and shale of the Ashcroft Formation. At the southeastern end of the Rattlesnake Hills, the basal conglomerate of the Ashcroft Formation unconformably overlies red volcaniclastics of the Nicola Group (Smith et al., 1998). Small outliers of Eocene volcanic rocks locally cap the hills.

118.9 km Lowest part of hill on north (right) side of highway is underlain by the Ashcroft Formation and the upper part by Kamloops Group.

120.4 km Descend hill; outcrop on left of Cache Cheek chert.

Cache Creek Complex (and Terrane)

The Cache Creek Complex is the remains of an early Mesozoic accretionary complex that probably accompanied the early Mesozoic Nicola and Rossland arcs, and thus locates the western edge of the early Mesozoic North American plate. It extends relatively continuously along the central part of the Canadian Cordillera from south of Cache Creek (latitude 50°30′N) to the southernmost Yukon (latitude 60°30′N). North of latitude ~54°N it separates the Stikine and Quesnel arc-dominated terranes, which have similar late Paleozoic and early Mesozoic faunas (Fig. 7). Here, in its type area, the Cache Creek Complex consists of three facies belts that differ somewhat in lithology and age (Fig. 3-5).

(1) Easternmost is melange (Stop 3-2) with a matrix of disrupted graphitic argillite and radiolarian chert containing blocks of chert, carbonate, basalt (in places with pillows), local volcaniclastics, gabbro and serpentinite. Ages of fossils (conodonts and fusulinids) in the blocks range from Middle Pennsylvanian to Early Permian, whereas those in the matrix (radiolarians and conodonts) are Early Permian to Late Triassic.

(2) A central belt of massive, shallow water carbonate called the Marble Canyon Formation (Stop 3-4) is associated with minor amounts of chert, argillite and mafic volcanics and contains fossils (fusulinids and conodonts) that range in age from Middle Permian to Late Triassic.

(3) A western belt of argillite, chert, volcaniclastics and minor carbonate contains fossils (conodonts and radiolarians) that range in age from Early Triassic to Early Jurassic.

Orchard and Danner (1991b) provide faunal lists and references for the Cache Creek Complex.

120.7 km Stop lights at junction of Highway 1, which here turns south (left), with Highway 17 to the north, in the village of Cache Creek. Continue straight on across junction to small road that crosses Bonaparte River.

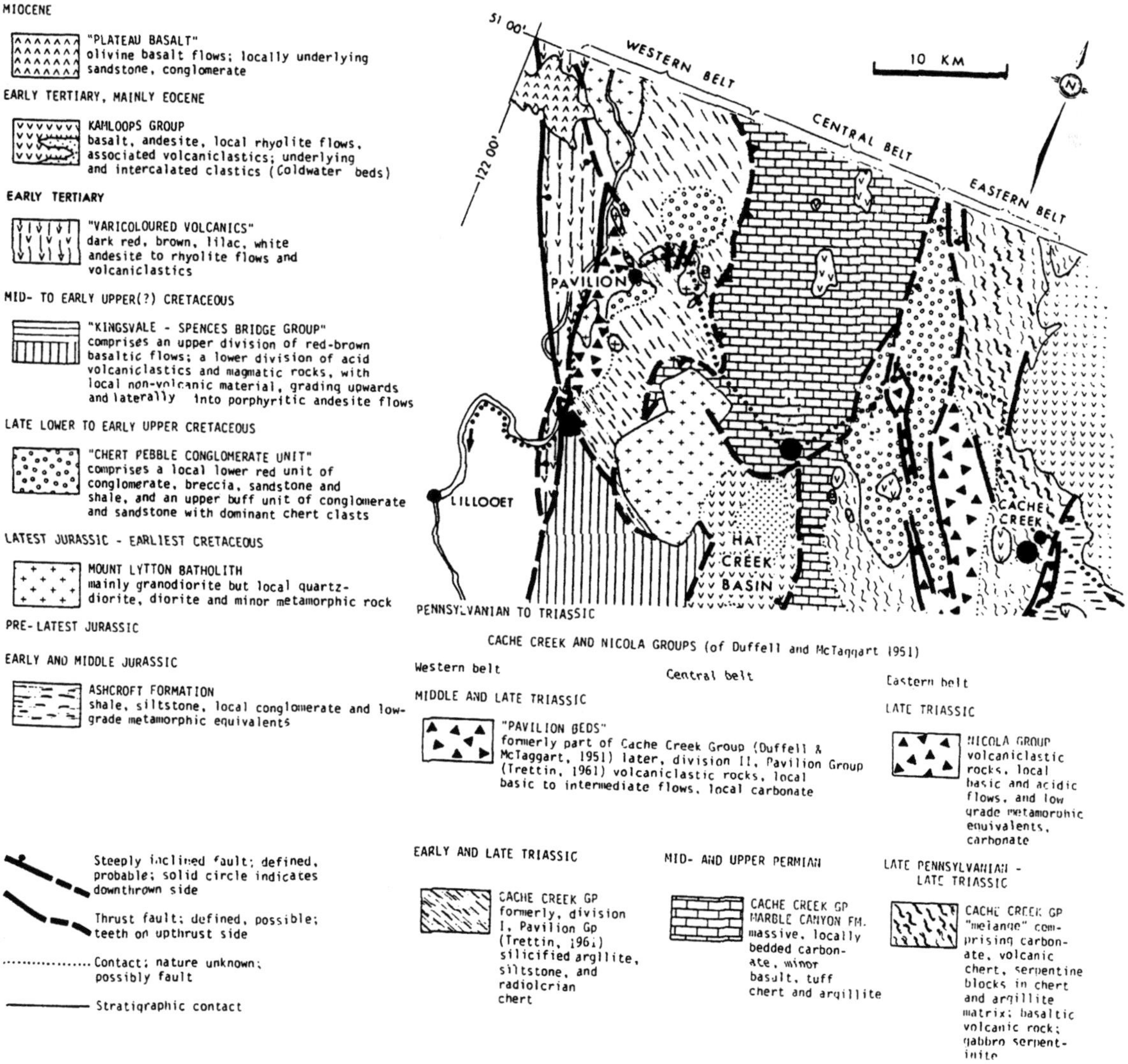

Figure 3-5. Cache Creek Complex; route shown by dotted line, stops by heavy black dots. Note: the "chert-pebble conglomerate" unit is now known to probably overlie the latest Early Cretaceous Spences Bridge Group, although this relationship is only known from drill core in the Hat Creek basin.

Stop 3-2 Cache Creek melange

120.8 km Cache Creek melange is exposed in long roadcut in hillside on the west side of the village of Cache Creek. The nature of the recessive matrix of the melange belt is generally only seen in artificial exposures such as this roadcut. At the north end of the roadcut is a block of Middle Pennsylvanian limestone in a matrix of cherty argillite. Continue south along the outcrop, through pervasively faulted, broken, contorted gypsiferous argillite, cherty argillite and radiolarian chert, which contains blocks of more massive chert and carbonate. At the southernmost end of the cut, near Highway 1, are thin beds of chert separated by argillaceous laminae called *ribbon chert.* The chert here contains abundant Late Triassic radiolarians, which are best seen on a fresh surface as small (≤1 mm) dark spots (crystalline quartz infills the hollow radiolarian tests) within the pale grey, translucent chert matrix. **Be careful when hammering splintery rock!**

The chaotic structural style seen in the roadcut is typical of that in parts of many accretionary complexes. It is unknown here just how much of the deformation was the result of the process of accretion and just how much was the result of later deformation (e.g. eastward thrusting of Cache Creek over Quesnellia).

From this locality rejoin Highway 1, and drive south; Cache Creek rocks are exposed on both sides of the highway. At ~2.4 km south of the southern boundary of the village of Cache Creek turn left (east), and take Highway 97C to the town of Ashcroft. The Bonaparte River, concealed in the valley the left, flows into the Thompson River just above the town of Ashcroft. Descend valley, cross the bridge over the Thompson River, and continue around Ashcroft on Highway 97C to Highland Valley and Logan Lake. Follow the road up the east side of the Thompson River valley to a point where the highway turns sharply east (left) uphill, ~2.8 km south of the Thompson River bridge.

Stop 3-3 Cache Creek-Quesnellian structural relationship

134.0 km The view of the Thompson River valley near Ashcroft displays the relationship between Cache Creek Complex (or terrane) and Nicola rocks (Quesnel terrane). Facing north (back towards Ashcroft), to the northeast (half right), the Early and Middle Jurassic (Pleinsbachian to Callovian) Ashcroft Formation stratigraphically overlies the Late Triassic Nicola Group and the Guichon Creek Batholith (Figs. 3-2A, 3-6). The stratigraphic contact between the Ashcroft and Nicola is exposed at the southeast end of the Rattlesnake Hills, in the central middle distance. The valley of the Thompson River is eroded in folded and faulted dark argillite, siltstone and sandstone of the Ashcroft Formation, much of which appears to be a deep water facies. These rocks are exposed in gullies on right hand side of Highway 97C just before the town of Ashcroft, and in river, road and railway cuttings on the west side of the Thompson River. To the northwest (half left), the top of the hill above the Thompson River features an inverted stratigraphic contact, with Ashcroft strata below the middle Norian clastics of the Nicola Group which form the top of the hill (Fig. 3-6). To the west, near Highway 1 and forming the hills up to the western skyline, are rocks of the Cache Creek Complex.

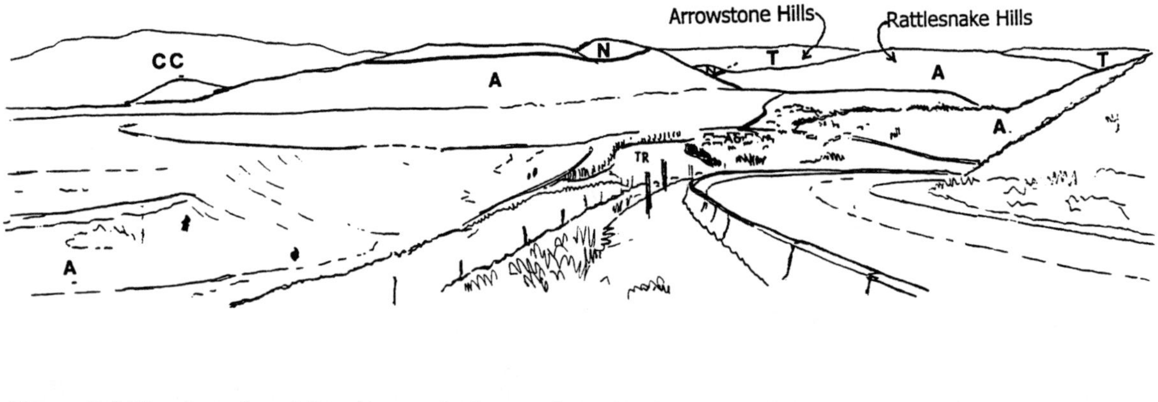

Figure 3-6. View from Stop 3-3 looking north, showing the scale of the Ashcroft syncline; overturned Nicola and Ashcroft contact exposed near crest of hill to northwest; right-way-up stratigraphic contact exposed on east end of Rattlesnake Hills, just below T on right. A = Ashcroft Formation; CC = Cache Creek Complex; N = Nicola Group; T = Kamloops Group; AS = town of Ashcroft; TR = Thompson River.

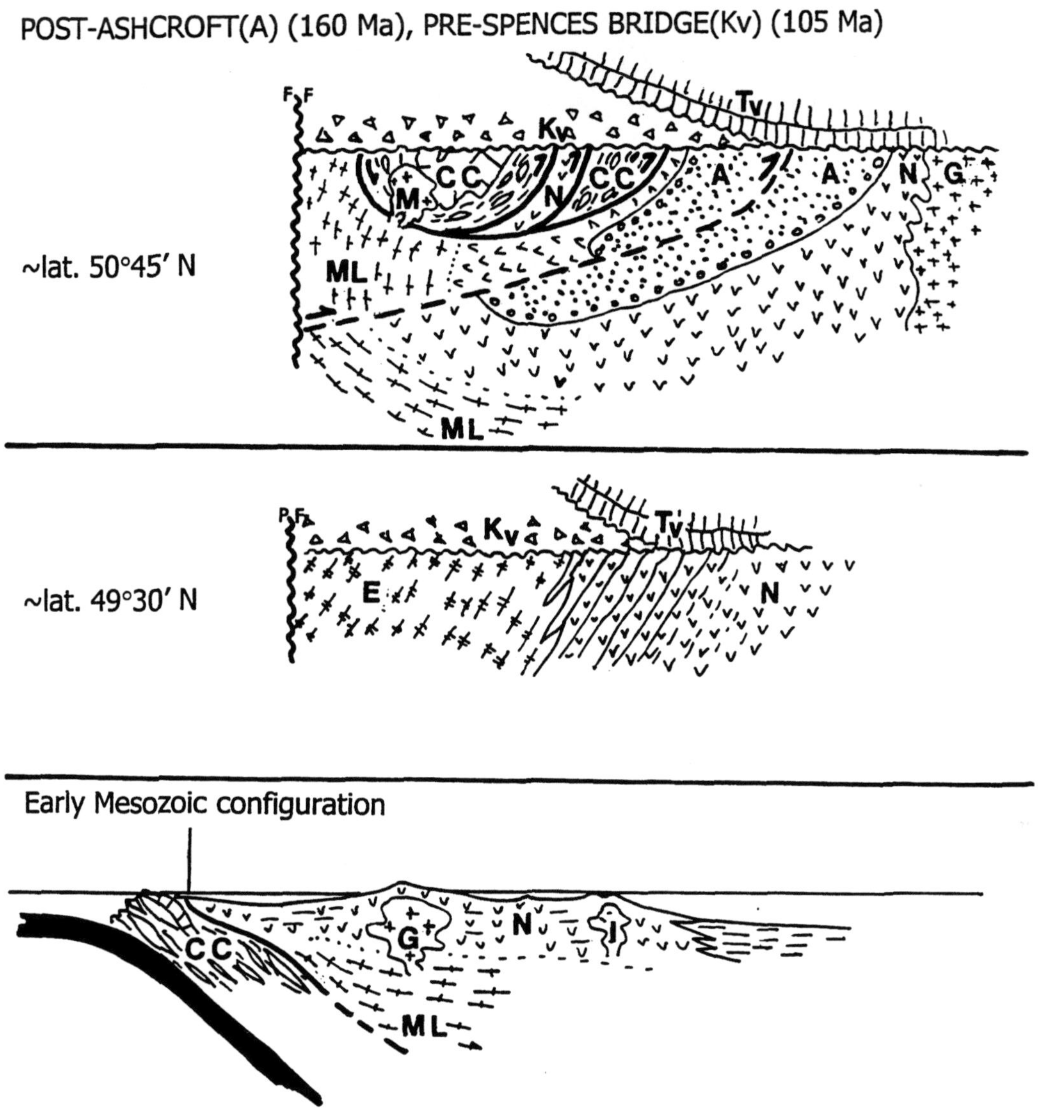

Figure 3-7. Cartoons illustrating vertical configuration of Mesozoic structures on the western edge of Quesnel terrane (west on left).

Top section. This section, drawn near Ashcroft (near latitude 50°45′N) shows the Ashcroft syncline in section (after Travers, 1978). The Cache Creek Complex, in places imbricated with the Nicola Group, is thrust eastwards over the overturned Nicola Group and Ashcroft Formation; to the east, the right-way-up Ashcroft Formation stratigraphically overlies the Nicola Group and the Guichon Creek Batholith.

Middle section. This section is drawn near latitude 49°30′N; the fabric in the ~140 Ma Eagle Plutonic Complex is concordant with that in the westernmost Nicola Group. The structure is along strike from the Ashcroft syncline (see Fig. 3-2B) and may be equivalent to it, but at a deeper structural level than that exposed in the Ashcroft syncline, along strike near latitude 50°45′N (refer to Figure 3-2B for the spatial distribution) and may be equivalent to it but at a deeper structural level. Both the Eagle and Mount Lytton complexes represent the uplifted and deeply eroded westernmost edge of Quesnel terrane.

Bottom section. This cartoon shows probable original configuration of the early Mesozoic Quesnellian arc.

A = Ashcroft Formation; CC = Cache Creek Complex; E = Eagle Plutonic Complex (JK); G = Guichon Creek batholith (UTr); I = Iron Mask Batholith (UTr); Kv = Spences Bridge Group; M = Mount Martley pluton (JK); ML = Mount Lytton Complex (PTr); N = Nicola Group (UTr); Tv = Kamloops Group; FF = Fraser Fault; PF = Pasayten Fault.

Travers (1982, 1987) suggested from these relationships that the Ashcroft Formation forms the core of an eastward-overturned, thrust faulted syncline, above which the Cache Creek Complex was thrust eastward over the western edge of Quesnel terrane (section at latitude 50°45′N on Fig. 3-7). His argument was supported by the juxtaposition of shallow water and deep water Ashcroft facies across the faulted syncline.

Travers (1982) felt that deformation was during or soon after Ashcroft deposition in late Middle Jurassic time, basing his argument on interpretation of features indicating soft sediment deformation. Later, Greig (1991) showed that the Nicola Group at the western margin of Quesnellia about 100 km south of here (section at latitude 49°30′N on Fig. 3-7) was deformed with a predominantly west-dipping cleavage, and that the deformation was associated with intrusion of gneissic granodiorite of the Eagle Plutonic Complex of latest Jurassic age (157–148 Ma). Regional mapping (Fig. 3-2B; Monger, 1989) shows that the deformed rocks near the Eagle Complex are on trend and congruent with those of the Ashcroft syncline, although formed at greater depth. Thus, the deformation near Ashcroft may be latest Jurassic. The deformation is late Early Cretaceous because continental arc volcanics of the Spences Bridge Group (dated at 105 Ma) overlap both the structures and Quesnel and Cache terranes (Fig. 3-2B).

The Ashcroft Formation not known to definitely lie stratigraphically on Cache Creek terrane, although (in three places) it is imbricated with the eastern, melange belt of the Cache Creek. The Ashcroft Formation is probably time-equivalent to the "Early-Middle" Jurassic of part of the western belt of Cache Creek.

Return to Cache Creek by same route.

134.6 km Tightly folded dark grey argillite and siltstone of the Lower and Middle Jurassic Ashcroft Formation is exposed in small canyon east of the highway, southeast of Ashcroft.

136.6 km Entering Ashcroft; rail cars with yellow covers (if present) contain concentrate from the Highland Valley porphyry copper mine, in the Guichon Creek batholith.

136.6–136.9 km Town of Ashcroft and bridge over Thompson River. Roadcuts west of river are in sandstone and shale of the Ashcroft Formation.

138.9 km Rattlesnake Hills across Bonaparte River valley to north (right) are underlain mainly by the Ashcroft Formation, which here ranges in age from Pliensbachian to Bajocian (~190–170 Ma), although Callovian (~157 Ma) fossils are known from just south of Thompson River (Smith et al., 1998). At the southeast end of the hills, the right-way-up stratigraphic contact between the Ashcroft Formation and the underlying Nicola Group is exposed. The knoll at the western end of the Rattlesnake Hills is underlain by overturned Nicola Group, above the Ashcroft Formation.

140.1–141.8 km Roadcuts on left in contorted dark grey argillite and sandstone of the Ashcroft Formation. Near top of hill above this, on left, the inverted stratigraphic contact between Ashcroft and Nicola strata is exposed.

142.0 km Bluff across Bonaparte River valley to north (right) is Nicola Group, at westernmost end of Rattlesnake Hills.

142.7 km A touchable terrane boundary! View on right is up the Bonaparte River valley. Contorted chert of the Cache Creek Complex low down to west (left) is in structural contact with volcaniclastic rocks of the Nicola Group to the east (right). The contact is a fault marked by greenish serpentinite. The age of the fault is not known, but is unlikely to be the original contact and may be Jura-Cretaceous and related to the Cache Creek overthrusting discussed at Stop 3-3.

144.0 km Junction of Highway 99C with Highway 1. Turn north along highway.

144.3–144.5 km Roadcuts in Cache Creek argillite on left. View to east (right) across the Bonaparte Valley shows serpentinite low down, above which are bluffs of middle Norian Nicola Group volcaniclastics. Travers (1982) found chert and greenstone clasts in the Nicola above the serpentinite and suggested that they were eroded from the uplifted outer arc ridge of the Cache Creek accretionary complex that accompanied the Nicola arc.

Towards the north end of the exposure is a large block of brownish weathering Eocene Kamloops Group flows overlying tuff and shale that slid downhill on top of Quaternary gravel.

145.3 km Knoll of Cache Creek chert and argillite on right.

146.4 km South end of village of Cache Creek, with outcrop of bedded radiolarian chert (south end of Stop 3-2) on left.

Overnight in Cache Creek.

WESTERN INTERMONTANE BELT

ROAD LOG — CACHE CREEK TO FRASER RIVER

147.1 km Proceed north on Highway 97.

147.8 km Cross bridge over Bonaparte River. Low down to the east (right), white weathering Eocene felsic intrusions are probably related to the Kamloops Group, which underlie the Arrowstone Hills, northeast of Cache Creek. To west (left), the hillside is underlain by Cache Creek melange; resistant blocks (*not* glacial erratics) stand out from the recessive covered matrix.

149.1 km To the east (right) brown bluffs in the lower valley side are pillow basalt that is a large lens within Cache Creek melange. The upper valley side is underlain by Eocene volcanic rocks of the Kamloops Group.

151.3–151.6 km To west (left) of Highway 97 are blocks of mainly chert and limestone in Cache Creek melange. At least one block is ignimbrite containing blue amphibole; ignimbrite is locally present in the western belt of the Nicola Group.

151.9 km Village of Bonaparte; at 153.3 km cross bridge over Bonaparte River.

153.8 km Low on hillside to west (left) are outcrops of conglomerate with granitic clasts and sandstone. The conglomerate is undated but is correlated on lithology with the Ashcroft Formation, which here is either faulted or folded into the Cache Creek.

154.6 km To east (right) on lower slopes, melange blocks resemble glacial erratics; above are massive, nearly flat-lying Eocene volcanic rocks of the Kamloops Group, that northwards gradually descend almost to level of highway.

158.2 km Junction of Highway 97 and Highway 99. Turn west (left) onto Highway 99 to Lillooet. At 158.7 km, recross Bonaparte River and at Carquile (158.9 km) cross a farm road which in the 1860s was the old Cariboo wagon road to the placer goldfields at Barkerville, some 300 km north of here. The large building among the ranch buildings south of the highway was once a hotel on the old road.

159.4–160.3 km Roadcuts in melange expose a large block of fine-grained amphibolite and blocks of crumpled chert in a mainly graphitic argillite matrix.

164.5 km Bluffs on north side of Highway 99 are composed of sandstone, conglomerate and shale palynologically dated as mid-Cretaceous (Albian-Cenomanian). The beds appear to have been deposited by southeast-flowing braided streams and contain a high proportion of chert grains that probably were derived from the Cache Creek Complex. At 166.6 km, similar rocks underlie much of the hillside to the south (left). The rocks form eastern limb of a faulted anticline, which is cored by possible Nicola Group volcanic rocks passed at 167.7 km.

Continental sandstones and conglomerate with a similar high content of chert clasts and of the same age are present in the eastern Coast Belt, and there overlie Methow and Bridge River terranes. They thus may constitute a mid-Cretaceous *overlap assemblage* between terranes of Intermontane and Coast belts.

167.1–172.2 km Roadcuts mainly along old road on north (right) of felsic volcaniclastics that locally contain β-quartz crystals. The rocks are undated here, but are correlated on lithology with some in the Late and locally Middle Triassic western belt of the Nicola Group.

168.5–172.2 km Roadcuts on north (right) side of highway in sandstone, shale and conglomerate on western limb of anticline. At 169.7 km a log is present in the sandstone. At 170.9 km the sandstone turns reddish, and between 171.9–172.2 km is red fanglomerate that in places contains clasts of carbonate bearing fusulinids similar to those in the Cache Creek Complex at Stop 3-4.

173.8 km Roadcut in graphitic argillite of the Cache Creek Complex.

174.0 km Roadcuts in carbonate of the Cache Creek Complex.

Hat Creek "graben"

To south (left) is a view of the Hat Creek valley. Beyond (to the southwest) the high ridges are capped by late Albian (~105 Ma) continental volcanics of the Spences Bridge Group unconformably overlying the western belt of the Cache Creek Complex. The lower part of the Hat Creek valley is underlain by poorly exposed early Tertiary clastics containing a ~300 m thick, low-rank, coal "seam." The coal occurs in a narrow graben-like structure, which probably formed by transpressional (strike-slip and reverse) fault movements, so that the coal became structurally thickened. Drill core data from the vicinity of the Hat Creek coal deposit area suggest that there the Spences Bridge Group is stratigraphically overlain by the mid-Cretaceous chert-rich clastics (Read, 1990).

176.2–177.0 km Carbonate on north (right) side of

Highway 99 belongs to the Marble Canyon Formation, which forms the central belt of the Cache Creek Complex.

Stop 3-4 Fusulinids in the Marble Canyon Formation, Cache Creek Complex

178.2 km Stop near yellow sign on right, just east of the junction with the Hat Creek road to the south. Walk back for about 200 m to a coquina of fusulinid tests in massive limestone of the Marble Canyon Formation in the roadcut on the north side of Highway 99.

The Marble Canyon Formation constitutes the central belt of the southern part of the Cache Creek Complex, and is mainly carbonate which ranges in age from Middle Permian to Late Triassic (Beyers and Orchard, 1991). The roadcut near the turnoff to Hat Creek contains a coquina with the large verbeekinid fusulinid (Fig. 3-8) that for ~50 years was called *Yabeina.*

This genus, and other associated fusulinacean, conodonts, and rare corals and brachiopods found in the Cache Creek Complex, are of great importance in Cordilleran paleobiogeographic reconstructions as they appear to be truly *exotic* with respect to the North American craton.

The following paragraph, modified mostly from Danner (*in* Smith, 1991, p. 170–171) summarizes the chequered history of generic names applied to this fossil. It was first collected near here in the early 1870s and misidentified as *Loftusia,* a Cretaceous-Eocene foraminiferan, although the assigned age was ignored by G.M. Dawson who geologically mapped the rocks in this area in the later 1870s and considered them to be Carboniferous. Similar fossils were known by the late 1800s in Indonesia, although the genus *Yabeina* was first described in the early 1900s in Indochina. It belongs to the verbeekinid group of fusulinids, which is widespread in eastern and southeastern Asia. *Yabeina* was first identified as such in this area by Thompson and Wheeler (1942), and was thought by them to represent the youngest marine Permian in North America. Since then, it has been restudied and renamed *Lepidolina* by Japanese paleontologists, and most recently called *Colania* by Rui and Nassichuk (Fig. 3-8; 1996). In the Asian standard biostratigraphic sections, *Colania* is slightly older than either *Yabeina* or *Lepidolina.*

In 1968, *Yabeina* (now renamed *Colania* by Rui and Nassichuk, 1996) was found in the Cache Creek Complex near Atlin in northwestern British Columbia, where it is associated with five different genera of ammonoids (Ross and Nassichuk, 1970; Nassichuk, 1977). Unlike the former *Yabeina,* the ammonoids also occur in the standard North American biostratigraphic sections of the Permian Basin of Texas and their presence permitted, for the first time, firm correlation between the standard Asian and North American Permian biostratigraphic sections. In Texas the ammonoids are of Wordian age (formerly early Late Permian but now Middle Permian), and that correlation showed that the age of the "Atlin *Yabeina*" was older in the North American biostratigraphic record than suspected by Thompson and Wheeler (1942).

About 200 km south of Atlin, a succession of different Permian fusulinid faunas dominated by schwagerinids was described by Pitcher (1960) in carbonate of a different stratigraphic sequence, which now is recognized as belonging to Stikinia. The Stikinian faunas could be readily correlated with those of the standard North American biostratigraphic successions. The youngest faunas present there were of similar, Wordian, age to the *Yabeina* near Atlin. The juxtaposition of the two different faunas of the same age, plus the east Asian and Paleo-Tethyan distribution of most verbeekinid fusulinids, plus the identification of the rocks hosting *Yabeina* as having formed in an oceanic setting located east of coeval arc-related strata (in what now are called Stikine, Alexander and Wrangellia terranes), provided some of the first reasons for believing that cryptic but very large amounts

Figure 3-8. Axial section of the verbeekinid fusulinid *Colania columbiana,* scale ×10. This figure is included to show the internal structure of the fusulinids in the coquina at Stop 3-4. The specimen illustrated here is from a location in the Cache Creek Complex near Atlin, northwestern B.C., although it is the same species as that in the Marble Canyon Formation (from Rui and Nassichuk, 1996).

of lateral displacement were concealed in the geology of the western Canadian Cordillera (Monger and Ross, 1971).

The association of massive shallow water carbonate with lithologies commonly recognized as having formed in oceanic settings (basalt, radiolarian chert, ultramafic rocks) is a puzzling characteristic of the Cache Creek Complex. The original natures of the contacts between central and flanking belts in the southern Cache Creek Complex are not clear, and at least in part are faults. However, in the Cache Creek terrane near Atlin in northwesternmost British Columbia, Late Mississippian to Middle Permian carbonate overlies Late Mississippian mafic volcanics with gradational contact. There, massive carbonate features algal-fusulinid-crinoidal and locally oolitic reef facies, and local lagoonal facies with mud-cracks, all deposited in very shallow water. The carbonate probably was an atoll founded upon a seamount (Monger, 1977).

Note: the Cache Creek Complex is only partly exotic. The association of carbonate with basalt and radiolarian chert seen in the Cache Creek Complex is not uncommon in older accretionary complexes in the circum-Pacific margin (e.g. Akioshi limestone of Japan; Madre de Dios complex of southwestern Chile). Coeval faunas many of these carbonates are similar, and the carbonates are probably remnants of atolls that probably formed at low latitudes in western Panthalassa or Paleo-Tethys, and later were dispersed by plate movements. The carbonates bearing the Paleo-Tethyan faunas in the Cache Creek were carried in later Permian, Triassic and earliest Jurassic time on the oceanic conveyer belt into the North American plate margin and preserved in the accretionary complex complementary to the early Mesozoic Quesnellian (and Stikinian) arc. The Harper Ranch Group underlying the Quesnellian (Nicola) arc contains a different but coeval Permian fauna with mid-continent-Andean affinities that probably lived at low latitudes near the eastern margin of Panthalassa, and later was displaced northwards relative to the craton. Just as the Franciscan Complex of California as a whole is not exotic with respect to the Sierra Nevadan arc but components (e.g. Laytonville limestone) are, so the Cache Creek Complex is not exotic with respect to the Quesnellian (and Stikinian) arc, although components like the Marble Canyon Formation are exotic.

178.6 km Road to Hat Creek turns off to south (left) of highway.

179.2 km Entering western end of Marble Canyon; roadcuts in limestone.

179.9–187.8 km Bluffs and cliffs on both sides of Highway 99 are underlain by massive carbonate of the Marble Canyon Formation. In this area, much of the carbonate is finely crystalline, in part because it is intruded by the Late Jurassic Mount Martley granodiorite pluton, which lies just south of the canyon. At 182.1 km pass Continental Lime plant. At 186.3 km, near Crown Lakes, the north side of the valley has a relief of ~700 m, all of limestone. Locally, the south side of the valley here is Mount Martley granodiorite.

187.8 km Pavilion Lake on left.

188.5 km Roadcut on north (right) side of Highway 99 is in Lower Triassic dark cherty argillite with limestone lenses and nodules. These rocks are part of the *western belt* of the Cache Creek Complex. Conodonts extracted from them are Early Triassic (Smithian, Dienerian; ~245–248 Ma; Orchard and Danner, 1991b). Known Early Triassic strata are very rare in the western Cordillera.

189.2–191.3 km Roadcuts in cherty argillite; western exposures contain Late Triassic age.

192.6–193.1 km Roadcuts in a granodiorite stock that is probably an apophysis of the Mount Martley pluton, and is K-Ar dated at 141 Ma (Preto et al., 1979). The approximate western limit of the body is crossed at ~196 km. Westward from 194.1 km, the reddish stains visible on the right valley wall are related to paleosols below the mid-Cretaceous chert-clast sandstone. At the top of valley wall in places, flat-lying beds of this unit overlie the Cache Creek Complex.

196.4 km View ahead of the Pavilion slide, which descends from southwest (left) side of valley. Beyond, the high peaks are in the Coast Mountains.

198.5 km Small roadcut on right in Cache Creek cherty argillite.

198.8–200.9 km Highway 99 crosses the toe of Pavilion earthflow, which is stable in its lower part, but still (1982) moving very slowly in its upper part. The reddish soil is characteristic of the surface below the mid-Cretaceous strata in this region. The village of Pavilion is located north of Highway 99 and below the toe of the slide.

201.2 km Site of Pavilion general store, which opened in 1872 and burned down a few years ago.

201.9–202.9 km The valley walls are composed of cherty argillite and local volcaniclastics, lithologies that are typical of the western belt of Cache Creek

Complex. On north (right) side of the valley, olistolithic(?) limestone blocks are present. The age of the western belt of Cache Creek ranges from Early Triassic to Early and possibly Middle Jurassic (Cordey et al., 1987).

202.8 km Enter the valley of the Fraser River, which the field trip follows as far as the town of Lillooet. Here, the western belt of Cache Creek is separated by the Fraser Fault from red continental volcanics (U-Pb dated at 104 Ma; M.L. Haskin, pers. comm. 1999) correlated with the Spences Bridge Group, which are exposed in the lower canyon/valley walls. The higher western valley walls, cresting in the Camelsfoot Range, are underlain by late Early Cretaceous (mainly Aptian-Albian, ~120–100 Ma) marine and nonmarine sandstone, shale and conglomerate of the Jackass Mountain Group, which is considered to be the uppermost unit of Methow terrane.

The Spences Bridge and Jackass Mountain groups are juxtaposed here across the north northeast-trending Slok Creek Fault, which diverges to the north northwest from the more northerly trending Fraser Fault. Note that the Slok Creek Fault must carry some of the ≥100 km of dextral displacement that is present farther south on the Fraser Fault (discussed at Stop 3-5), as Spences Bridge rocks occur on the east side of the Fraser Fault only ~10 km south of here.

203.9–206.5 km Bluffs on east (left) of Highway 99 are mainly in argillite of the western belt of Cache Creek Complex, which in this area is intruded by small bodies of diorite and granodiorite, possibly related to the Mount Martley stock. The west side of the valley containing Fraser River consists of Jackass Mountain Group.

207.0 km Brownish weathering Eocene volcanics, correlated with Kamloops Group, on both sides of highway.

208.9 km Cross British Columbia Rail track. Cache Creek Complex to left.

213.2–213.3 km Roadcuts on east (left) side of highway in Spences Bridge volcanics.

Stop 3-5 The Fraser Fault system

214.9 km Viewpoint to discuss the Fraser Fault system.

(1) In a view to the northeast (Fig. 3-9; above Highway 99), are mainly rocks of the Cache Creek Complex, with the knoll of Eocene volcanics in the middle distance (passed at 207 km).

(2) To the southeast, the ridges are capped by continental volcanics of the late Early Cretaceous

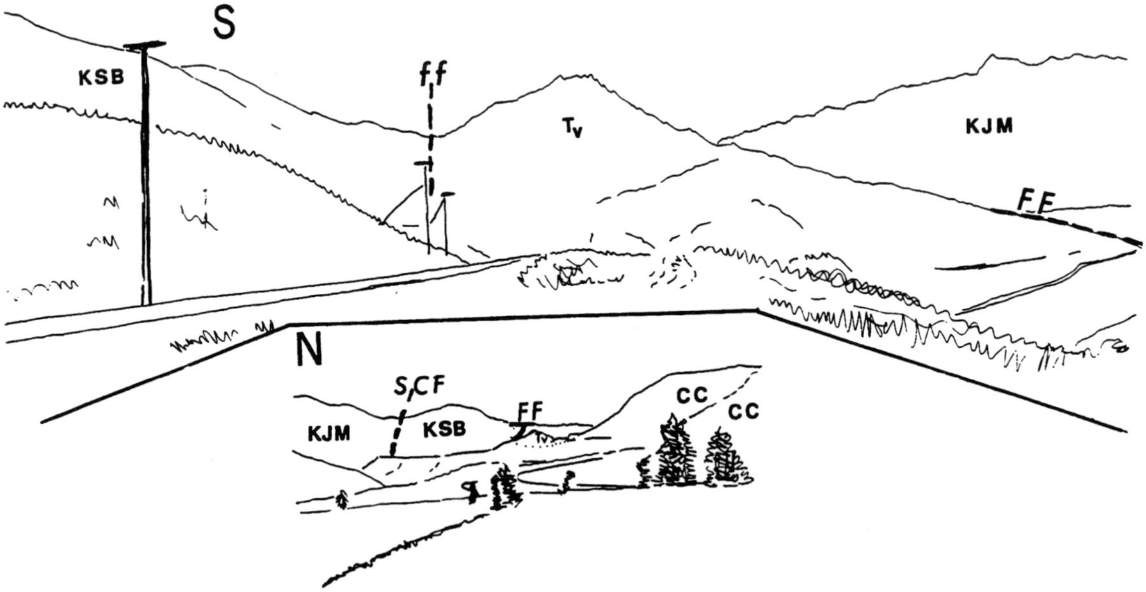

Figure 3-9. Two views of the area near the Fraser Fault from Stop 3-5, looking south (S) and north(N).
To south (S), FF = Fraser Fault follows Fountain Valley, which at this latitude is the boundary between Coast and Intermontane belts; ff = unnamed splay off Fraser Fault; horse of Eocene volcanic rocks (Tv) between the two faults; KJM = Jackass Mountain Group forms Fountain Ridge on right; KSB = Spences Bridge Group forms mountain on left.
To north (N), Fraser Fault is capped by 2 Ma flow of basalt; SCF = Slok Creek Fault; CC = Cache Creek Complex to right.

Spences Bridge Group, which here stratigraphically overlie the Cache Creek Complex (and elsewhere overlie the Nicola Group; Figs. 3-2A, B). The rock fragments in the colluvium at Stop 3-5 are mainly derived from the Spences Bridge higher up the slope.

(3) To the south (Fig. 3-9), and forming the canyon wall to the west across Fraser River, is a horse of Eocene (~50 Ma) volcanics within the Fraser Fault system.

(4) To the southwest, Fountain Valley follows the trace of the main strand of the Fraser Fault. Rugged Fountain Ridge on the west side of Fountain Valley is underlain by late Early Cretaceous marine and non-marine sandstone, shale and conglomerate of the Jackass Mountain Group.

(5) To the northwest, the west side of the valley is underlain by Jackass Mountain Group.

(6) To the north, the Jackass Mountain Group is separated by the valley containing the trace of the north northwest-trending Slok Creek Fault from the Spences Bridge Group, which forms the low hill. The flat surface on the skyline is underlain by flows of the Late Miocene-Pliocene Chilcotin basalt (here ~2 Ma), that overlap the Fraser Fault, which separates the Spences Bridge Group on the west from the Cache Creek Complex on the east (Fig. 3-9).

The Fraser Fault, and its southerly extension in Washington State called the Straight Creek Fault, is a major dextral strike-slip fault that trends north-south (Figs. 1, 3-10). The fault is of Middle to Late

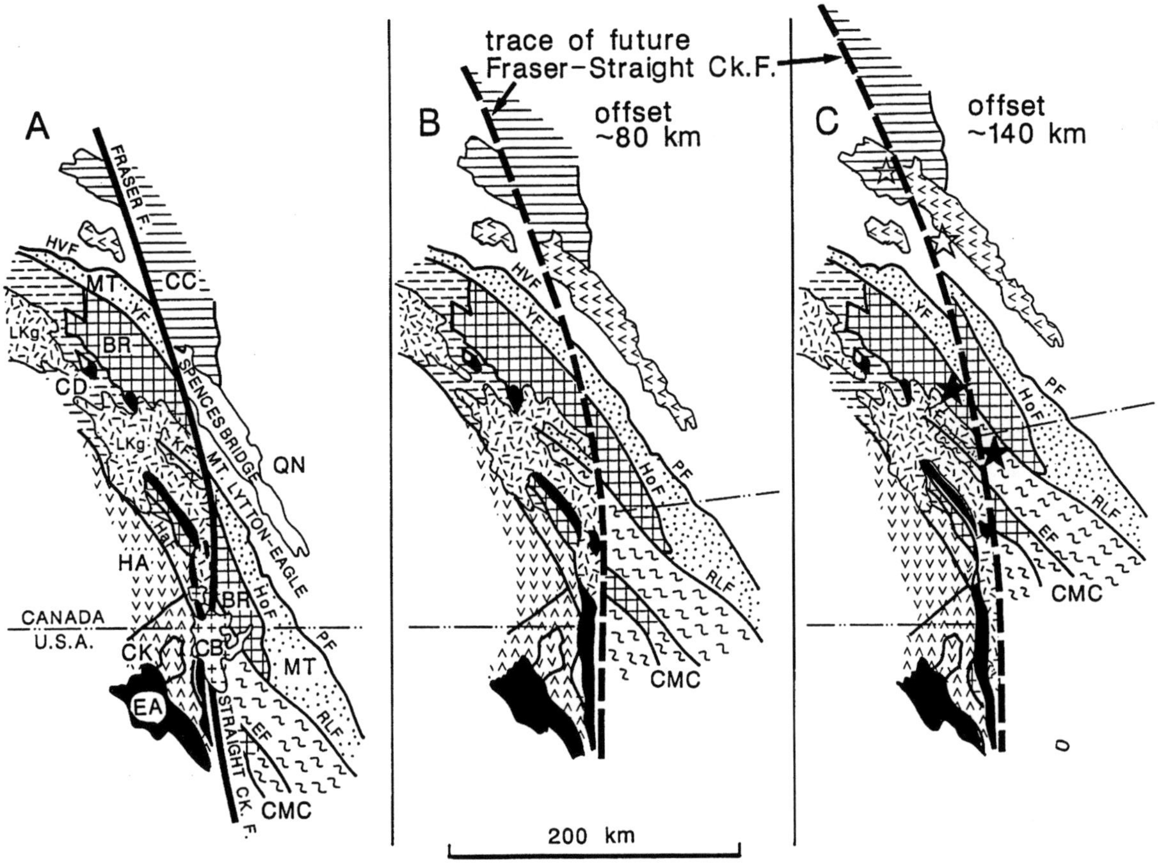

Figure 3-10. Fraser-Straight Creek Fault (figure from Monger and Journeay, 1994a).
A. Present map pattern; terrane abbreviations as in Fig. 3-7 except that EA (Easton terrane) is included in Bridge River terrane in Canada; it probably corresponds with parts of Cayoosh assemblage and Chism Creek schist, discussed in the Coast Belt section; CB = Chilliwack batholith; CMC = Cascade metamorphic core; LKg = Late Cretaceous granitic rocks. Fault abbreviations: EF = Entiat Fault; HaF = Harrison Fault; HoF = Hozameen Fault; HVF = Hungry Valley Fault; KF = Kwoiek Fault; PF = Pasayten Fault; RLF = Ross Lake Fault; YF = Yalakom Fault.
B. Restored dextral offset of 80–100 km on Fraser-Straight Creek fault system.
C. Restored dextral offset of ~140 km on Fraser-Straight Creek Fault; black stars are offset early Tertiary granitic rocks; open stars are offset Permian granitic rocks (Farwell Canyon west of Fraser Fault; northern Mount Lytton Complex on east).

Eocene age (47–35 Ma). The maximum age comes from a narrow north northwest-trending belt of Eocene (~47 Ma) intrusions offset by the fault. The minimum age is established by the latest Eocene-Oligocene (~35–25 Ma) Chilliwack batholith which intrudes the fault south of the town of Hope (~160 km south of here) and is not offset across it (3-10A).

Estimates of the amount of dextral offset across the Fraser Fault range from ~80 km to 190 km, depending upon criteria used. The north-trending fault offsets rocks and structures that follow the north northeast-trending structural grain established in Late Cretaceous-early Tertiary time. Although the fault was first mapped in the late 1800s in Canada, Misch (1966) and his students in Washington were probably the first to identify its southerly continuation, the Straight Creek Fault, as a right-lateral strike-slip fault. Later, Misch (1977) suggested the fault had a displacement of ~190 km, a number based on offset of the lithologically similar Settler and Chiwaukum schists. Subsequently, workers in Canada (Mathews and Rouse, 1984; Kleinspehn, 1985; Monger *in* Price et al., 1985) suggested amounts of displacement of between 80–110 km. The figures were based on matching similar rock packages (terranes) and their bounding faults across the Fraser Fault (Fig. 3-10B). More recently, Coleman and Parrish (1991) studied structures associated with the belt of Eocene intrusions. They matched the Mission Ridge Fault (which lies west of the town of Lillooet, ~12 km southwest of Stop 3-5) with the Ross Lake Fault (southwest of the town of Hope, ~160 km to south), to give perhaps a "best" estimate of ~130–150 km of offset (Fig. 3-10C). Restoration of this amount of offset restores the townsite of Lillooet to a position near Hope. In addition, this figure is comparable with the amount of separation of Permian granitic rocks at the north end of the Mount Lytton Complex and east of Fraser Fault with the Permian Farwell Canyon Pluton to the west of the fault (Fig. 3-10C; Friedman and van der Heyden, 1992).

Note that with 130–150 km of offset, parts of Bridge River and Methow terranes do not match up across the Fraser Fault (Fig. 3-10C). This discrepancy can be accounted for by the lateral migration of terrane boundaries with differential uplift and erosion. Interpretation of structural and seismic data suggests the terranes are disposed in relatively thin (≤10 km thick) thrust sheets, and (south of here) rocks west of the Fraser Fault are of generally higher metamorphic grade than the equivalent rocks east of it.

215.7 km Bridge over British Columbia Rail track.

216.4–216.7 km Roadcut and railway cut in red-grey Eocene volcanics.

217.4 km Fraser Fault near Fountain Valley marks boundary between Omineca and Coast belts. Highway 99 crosses the main strand of the Fraser Fault on the west side of the horse of Tertiary volcanics. This is 70.3 km from the junction of Highways 1 and 97 in Cache Creek and 1 km east of the B.C. Rail underpass (ahead). It becomes 0 km for Section 4: Coast Belt.

Reset odometer to 0 km.

4. COAST BELT

FRASER RIVER TO VANCOUVER

INTRODUCTION

The Fraser Fault marks the eastern boundary of the Coast Belt where crossed on Highway 99. This place is also the effective location of the Middle Jurassic-Early Cretaceous continental margin (Fig. 4-1), which had migrated oceanward from the position it occupied in pre-Jurassic time between Revelstoke and Sicamous as the result of accretion of terranes in the Intermontane Belt the ancient continental margin.

The morphogeological Coast Belt extends for nearly 1700 km (between latitudes 47°30′N and 62°N) and is between 100–200 km wide (Fig. 4a). It includes the physiographic *Coast Mountains,* west (and north) of Fraser River, and their southern extension the *Cascade Mountains,* east (and south) of Fraser River between the town of Lytton and the International Boundary at latitude 49°N (Fig. 2). South of the International Boundary, the name *North Cascade Ranges* is used.

The Coast Mountains are composed of mainly hard (granitic) rock, a feature which in combination with the high precipitation in their western part and sculpting by alpine glaciation, results in their very rugged topography. Where crossed on Highway 99,

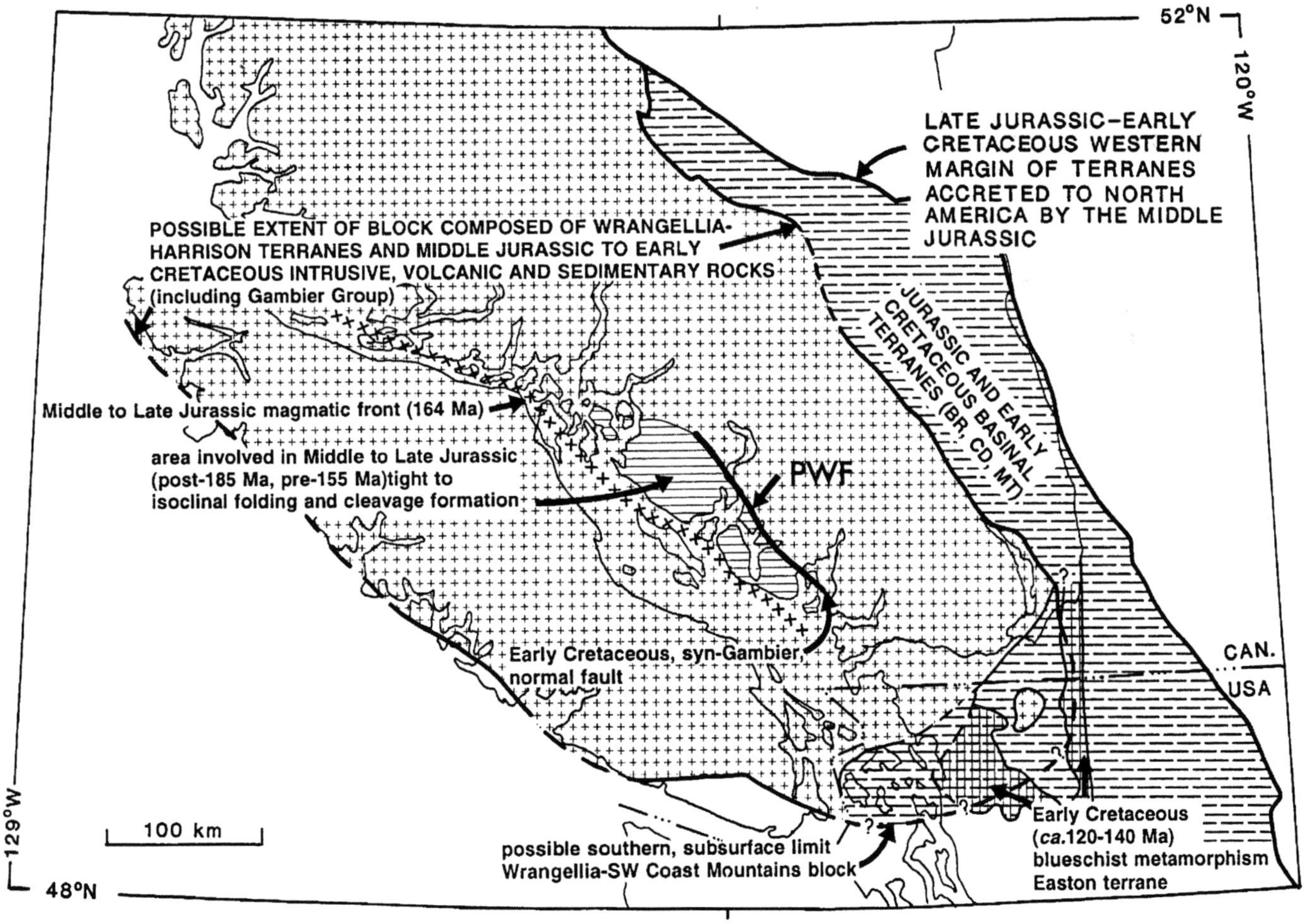

Figure 4-1. Distribution of pre-mid Cretaceous (≥95 Ma) features in the southern Coast Belt. These features existed prior to accretion of Wrangellia and the southwestern Coast Belt block (Gravina-Gambier arc founded on Wrangellia (and Alexander) terranes (from Monger and Journeay, 1994a, b).

the southeastern Coast Mountains contain many peaks whose summits approach 3000 m in elevation, whereas the southwestern Coast Mountains are generally lower (~2000 m) with only a few peaks approaching 3000 m, and descend to sea level.

The present topography of the southern Coast Mountains reflects differential uplift in the last 10 million years (Parrish, 1983). This conclusion is based on fission track studies, on a westward increase in elevation of a pre-Late Miocene erosion surface that can be traced from the Interior Plateau into the Coast Mountains, and on a Late Miocene change from wet to dry floras in the Interior Plateau, which today is in the rain shadow of the Coast Mountains.

Belt Boundaries

The eastern boundary of the southern Coast Belt is delineated by a series of linked faults, all of which were active in the early Tertiary. Northwest of where crossed by Highway 99, the boundary is the northwest-trending Hungry Valley Fault. Southward from that point to the town of Lytton, ~70 km to the south, it is the north-trending Fraser Fault. South southeastward from Lytton across the International Boundary, it is the Pasayten Fault (Fig. 1). These three faults differ in character. The Hungry Valley Fault is of probable Eocene age and is a southwest-side-up reverse, probable restraining bend fault related to Late Eocene dextral strike-slip movement on the Fraser Fault. The Pasayten Fault was active as a sinistral fault in the late Early Cretaceous (105 Ma), and also is an early Tertiary, east-side-up reverse fault.

The western boundary of the southern Coast Belt is mostly submerged beneath the east side of the Strait of Georgia, which separates the mainland from Vancouver Island. Where exposed on islands at the north end of the Strait and at one locality on the northeast coast of Vancouver Island, the boundary is intrusive—a "magmatic front"—along which Middle Jurassic (~170 Ma) granitic rocks intrude Wrangellian stratigraphy with its older "terrane specific" plutons (Fig. 4-1; Nelson, 1979; Monger and McNicoll, 1993).

Characteristics

The Coast Belt is mainly (80–85%) granitic rock, a composition which gave rise to the older but more-or-less synonymous terms Coast Range Batholith and Coast Plutonic Complex. The remaining rock types are variably metamorphosed Paleozoic through Mesozoic volcanic and sedimentary strata and Tertiary through Holocene volcanic rock.

(1) In composition, the granitic rocks are calc-alkalic, and mainly quartz diorite to granodiorite with low potassium content and juvenile isotope geochemistry (ε_{Nd} = +2.4 to +8.0; Sr_i = 0.703 to 0.7042; Roddick, 1983; Cui and Russell, 1995; Friedman et al., 1995).

(2) Ages of granitic rocks in the southern Coast Belt are mostly Middle Jurassic (~170 Ma) to early Tertiary (45 Ma) but locally are younger. Older (mainly ≥95 Ma) granitic rocks are in the southwestern Coast Belt, and younger (mainly ≤95 Ma) rocks in southeastern Coast Belt (Fig. 4-2; Friedman and Armstrong, 1995; Friedman et al., 1995). The youngest plutons (35–8 Ma) are the northernmost, deeply eroded, intrusive parts of the still-active Cascade Magmatic Arc (Fig. 3).

(3) The subordinate, variably metamorphosed sedimentary and volcanic rocks occur mostly as septa and fault-bounded slivers within the plutonic rocks. Southeastern Coast Belt contains Bridge River terrane and Methow terrane and its possible correlative Cadwallader terrane, all of which were founded on late Paleozoic oceanic basements, and which in Late Jurassic and Early Cretaceous time received marine clastic sediments (Figs. 4-1, 4-3, 4-4). Southwestern Coast Belt contains the eastward extension of the Devonian-Middle Jurassic Wrangellia terrane on Vancouver Island, and Middle Jurassic to mid-Cretaceous volcanic and sedimentary rocks that are largely coeval with the granitic rocks of southwestern Coast Belt (Figs. 4-1, 4-3, 4-4).

(4) The dominant structures in the southern Coast Belt are crust-penetrating systems of mainly west-vergent and locally east-vergent contractional faults of Late Cretaceous-early Tertiary age, which root mainly in the southeastern Coast Belt (Figs. 5, 4-5, 4-6; Varsek et al., 1993; Journeay and Friedman, 1993). Late Cretaceous and early Tertiary strike-slip faults occur mainly near the southeastern boundary of the Coast Belt, and early Tertiary extensional faults are recognized locally along the east side (Coleman and Parrish, 1991; Schiarizza et al., 1997).

(5) The metamorphic grade of the southern Coast Belt is predominantly greenschist and amphibolite (Fig. 4b). The highest metamorphic grades (7.4–9.6 kb; 675–750°C) are associated with the contractional deformation and Late Cretaceous-early Tertiary plutonism in the southeastern Coast Belt south of latitude 50°N (in the Cascade metamorphic core and the southeasternmost Coast Mountains east of Harrison Lake). The metamorphic grade decreases along the regional strike from that area to the north-northwest,

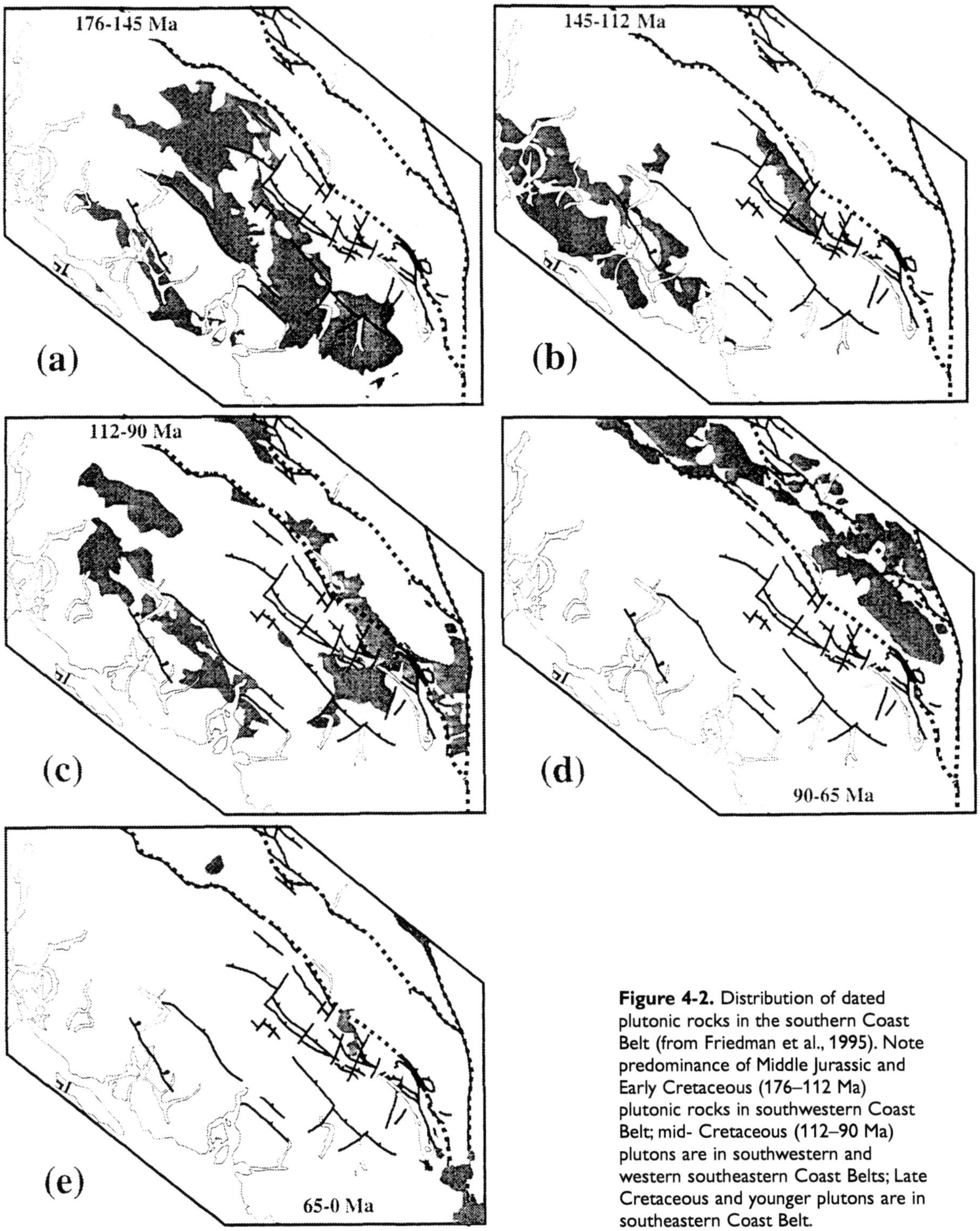

Figure 4-2. Distribution of dated plutonic rocks in the southern Coast Belt (from Friedman et al., 1995). Note predominance of Middle Jurassic and Early Cretaceous (176–112 Ma) plutonic rocks in southwestern Coast Belt; mid- Cretaceous (112–90 Ma) plutons are in southwestern and western southeastern Coast Belts; Late Cretaceous and younger plutons are in southeastern Coast Belt.

and where traversed on Highway 99 is mainly of greenschist grade. Still farther north-northwest along strike, the rocks are mostly of subgreenschist grade, and locally contain blueschist (Read et al., 1991; Journeay, *in* Monger and Journeay, 1994a; Schiarizza et al., 1997). In the southwestern Coast Belt, the metamorphic grade is mainly greenschist and amphibolite, and the metamorphism there appears to be mainly related to heat supplied by the very abundant granitic intrusions.

Two-Part Division of the Southern Coast Belt

In 1930, C.H. Crickmay recognized the division of the southern Coast Belt into two parts (Crickmay, 1930). The boundary he drew between them was near the Pemberton-Harrison Lake valley, which is located near the middle of the southern Coast Belt about 90 km west southwest of Stop 3-5, and crossed on Highway 99 at ~105 km. The western part, his "Coast Range," was dominated by granitic rocks and "Jurasside" structures, and disappeared beneath Tertiary and Quaternary sediments of the Fraser Lowland, southeast of the city of Vancouver. The eastern part, his "Cascade Range," contained rock units and structures that he traced southwards across the Fraser River into the North Cascade Ranges of Washington State. It featured "Laramide" folds and faults that wrapped around the "buttress" of his Coast Range. In general, Crickmay's perspicacious view has held up and can be explained in terms of modern mobilistic tectonic concepts (Figs. 4-1, 4-5).

The southwestern Coast Belt consists mainly of granitic rocks ranging in age from Middle Jurassic to mid-Cretaceous (~170–90 Ma; Fig. 4-2; Friedman and Armstrong, 1995; Friedman et al., 1995). Together with coeval volcanic and sedimentary stra-

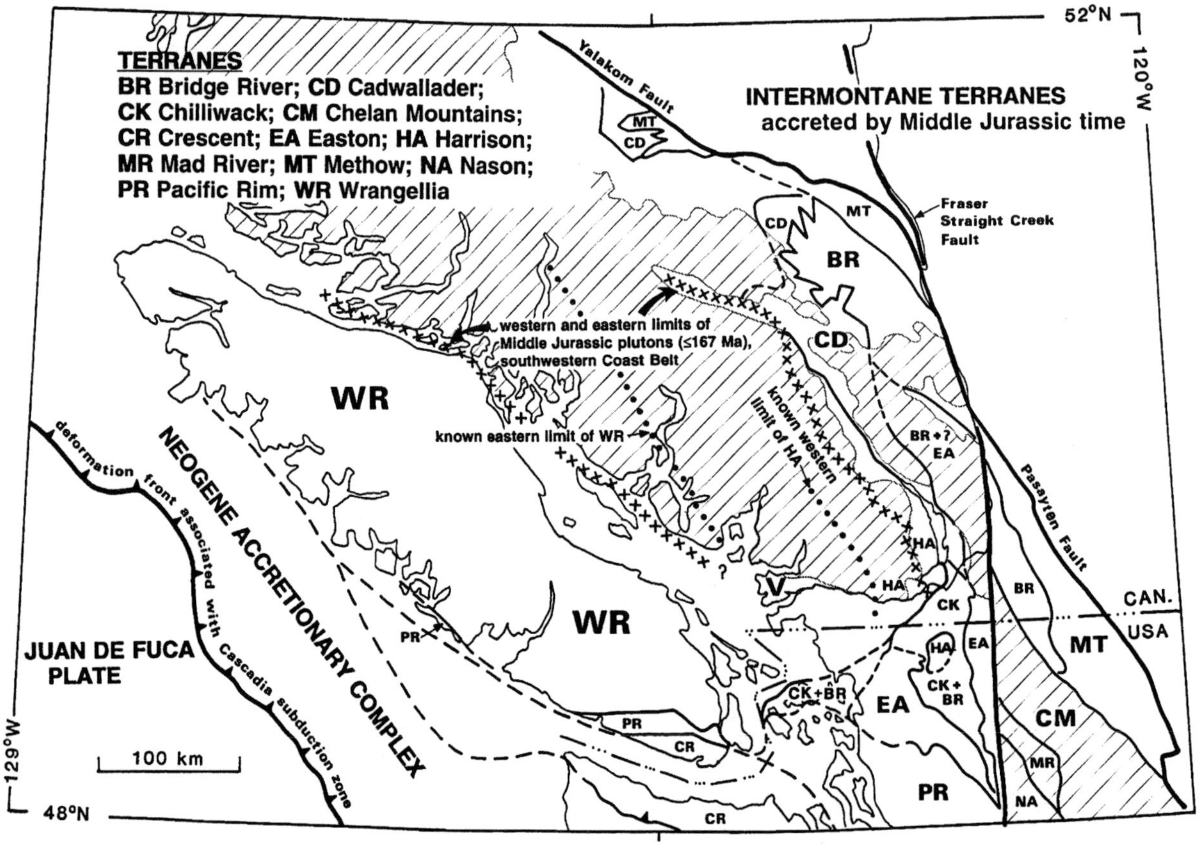

Figure 4-3. Distribution of terranes in the southern Coast Belt region. The Yalakom, (Fraser) and Pasayten faults on the east, and the Neogene accretionary complex on the west, give the limits of new, specifically Cordilleran, crust formed since the Early Cretaceous. Hachured areas are dominated by granitic and high-grade metamorphic rocks, within which the terranes occur as variably metamorphosed septa (from Monger and Journeay, 1994a, b). Schiarizza et al. (1997) consider that Methow and Cadwallader terranes are probably different parts of the same arc-basin system, both founded on Permian oceanic lithosphere.

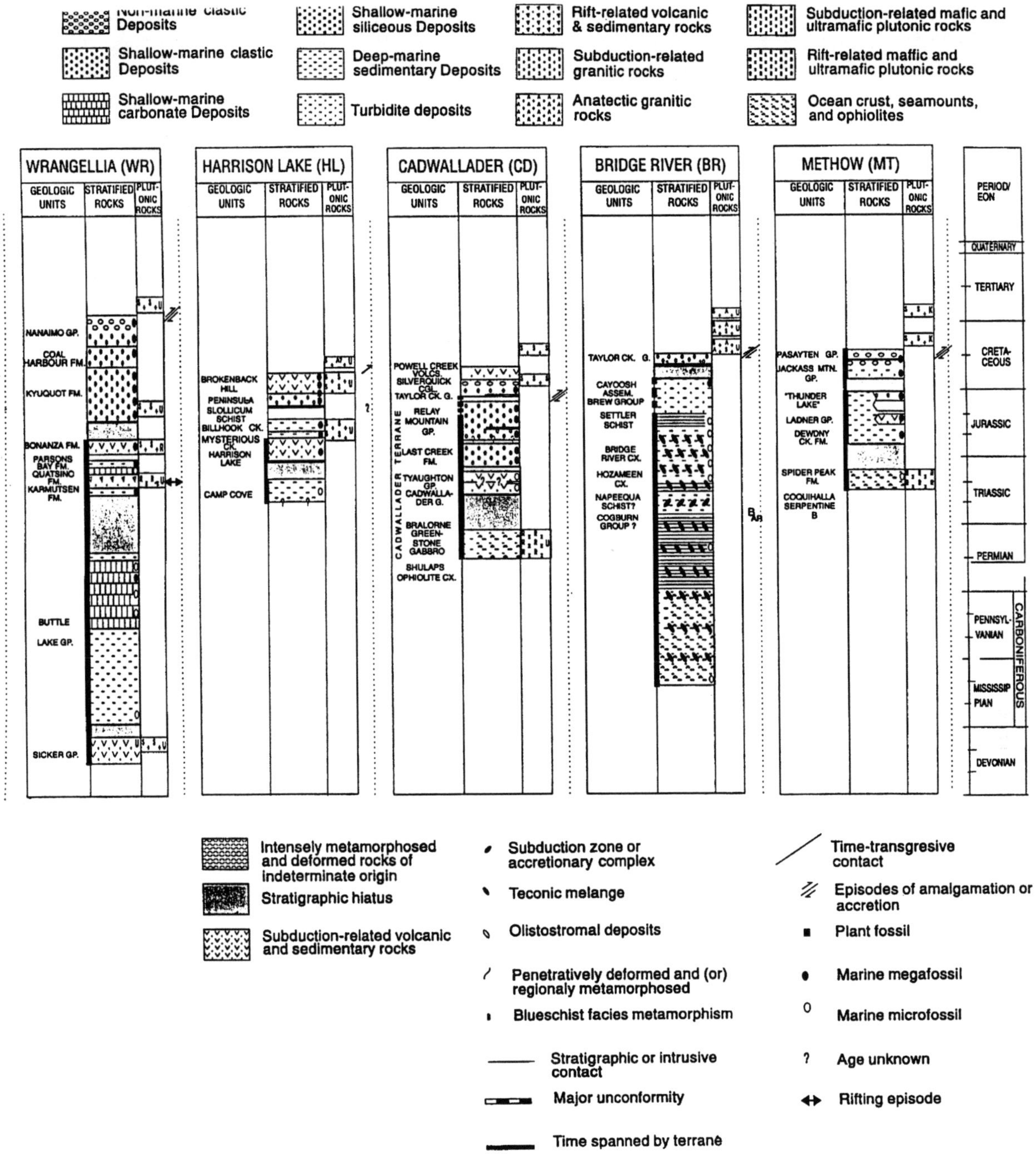

Figure 4-4. Terranes of the southern Coast Belt region; the recent correlation of Methow and Cadwallader terranes made by Schiarizza et al. (1997) suggests that ocean floor rocks at the base of Methow terrane (Spider Peak, Coquihalla; herein dated by Triassic conodonts in chert breccia unconformably above the pillow basalt, are more probably of Permian age. From Monger and Journeay (1994a, b).

ta, the granitic rocks belong to what has been called by Monger and Nokleberg (1996) the *Gravina-Gambier arc,* which north of latitude ~54°N lies entirely west of the western margin of the Coast Belt, within the Insular Belt (Fig. 8). The granitic rocks in southwestern Coast Belt intrude stratified volcanic and sedimentary sequences that range in (known) age from Middle Triassic through Early Cretaceous. Triassic and Early Jurassic *Wrangellian* strata can be traced from Vancouver Island into the western Coast Belt (Fig. 4-3). Near Harrison Lake, a Middle Triassic through Early Cretaceous succession has been called *Harrison Lake terrane* (Figs. 4-3, 4-4; Monger and Journeay, 1994b). The oldest specifically Coast Belt granites are Middle Jurassic (~170 Ma) and intrude Wrangellia on the west, and Harrison Lake terrane on the east. Geochemical studies suggest that Early to Middle Jurassic volcanic rocks of both terranes are correlative (Mahoney and DeBari, 1995), and arguably by late Early Jurassic time (~180 Ma), and probably by Middle Jurassic time (~170 Ma), Wrangellia extended at least as far eastward as the Pemberton-Harrison Lake valley.

Deformation in the southwestern Coast Belt is concentrated in widely spaced (±10 km) shear zones associated with mostly east-side-up reverse faults of early Late Cretaceous age (~95–90 Ma; Fig. 4-5). Between the shear zones, the rocks are relatively little deformed (Monger and Journeay, 1994a). As first recognized by Crickmay (1930), it appears that rocks of the southwestern Coast Belt and Wrangellia acted as a semi-rigid *foreland* (Crickmay's "buttress") to the Late Cretaceous thrust faults rooting in southeastern Coast Belt. The foreland received Late Cretaceous clastics that were eroded from the tectonically uplifted rocks to the east and deposited to form the marine and non-marine "molasse" of the Nanaimo Group (Mustard, 1994).

The southeastern Coast Belt has a lower percentage of granitic rocks than southwestern Coast Belt. These are distributed in four north northwest-trending plutonic belts that from southwest to northeast decrease in age from mid-Cretaceous (≤95 Ma) to early Tertiary (≥45 Ma). They intrude stratified rocks

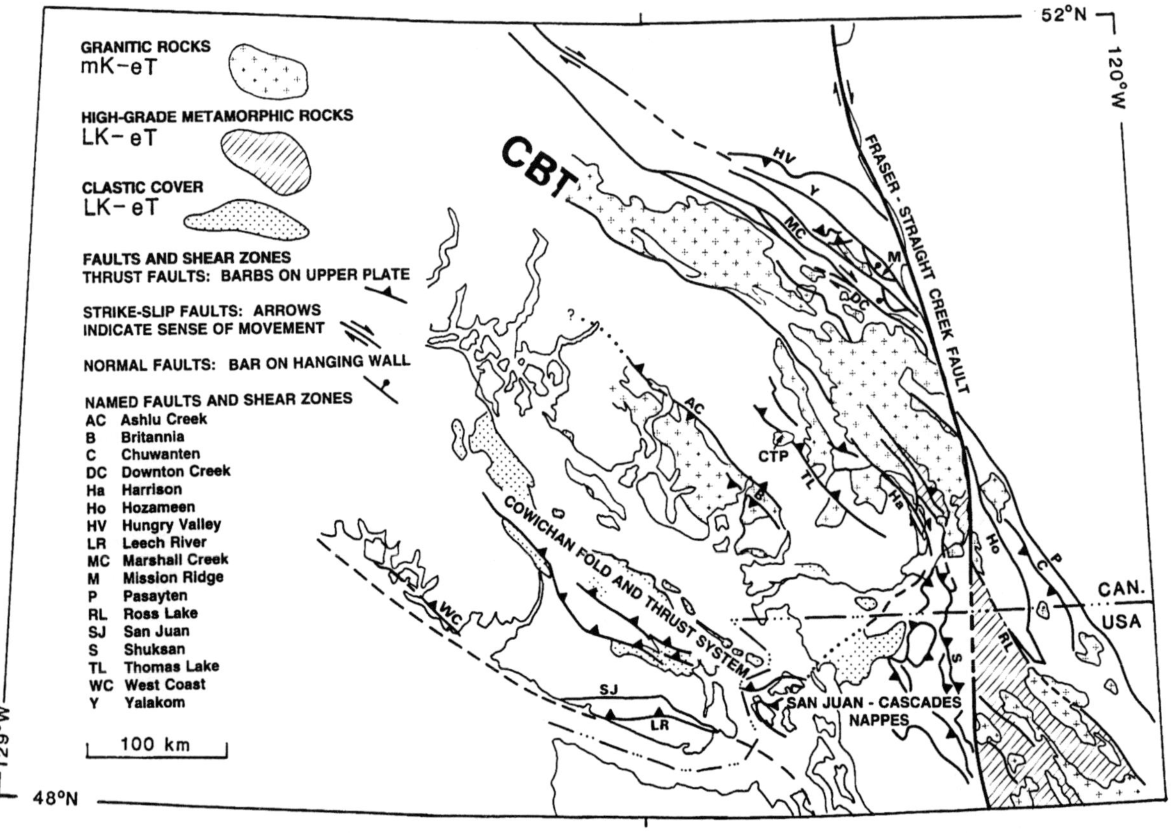

Figure 4-5. Distribution of mid-Cretaceous to early Tertiary (95–40 Ma) features of the southern Coast Belt, that were formed during the major period of deformation, metamorphism, crustal thickening and uplift. CTP = Castle Towers Pluton; CBT = approximate location of Coast Belt thrust system (from Monger and Journeay, 1994a, b).

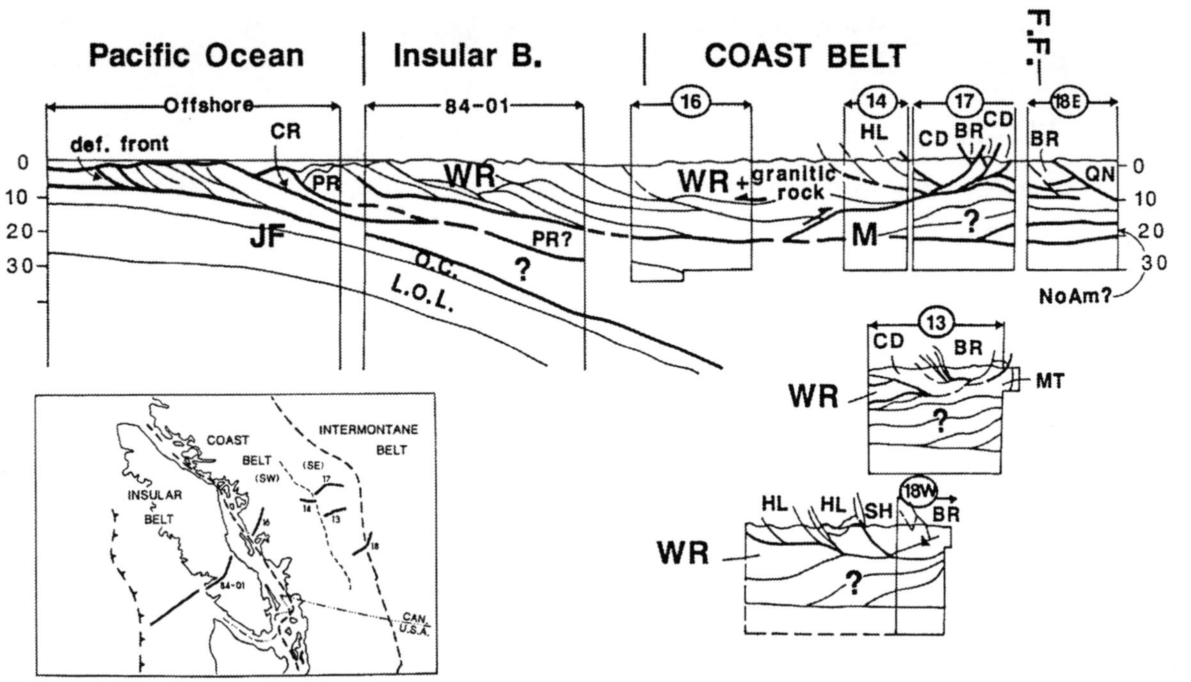

Figure 4-6. Crustal cross section across southern Coast and Insular belts, based on surface geological mapping and Lithoprobe deep seismic reflection profiles (positions of lines 13-18 and 84-01 shown on inset map). The section, drawn by J.M. Journeay, shows the possible vertical distribution of terranes, and their complex interleaving, and differs in detail from those of Varsek et al., 1993, and Clowes et al., 1995; Fig. 5, after Journeay). No granitic rocks are shown; terranes are lettered as in Fig. 4-3, except WR + granitic rock denotes the dominant granitic rocks emplaced in Wrangellia in southwestern Coast Belt. M = reflection Moho; O.C. = oceanic crust, L.O.L. = lower oceanic lithosphere, JF = Juan de Fuca Plate (from Monger and Journeay, 1994a, b).

belonging to *Methow, Cadwallader* and *Bridge River* terranes, all of which were founded on late Paleozoic oceanic crust and all of which remained marine until mid-Cretaceous time (Figs. 4-3, 4-4). Recently, it has been suggested by Schiarizza et al. (1997) that Methow and Cadwallader terranes are different parts of the same arc-basin system founded on Permian oceanic lithosphere, with the distinctive Triassic arc sequence of Cadwallader terrane presumably removed by erosion from Methow terrane. Bridge River terrane is a chert-rich accretionary complex with local Triassic blueschist, and contains radiolarian chert facies ranging in age from Mississippian through Middle and possibly Late Jurassic (Callovian-Oxfordian, ~155 Ma; Cordey and Schiarizza, 1993; Schiarizza et al., 1997). It is partly overlain by, and may partly incorporate, a clastic sequence called Cayoosh assemblage that is at least as young as early Early Cretaceous (Neocomian). This clastic sequence may be common to both Bridge River and Methow terranes (Journeay, *in* Monger and Journeay, 1994a). Discontinuous outcrop areas of mid-Cretaceous (Albian-Cenomanian) marine and non-marine clastics containing abundant chert grains overlie the Bridge River terrane. They are compositionally similar and age equivalent to strata in isolated outcrop areas in the Intermontane Belt (passed earlier on Day 3 at ~164–172 and 194 km), a correlation which hints at a tie between strata in the Coast and Intermontane belts by mid-Cretaceous time.

Rocks of southeastern Coast Belt are typically penetratively deformed (in contrast to those of southwestern Coast Belt) with folds associated with thrust, strike-slip, and normal faults. Westerly-directed thrust faults of the *Coast Belt thrust system,* which mostly lie just east of the Harrison Lake-Pemberton valley, carry rocks of southeastern Coast Belt over those of the southwestern Coast Belt (Journeay and Friedman, 1993). The deformation was accompanied by regional metamorphism up to high grades, and intrusion by mid-Cretaceous through early Tertiary granitic rocks. P-T-t considerations suggest that the deformation was accompanied by regional uplift of 10–15 km between 95–85 Ma (Journeay, *in* Monger and Journeay, 1994a). Clasts eroded from the uprising Coast Belt were deposited to the west to form the marine and non-marine Nanaimo Group (Mustard, 1994). The southeastern Coast Belt contains the "*suture*" between terranes of the Intermontane Belt, whose terranes were accreted in the Jurassic to the ancient continental margin, and terranes of the southwestern Coast and Insular belts.

Conflicting Paleogeographic Hypotheses

Although the material pertaining to the Coast Belt presented above is generally accepted, a current debate centres upon the possible late Mesozoic paleogeographic configurations of the rocks in the southern Coast Belt. Currently, there are three main hypotheses.

(1) Interpretation of paleomagnetic data suggest that the Late Cretaceous (~85 Ma) Powell Creek volcanic rocks in the *easternmost* southern Coast Belt formed ~3500 km south of their present locations relative to the craton (Wynne et al., 1995). Several other Coast Belt paleomagnetic studies indicate Cretaceous paleopoles that are different from those of cratonal North America, but they are from plutons whose paleohorizontal attitude is unknown or at best debated. However, the Powell Creek data set is from stratified rocks in which it appears that the paleohorizontal attitude can be established, and so it provides the strongest foundation for the long-lived Baja B.C. hypothesis.

The Powell Creek volcanics in Churn Creek, which is located ~100 km northwest of where Highway 99 crosses the eastern boundary of the Coast Belt, are continental and stratigraphically overlie continental Albian-Cenomanian chert-rich clastics. In turn, the clastics overlie continental mid-Cretaceous (105 Ma) volcanics long correlated with the Spences Bridge volcanics of the southern Intermontane Belt (Monger and Price, 1996; Mahoney et al., 1996). The paleomagnetic results from the type Spences Bridge suggest ~1100 km of south-to-north displacement with respect to the craton (Irving et al., 1995). Both paleomagnetic interpretations, when set within the above stratigraphy, imply that *continental* rocks in the southern Coast Belt *and* southwesternmost Intermontane Belt moved southward with respect to the craton for ~2400 km between 105 and 85 Ma, and then northward for ~3500 km between 85 and 50 Ma.

One problem with this hypothesis is that post-85 Ma offsets of geological markers across *recognized* strike-slip faults in the southwestern Canadian Cordillera only amount to ~300 km. Another is that Late Cretaceous marine faunas of the Nanaimo Group on Vancouver Island appear to be in their expected northerly latitudinal position relative to other west coast faunas (Elder and Saul, 1993). Just possibly an explanation for the conflict between the paleomagnetic interpretation on the one hand, and geological and paleontological interpretations on the other, may be found in the recent suggestion by Sager and Koppers (2000) that there was a rapid Late Cretaceous *true* polar wander event, which has not been identified in the North American apparent wander path.

The second and third hypotheses involve different times of arrival of rocks in the southwestern Coast Belt and Insular belts prior to the mid-Cretaceous collapse and inversion of the basin(s) floored by Bridge River and Methow-Cadwallader terranes (Fig. 4-3).

(2) Monger et al. (1994) suggested that the rocks in southwestern Coast Belt arrived just before their mid-Cretaceous accretion to the Early Cretaceous continental margin (Fig. 4-7). They speculated that the volcanic and granitic rocks of the Gravina-Gambier arc in southwestern Coast Belt, and the continental Spences Bridge arc in southeastern Intermontane Belt were different segments of a *single* Early Cretaceous North American plate margin arc. That arc was cut by a hypothetical sinistral fault in Early Cretaceous time, and the northwestern, Gravina-Gambier segment was displaced southward by a minimum of ~800 km to lie oceanward of the Spences Bridge segment. The basin founded on the Bridge River accretionary complex was trapped between the two arc segments. Their view was supported by scattered structural evidence for orogen-parallel displacement with a sinistral component on structures of late Early Cretaceous age in the southeastern Coast Belt (e.g. Pasayten Fault, ~105 Ma, Hurlow, 1993). It is in accord with oceanic plate reconstructions which permit oblique subduction with a sinistral arc-parallel component at ≥105 Ma (Engebretson et al., 1985; D.C. Engebretson and K. Kelley, pers. comm. 1993), and with structural patterns in and near the eastern Coast Belt in northwestern British Columbia which indicate Cretaceous sinistral convergence (Evenchick, 1997).

(3) In contrast, Schiarizza et al. (1997) proposed that the western block arrived oceanward of Bridge River-Methow-Cadwallader terranes earlier, possibly as early as late Middle Jurassic time. Their proposition was based on the presence of the late Middle Jurassic-Early Cretaceous clastic rocks that overlap Bridge River, Methow and Cadwallader terranes (Brew and Relay Mountain Groups, Cayoosh assemblage) that possibly (and probably by Albian time) were sourced from both east and west. Their reconstruction appears to demand *two* Late Jurassic-Early Cretaceous arcs, separated by an amagmatic basin.

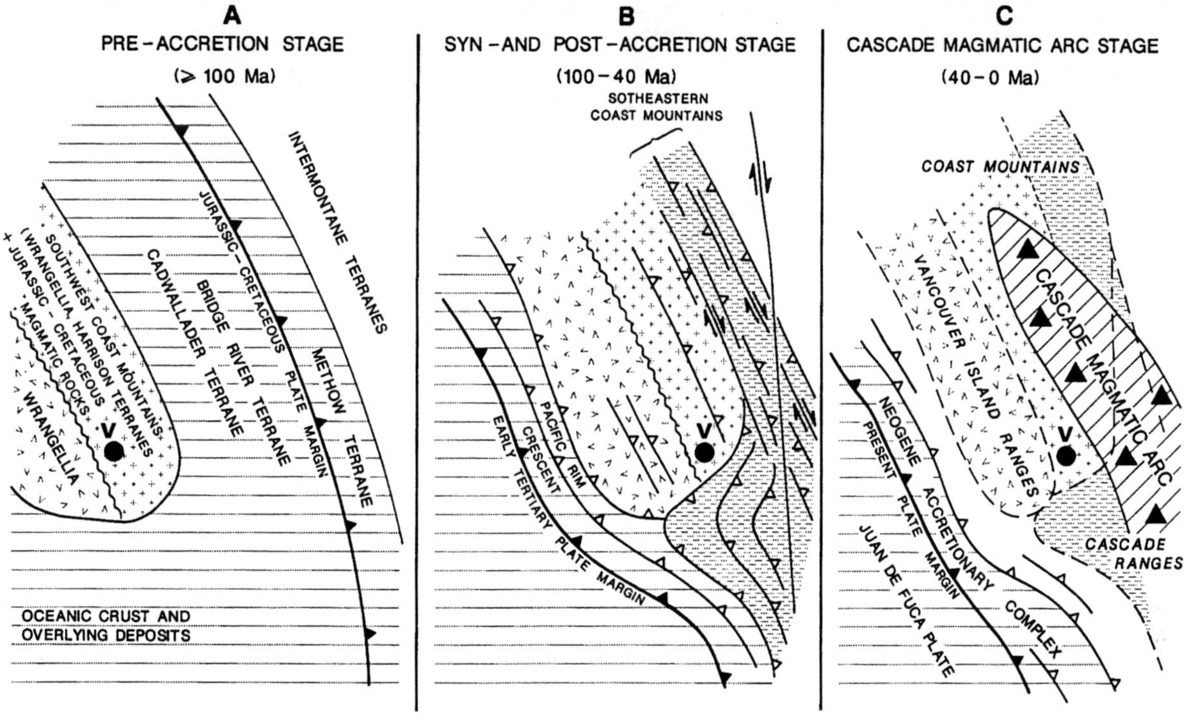

Figure 4-7. A model of the tectonic evolution of the southern Coast Belt; V = denotes city of Vancouver (from Monger and Journeay, 1994b).
A. Shows basinal terranes (Bridge River, Methow, Cadwallader; according to Schiarizza et al., 1997, Methow = Cadwallader, if so, Cadwallader would have lain east of Bridge River). The western Wrangellia-southwestern Coast Mountains (Gravina-Gambier) block possibly was emplaced by intra-plate, sinistral, orogen-parallel faults in Early Cretaceous time.
B. Shows syn- and post-accretionary stage, the interval when the structural and metamorphic features formed that give integrity to the entire Coast Belt.
C. Approximate extent of the Late Eocene-Quaternary Cascade magmatic arc.

Some Coast Belt Sources

Much of the material covering the section along Highway 99 between Vancouver and Lillooet is modified from a more detailed guidebook by Monger and Journeay (1994a), in which Monger was largely responsible for the southwestern Coast Belt and Journeay for the southeastern Coast Belt. Both authors owe a great debt to J.A. Roddick, who did much of the initial mapping in the region and who in 1985 led the Coast Belt part of the Trans-Cordilleran field trip (Price et al., 1985). Since 1985, new mapping by the Geological Survey of Canada (e.g. Journeay, 1993) and the British Columbia Geological Survey Branch (Schiarizza et al., 1997), as well as Lithoprobe seismic refraction and reflection profiles across the region (e.g. Clowes et al., 1995), and extensive U-Pb dating related to these projects (e.g. Friedman and Armstrong, 1995), contribute to the view presented herein.

FRASER RIVER TO PEMBERTON — SOUTHEASTERN COAST BELT

ROAD LOG

0.5 km Roadcut in Jackass Mountain strata on left.

1.0 km Underpass below British Columbia Rail tracks.

1.9 km Junction of Highway 99 with Fountain Valley road to the left. South of here, Fountain Valley follows the trace of the Fraser Fault, which rejoins the Fraser River ~25 km to the south.

4.7–6.3 km Roadcuts in shale, and sandstone of the Jackass Mountain Group. At 5.0 km the knoll on north (right) of Highway 99 is composed of conglomerate. Massive beds of buff-green sandstone interbedded with dark grey shale form the valley wall across the Fraser River.

Jackass Mountain Group and Methow Terrane

The Jackass Mountain Group is later Early Cretaceous (Barremian-Albian; 125–100 Ma) in age. Current direction measurements and clast compositions in the Jackass Mountain Group show that the provenance lay to the east (Kleinspehn, 1985). The clasts include feldspar and quartz derived from granitic rock (a U-Pb date on a granitic boulder is 156 Ma; O'Brien et al., 1992), volcanic rock, argillite and chert. The Jackass Mountain Group probably was deposited mostly in submarine fan complexes in a fore-arc setting. The related arc is probably (at least in part) the Albian (105 Ma) Spences Bridge Group and coeval granitic rocks in southwestern Intermontane Belt (Fig. 3-2B).

South of the town of Lytton, the Jackass Mountain Group is separated from Spences Bridge Group by the Pasayten Fault. The fault apparently moved sinistrally at 105 Ma, with ~20 km of offset (Hurlow, 1993), but also has Tertiary reverse movement. East of the fault, the Spences Bridge Group stratigraphically overlies Nicola and Cache Creek strata, as well as a narrow welt of plutonic and metamorphic rock (Mount Lytton and Eagle Plutonic complexes; Figs. 3-2A, B) that marks the uplifted and eroded southwestern edge of Quesnel terrane. Uplift was in the late Early Cretaceous, concomitantly with extrusion of the overlying Spences Bridge volcanics. Clasts in the Jackass Mountain Group are compatible with such a varied source terrane.

The Jackass Mountain Group is the uppermost component of Methow terrane (Figs. 4-3, 4-4). It unconformably overlies generally finer grained Jura-Cretaceous(?) clastics, and fine-grained clastics with locally intercalated volcanics (Dewdney Creek Formation) of the Jurassic Ladner Group. In turn, this was deposited unconformably on pillow basalt with MORB geochemistry of the Spider Peak Formation of pre-Triassic and possibly Permian age. The latter is associated with alpine-type ultramafic rocks and gabbro. The Methow terrane thus appears to be founded wholly or partly on oceanic lithosphere. In contrast with the Intermontane Belt, it remained marine until latest Early Cretaceous time.

6.9 km The view ahead is into the eastern Coast Mountains, whose summit heights in this area approach 3000 m. Below, across Fraser River, is the northwest-trending Bridge River valley. Rocks exposed at the confluence of Fraser and Bridge rivers are volcaniclastics, containing Middle Jurassic ammonites, and are correlated with the Dewdney Creek Formation of the Ladner Group of Methow terrane.

Yalakom Fault

About 10 km northwest of here, the lower Bridge River and its tributary the Yalakom River follow the trace of the northwest-trending dextral strike-slip Yalakom Fault (Fig. 1), which was active in latest Cretaceous-earliest Tertiary time. The amount of dextral offset on the Yalakom Fault is estimated to be ~150 km by Kleinspehn (1985) based on offset facies in Early Cretaceous clastic strata, and ~115 km by Riddell et al. (1993) based on matching cutoffs of Late Cretaceous structures near the fault. Highway 99 crosses the Yalakom Fault at 17.1 km, south of the town of Lillooet.

8.4 km Siltstone and argillite by Highway 99 belong either a fine-grained facies of the Jackass Mountain Group or else to an underlying, undated Jura-Cretaceous sequence tentatively correlated with the Brew Group and Cayoosh assemblage (Monger and Journeay, 1994a).

11.7–13.7 km View to south along the valley of the Fraser River, which here follows the traces of the north northeast-trending Yalakom Fault system and other parallel faults. The Fraser River canyon here and to the south contains sand and gravel bars that were worked for placer gold in the mid- to late 1800s. The town of Lillooet across the river was important in the early settlement of British Columbia by people of European and Chinese origin, and in the late 1850s marked the end of the route from the coast across the Coast Mountains. A cairn in town marks "Mile 0" on the road to the Cariboo goldfields at Barkerville. Highway 99 crosses the Fraser River and continues through the steep-walled valley visible southwest of Lillooet.

15.7 km Turn west (right) on Highway 99 at its junction with Highway 12, which follows the Fraser River south to the town of Lytton, located at the confluence of Fraser and Thompson rivers.

16.2–16.5 km Cross the Twenty-three Camels Bridge over the Fraser River. The name comes from an attempt in the 1860s to use camels as pack animals in this arid region. The camels worked well but apparently horses and mules reacted strongly to their presence. This caused problems on narrow trails, and so the camels were turned loose; one survived for at least 20 years.

16.9 km Turn south (left) on Highway 99 to Pemberton and Whistler. To the east, across Fraser River, beds of Jackass Mountain sandstone, shale and local conglomerate form Fountain Ridge.

17.1 km Cross the bridge over Seton River. South of the bridge is a knoll of serpentinite, one of several located along the Yalakom Fault. The fault follows the Fraser River valley southward to merge with the more northerly-trending Fraser Fault ~30 km to the south. One possible equivalent, offset across the Fraser Fault, is the Ross Lake Fault; another is the Hozameen Fault (Fig. 3-10).

West of the fault, and underlying the knoll below the steep valley wall to the south (left) is a ~1 km wide fault-bound slice of disrupted, subgreenschist-grade chert, basalt, gabbro and serpentinite belonging to the Bridge River Complex.

Two Important Rock Units: Bridge River Complex and Cayoosh Assemblage

For much of the remainder of the trip across the southeastern Coast Mountains as far as ~70 km, Highway 99 crosses mainly greenschist-grade, foliated, late Paleozoic and Mesozoic rock units belonging either to the chert-rich Bridge River Complex or to the mainly clastic Cayoosh assemblage. Both units were involved in Late Cretaceous and early Tertiary deformation and both are intruded by Late Cretaceous or early Tertiary granitic rocks. They are shown on Fig. 4-4 as part of Bridge River terrane because in places the Cayoosh assemblage stratigraphically overlies the Bridge River. However, the Cayoosh assemblage also may correlate with Late Jurassic and Early Cretaceous strata of Methow (=? Cadwallader) terranes (e.g. Relay Mountain Group), and thus may be an overlap assemblage.

Bridge River Complex

The Bridge River Complex is best preserved and exposed ~50 km northwest of here in roadcuts near Carpenter Lake, in the dammed upper section of the Bridge River valley. There, it consists of tectonically disrupted radiolarian chert, pillowed basalt, mafic intrusive rock, argillite, sandstone, local serpentinite and rare limestone olistoliths. Radiolarians and conodonts show the complex ranges in age from Mississippian to late Middle Jurassic (Callovian, ~350–160 Ma) and possibly Late Jurassic (Oxfordian, ~155 Ma, Cordey and Schiarizza, 1993; F. Cordey, pers. comm. 1994), with Late Triassic-Early Jurassic forms being the most abundant. The time spanned by the Bridge River (~190 million years), and the pelagic nature of most of the sedimentary rocks within it, suggest that it is the remains of a large ocean basin. The metamorphism in the Carpenter Lake area is mostly of pumpellyite-prehnite grade, with local blueschist $^{40}Ar/^{39}Ar$ dated at 220–230 Ma (Middle to Late Triassic; Schiarizza et al., 1997). Structures within the Bridge River Complex are complex. Arguably, the initial disruptive deformation ("broken formation") took place within a long-lived accretionary complex, but superimposed on any structures formed during accretion are brittle Late Cretaceous-early Tertiary folds and thrust faults, strike-slip faults and at least locally, normal faults (Schiarizza et al., 1997).

Unfortunately, the Carpenter Lake route across the mountains is blocked by snow in winter and spring. With the exception of rocks just crossed near 17.5 km (between Yalakom and Mission Ridge faults), the Bridge River Complex near Highway 99 comprises actinolitic greenstone, metachert, phyllite, rare marble and talc schist, metamorphosed in greenschist or lower amphibolite facies. The intense Cretaceous and early Tertiary deformation of these rocks transposes and homogenises their earlier, probable highly disrupted, nature.

Cayoosh assemblage and correlative units

The Cayoosh assemblage of Journeay and Mahoney (1994; Fig. 4-8) is a sequence of interbedded turbiditic sandstone, in part arc volcanic-plutonic derived but including local quartzite, shale and local conglomerate with plutonic clasts, and rare volcanic horizons. The Cayoosh assemblage mostly is metamorphosed in greenschist facies and is penetratively deformed, but contains rare fossils showing its Jurassic-Cretaceous age. The Cayoosh assemblage can be correlated with a number of other clastic units in the southeastern Coast Belt. It closely resembles in its range of lithologies and metamorphic grade, the Brew Group of Duffell and McTaggart (1952), which is crossed on Highway 99 at ~25–40 km. It also can be correlated with less metamorphosed clastic rocks near Carpenter Lake, including the late Middle Jurassic to Early Cretaceous Relay Mountain Group, the Gun Lake and Downton Lake units, and the fossiliferous, latest Jurassic (Volgian; ~152–145 Ma) Truax Creek conglomerate (Schiarizza et al., 1997). A lithogical correlation is with parts of the Chism Creek Schist of Rusmore (1985), exposed near Duffey Lake (Stop 4-3). Earlier, Monger (1991) correlated the rocks now included in Cayoosh assemblage with amphibolite-grade metagreywacke, pelite and metabasalt of the Settler Schist, ~120 km to the south, and east of Harrison Lake. The Settler Schist overlies (structurally?) an undated assemblage of metachert, metabasalt, minor marble and ultramafic rock called the Cogburn Group, which is lithologically similar to the Bridge River Complex.

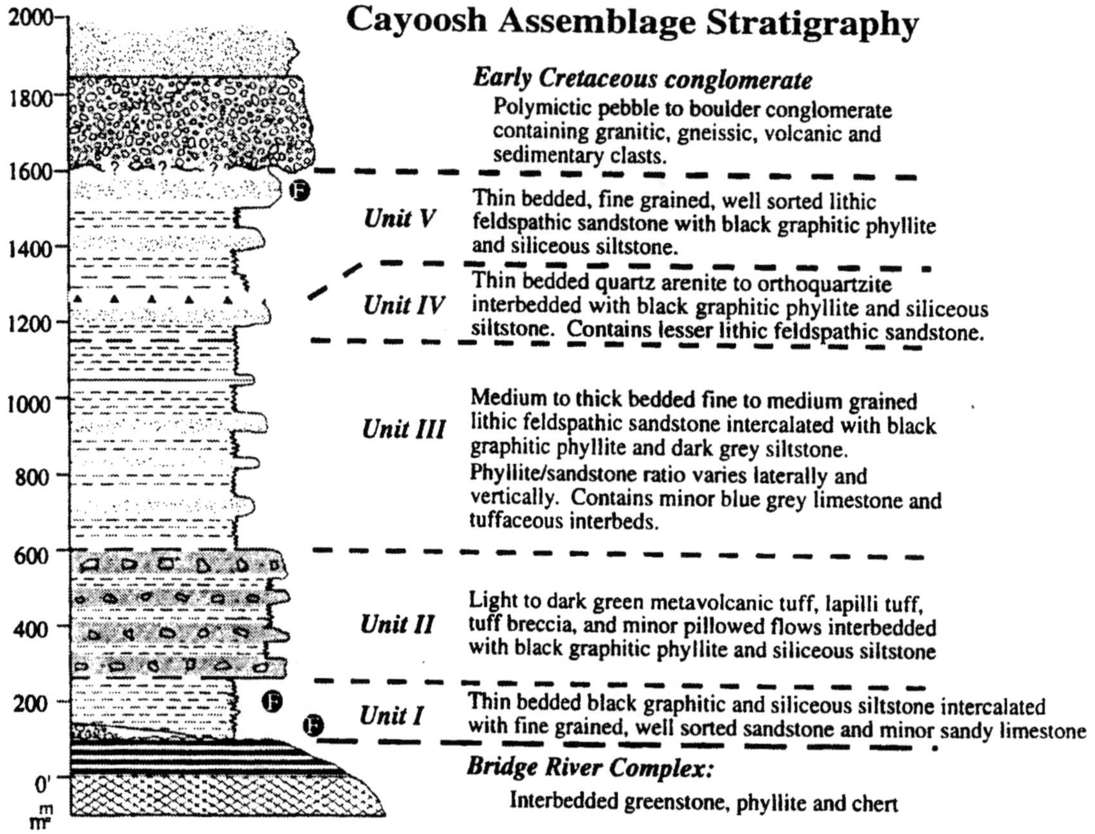

Figure 4-8. Schematic stratigraphic section of the Cayoosh assemblage, showing the range of lithologies; compiled from measured sections in the Mount Brew and Melvin Creek areas (from Journeay and Mahoney, 1994). The upper fossils (F) are Early Cretaceous (Neocomian) in age; the lower fossils are Early Jurassic. Note that the boundary between Bridge River and Cayoosh must be strongly time-transgressive as the Bridge River contains fossils in chert facies as young as Callovian (Middle Jurassic) and possibly as young as Oxfordian (Late Jurassic).

In addition, the Cayoosh assemblage also may be correlative with a sequence of siltstone and shale that underlies the late Early Cretaceous Jackass Mountain Group, on the east side of Fraser River, ~5 km southwest of Lillooet, that is included in Methow terrane. To the north, the Relay Mountain Group overlies Cadwallader terrane (P. Schiarizza, pers.com. 2000). Thus, the Cayoosh assemblage and correlative units may be an *overlap assemblage* on Bridge River, and Methow (=? Cadwallader) terranes.

Although many contacts between the Bridge River Complex and Cayoosh assemblage are faults, the Cayoosh assemblage (and correlatives) apparently *conformably* overlies the Bridge River Complex in places, and elswhere is a lateral facies equivalent (Journeay and Mahoney, 1994; Journeay *in* Monger and Journeay, 1994a, p. 45; Schiarizza et al., 1997). From the less metamorphosed rocks in the Carpenter Lake area ~60 km northwest of here, Schiarizza et al. (1997) report that sandstone in their Gun Lake unit is similar in composition to sandstone intercalated with chert and argillite of the underlying Bridge River Complex, and furthermore, that sandstone in the Bridge River is most abundant in places adjacent to the Gun Lake unit. They conclude from these observations that the contact between the two units probably is gradational. A change from a dominantly chert-rich to a dominantly clastic-rich facies between Early and Late Jurassic time is common in Cordilleran accretionary complexes.

The Cayoosh assemblage and correlative units presumably were deposited partly on the deformed Bridge River accretionary complex, although an angular unconformity between the two is nowhere recognized. Atypically, some rocks in the Cayoosh assemblage are highly disrupted, in the style characteristic of accretionary complexes. These relationships raise an important question. Were some rocks of Cayoosh assemblage deposited on the Bridge River ocean floor prior to subduction, as suggested by the probable gradational relationships between the two, whereas others (most?) were deposited on top of the Bridge River accretionary complex in a possible forearc basin?

As noted above, Monger (1991) correlated rocks later called Cayoosh assemblage with the Settler Schist (east of Harrison Lake). He also argued that the latter unit was equivalent to the lower grade Shuksan greenschist and Darrington phyllite (EA of Figure 4-3) in the western North Cascade Ranges, which contain Early Cretaceous (~130 Ma) blueschist. If that correlation is correct, it would discriminate between paleogeographic hypotheses 2 and 3 in the introduction to the Coast Belt.

17.1 km Cross the southeastern extension of the Mission Ridge Fault (Fig. 4-5; formerly called the Lillooet Fault), which here dips steeply to the northeast, with the steep western slope of the Fraser valley wall south of Highway 99 probably approximating the dip of the fault. Across Seton River (on the right) the location of the fault is marked by a notch on the northern skyline.

Mission Ridge Fault

The sliver of subgreenschist grade Bridge River rocks just crossed are juxtaposed across the Mission Ridge Fault with cliff-forming, upper greenschist grade siliceous phyllite, local biotite schist and amphibolite, which are intruded by numerous felsic sills and dykes of Eocene age (~47 Ma). The metamorphic rocks are derived in part from the Bridge River Complex and probably in part from the Cayoosh assemblage. These rocks form the steep walls of the valley containing Seton River.

Coleman and Parrish (1991) considered the Mission Ridge Fault to be a northeast-dipping Eocene normal fault that separates low grade (prehnite-pumpellyite facies) rocks of the Bridge River Complex and overlying mid- to Late Cretaceous chert pebble conglomerate in the hanging wall from upper greenschist-lowest amphibolite grade schist in the footwall. Where studied, ~20 km northwest of here, barometric discontinuities across the fault indicate a downdip displacement of 10–18 km. The fault cuts deformed rocks of the Mission Ridge pluton of Eocene age (47.5 ± 2 Ma), and to the southeast is truncated by the Fraser Fault.

~20 km Highway 99 crosses bridges over Seton River and canal, and climbs south wall of valley.

Stop 4-1 Eastern Coast Belt view

21.7 km Pull out to right into parking area above Seton Lake. To the north across Seton River, the cliffs at the southeast end of Mission Ridge are composed of high greenschist to lower amphibolite (common biotite, rare garnet) grade pelitic, quartzitic and amphibolitic schist derived from the Bridge River Complex and(?) Cayoosh assemblage. White felsic intrusions, in places concordant with the schistosity and with mylonitic fabrics, and elsewhere cross-cutting, are all probably of Eocene age. These rocks are in the exhumed lower plate of the Mission Ridge Fault.

To the northwest, the high peaks visible along the skyline (Cayoosh Range) are underlain by mainly subgreenschist grade rocks of the Bridge River Complex. To the southwest, Mount Brew (el. 9470 feet), is composed of greenschist grade, sandstone, pelite and local conglomerate of the Brew Group (= Cayoosh assemblage). These rocks are in the hanging wall of yet another northwest-trending normal fault, called the Marshall Creek Fault (Fig. 4-5), whose trace is marked by a notch in the ridge on the north side of the west end of Seton Lake.

Stop 4-1 also is of historical interest in that it contains the remains of aboriginal pit dwellings that are ~1200 years old. In addition, Seton Lake is the easternmost of the chain of waterways (Fraser River, and Harrison, Lillooet, Anderson, and Seton lakes) connected by trails that were used in the 1860s by travellers from the coast to Lillooet.

~23.1 km Cross Marshall Creek Fault. The Marshall Creek Fault extends northwestwards from its cut-off with the Fraser Fault for a strike distance of over 130 km. Where it crosses Seton Lake, Marshall Creek Fault comprises an older system of shallow, northeast-dipping shear zones, and a younger (Eocene?) steeply southwest-dipping brittle normal fault.

24.5–31.0 km Road ascends and enters the canyon of Cayoosh Creek. The roadcuts are in foliated metagreywacke and phyllitic siltstone of the *Brew Group*, cut by undated, probably Eocene, felsic intrusions.

Brew Group

To south (left) of road, across the canyon of Cayoosh Creek, similar rocks to those passed in the roadcuts, together with local conglomerate, make up much of Mount Brew (elevation 2887 m). Mount Brew is the type-area of the Brew Group (Duffell and McTaggart, 1952) and is correlative with the Cayoosh assemblage studied mainly west of this area by Journeay and Mahoney (1994; Fig. 4-8).

The Brew Group is a succession of metamorphosed, variably foliated, pelitic siltstone, volcanic-rich greywacke, quartzite (rare in the western Cordillera!) and local, matrix-supported conglomerate beds with granite, black chert, siltstone and argillite clasts. The sandstone units locally contain shale rip-up clasts, and the package as a whole appears to have been deposited in deep water by turbidity currents. The rocks cut by small felsic intrusions, and typically are of greenschist grade, and folded into recumbent folds, including a syncline with an overturned limb that is several kilometres long. Near the summit of Mount Brew, and near the stratigraphic top of the succession, calcareous siltstones contain Neocomian (early Early Cretaceous, 145–127 Ma) bivalves (*Buchia*, formerly *Aucella*) (Duffell and McTaggart, 1952).

In addition to other units discussed earlier under Cayoosh assemblage, the Brew Group resembles a sequence of siltstone and shale that stratigraphically underlies the late Early Cretaceous Jackass Mountain Group, exposed on the east side of Fraser River, ~5 km southeast of Lillooet, and part of Methow terrane. With the exception of rare belemnites, no fossils are known from this sequence, although a boulder of argillite with Early Cretaceous *Buchia* is "almost certainly from the lower part of this unit" according to Duffell and McTaggart (1952). Although this correlation supports linkage between Methow and Bridge River terranes by Early Cretaceous time, the two are separated here by the Yalakom Fault, with >100 km of right-lateral offset across it.

For the next ~40 km, the roadcuts along Highway 99 are mostly either in foliated, greenschist grade rocks of the Bridge River Complex or in the Cayoosh Assemblage. According to Journeay (*in* Monger and Journeay, 1994a) are disposed in a complex system of mainly southwest vergent, locally northeast vergent, thrust faults of Late Cretaceous age. In a few places, there is evidence for (late) normal faults and strike-slip faults.

31.0 km Lower bridge over Cayoosh Creek. The roadcut on the right hand side before the bridge is in isoclinally folded dark grey phyllite and light grey metasandstone, locally with calcareous siltstone and marble, similar to Lower Cretaceous rocks of the Brew Group exposed near the summit of Mount Brew. Shear bands and asymmetric folds are locally developed in discrete ductile shear zones, and record a history of both flattening and northwest directed rotational strain. The rocks are structurally separated by a low angle fault from overlying schistose greenstone, metachert and phyllite of the Bridge River Complex. It is not clear whether the fault is a west-vergent Late Cretaceous thrust, or a low-angle Tertiary normal fault.

31.2–39.7 km On the left, discontinuous roadcuts are in phyllite and siltstone of the Brew Group. On the right, across Cayoosh Creek, the cliffs (and recent large rock slide) feature massive foliated greenstone of the Bridge River Complex that structurally overlies the Brew Group. This area was an active mining camp in the late 1800s; an adit formerly was present by the road, but is now buried, and several small adits are present high on the cliffs across the creek.

42.7–43.0 km Interlayered greenstone, metachert and siliceous argillite of the Bridge River Complex.

45.1–47.1 km This southwest-trending part of Highway 99 follows the trace of a high-angle, southeast side up, brittle normal fault. It separates granodiorite with prominent flat-lying joints of the Boulder Creek Pluton to the left (southeast) from thinly laminated siltstone and argillites of the Cayoosh assemblage exposed across Cayoosh Creek to right. The pluton is part of the latest Cretaceous-earliest Tertiary (~63 Ma) Bendor plutonic suite, which is a northwest trending linear belt of mainly medium and coarse grained, unfoliated, hornblende biotite quartz diorite, granodiorite and local granite bodies. The Bendor suite is southwest of the youngest, Eocene, suite of granitic rocks near Stop 4-3.

46.0 km Cross Boulder Creek.

47.6 km Approximate position of Downton Creek Fault on Highway 99. According to Journeay (*in* Monger and Journeay, 1994a) the Downton Creek Fault (Figs. 4-9, 4-10) is a southwest-directed contractional fault that also records a late-stage history of dextral-slip displacement. It is well exposed to the northwest in the headwaters region of Downton Creek and along the shore of Anderson Lake. It is intruded by the latest Cretaceous-early Tertiary Boulder Creek pluton, and is conidered to have been active at about 77 Ma.

Stop 4-2 Transition between Bridge River Complex and Cayoosh Assemblage

48.3 km Stop just before bridge over Gott Creek. Roadcuts in well-foliated siliceous siltstone, metachert (pale grey), talc-chlorite schist, greenschist and (on old roadcut up around corner, above bridge) thin layers of marble. These tectonized greenschist grade rocks are an example of the transition in this area between the Bridge River Complex and hemipelagic siltstones, turbidites and volcaniclastic sandstone typical of the Cayoosh assemblage. The contact between this transitional facies and conformably overlying basal successions of the Cayoosh assemblage is gradational; the contact is mapped between the last chert-bearing facies (Bridge River) and the clastic succession (Cayoosh). Note that here any bedding is transposed, and any formerly disrupted bedding which is characteristic of the Bridge River in its sub-greenschist facies, is "homogenized."

49.0–49.5 km Foliated greenschist and phyllite of the Bridge River Complex.

49.6 km Cross Cayoosh Creek; here, metachert and greenstone are interleaved with phyllite and siliceous siltstone, near the boundary between Bridge River Complex and Cayoosh assemblage.

53.5 km On right, road to the "Arctic White" granite quarry, located in the Downton Creek pluton of the Bendor Plutonic suite (~63–57 Ma).

55.8 km Bridge across Cayoosh Creek; just east of bridge, metasandstone and pelite of the Cayoosh assemblage contain small andalusite crystals.

55.8–58.0 km Discontinuous outcrops of greenstone, greenschist, metachert, phyllite belong to the Bridge River Complex. The higher ridges south of Highway 99 are underlain by medium and coarse grained unfoliated granodiorite of the (third from northeast), Late Cretaceous Scuzzy plutonic suite, U-Pb dated at ±85 Ma. The contacts with the hosting Bridge River and Cayoosh rocks are sharp and marked by a narrow, Buchan-style metamorphic aureole containing post-kinematic porphyroblasts of biotite, garnet, hornblende, andalusite and staurolite.

58.6 km Cross Channel Creek bridge.

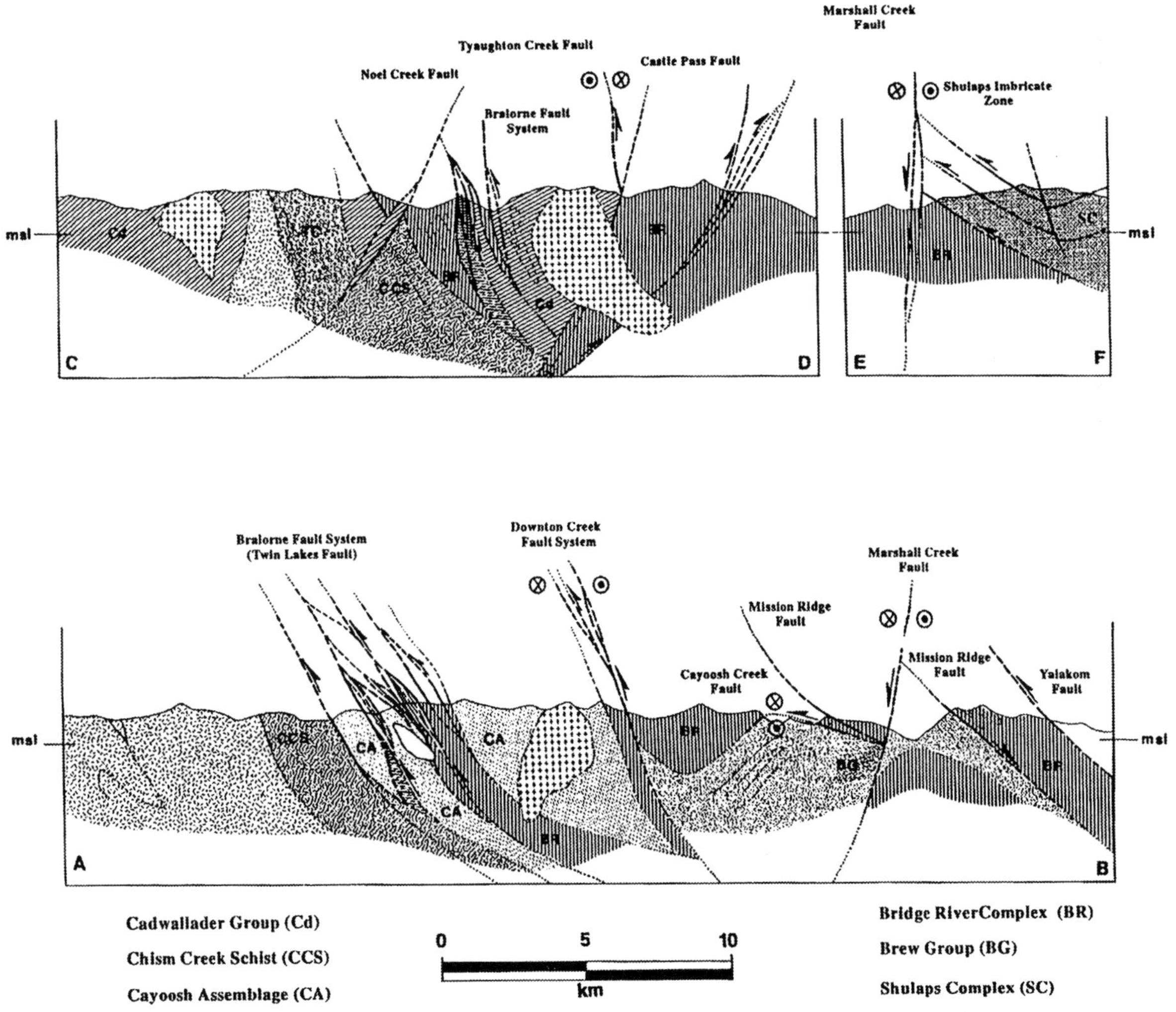

Figure 4-9. AB is a structural cross-section near Highway 99. CD, EF are drawn near Carpenter Lake to north. Compare with Figure 4-10 (Journeay, *in* Journeay and Monger, 1994a).

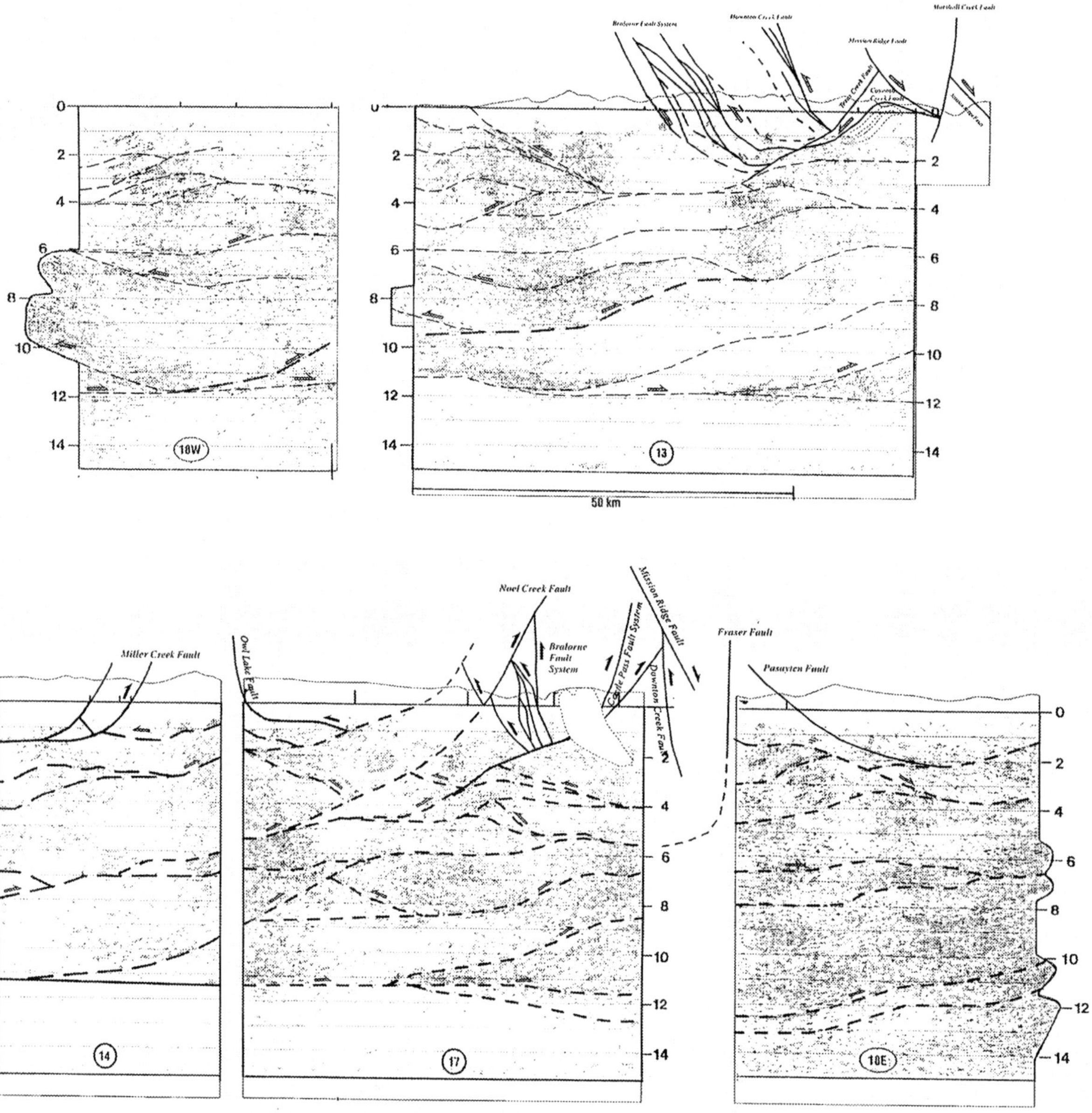

Figure 4-10. Integrated crustal cross-sections (vertical scale in seconds, two-way-travel time) based on geological mapping and Lithoprobe Southern Cordilleran seismic reflection sections 13, 14, 17, 18E, W; located on Figure 4-6. Cross-sections drawn by Journeay (in Monger and Journeay, 1994b) near Highway 99 (top) and near Carpenter Lake to the north (bottom).

59.0–61.5 km Roadcuts in metachert, greenstone, greenschist, and pelite of the Bridge River Complex are imbricated with slivers of phyllite and volcaniclastic sandstone of the Cayoosh assemblage and, near the western end of the series of roadcuts, minor serpentine and talc-carbonate schist.

Twin Lakes Fault System

Rocks in this stretch of road are involved in a system of northwest-trending, southwest-directed imbricate faults, called the *Twin Lakes fault system* (Figs. 4-9, 4-10). Although separated by plutons, this fault system appears connect with the Bralorne Fault to the northwest, and the Kwoiek Fault to the southeast. The system comprises southwest-directed thrust faults and left-lateral oblique slip faults (Journeay et al., 1992). Where crossed on Highway 99, the fault system marks the western limit of the Bridge River Complex. Individual faults on the Twin Lakes Fault system are marked by narrow zones that contain slivers of sheared ultramafic rock, and separate horses of greenstone, chert and clastic rocks derived from both Bridge River and Cayoosh rocks. The lower plate of the thrust system comprises polydeformed and more metamorphosed rocks called the Chism Creek Schist, which includes distinctive micaceous and phyllitic quartzite, phyllitic siltstone and marble, and is partly correlative with the Cayoosh assemblage. The fault system is cut by the Scuzzy Pluton, U-Pb dated at 85 Ma. Near Bralorne, to the northwest, shear-related mesothermal gold quartz veins formed between 91–86 Ma (Leitch, 1989). In the Harrison Lake area, congruent structures are known to have been active between 95–91 Ma (Journeay and Friedman, 1993).

Lithoprobe line 13 was run along this part of Highway 99. The east-dipping thrust faults and corresponding crustal reflections of the Bralorne-Twin Lakes-Kwoiek fault system can be traced down to ~3–4 seconds (~10–12 km), where they are apparently truncated by a system of sub-horizontal and west-dipping reflectors (Figs. 1-6, 4-10). On Line 14, north of this area, the west-dipping reflections can be traced up into shallow crustal levels (1 second) and these can be correlated with northeast-directed thrusts (of the Cadwallader-Castle Pass fault system), which cut the Bralorne-Twin Lakes-Kwoiek fault system (Journeay et al., 1992). Based on these observations, the west-dipping reflections at mid-crustal levels on Line 13 are interpreted to be part of the northeast-directed thrust system. The structures cut across the basal décollement of the Coast Belt thrust system (discussed below), and apparently root westward beneath Wrangellia to provide an example of *backthrusting* and *crustal delamination* in the Coast Belt.

62.2–64.0 km Discontinuous roadcuts along Highway 99 and along logging roads to northwest across Cayoosh Creek are in volcaniclastic sandstone and siltstone of the Cayoosh assemblage. This section is correlated with the upper clastic facies of the Cayoosh assemblage, and contains broken pelecypod (*Buchia?*) fragments that unfortunately are not diagnostic of age. The rocks are in an overturned panel near the base of the Twin Lakes fault zone.

64.7 km Bridge across Blowdown Creek approximately locates the footwall thrust of the Twin-Lakes fault zone, and separates hanging wall rocks of the lower Cayoosh assemblage from polydeformed footwall rocks of the Chism Creek Schist.

Chism Creek Schist

The term was coined by Rusmore (1985) for structurally interleaved amphibolite grade rocks of the eastern Cadwallader Range, ~100 km northwest of here. Chism Creek schist (*sensu stricto*) is not a stratigraphic unit, but rather a heterogenous tectonic package comprising imbricated slivers of bedded metachert, amphibolite, mafic and ultramafic rock, pelitic schist, micaceous and phyllitic quartzite and minor marble (sound familiar?). Within the Chism Creek schist, massive and thin bedded, locally disrupted quartzite, pelitic schist and graphitic phyllite are by far the most abundant rocks and probably are derived from the Cayoosh assemblage. Some quartzite associated with amphibolite may be metachert; and both probably are derived from the Bridge River Complex. Mafic and ultramafic rocks, and calcareous metasediments may be derived from Cadwallader (=Methow) terrane and its oceanic basement.

Pendants with a similar range of lithologies extend for 100+ km south southeastward, and probably include the Settler Schist exposed east of Harrison Lake (Monger, 1991). The belt is over 50 km wide across strike. In general, the metamorphic grade is higher than the Bridge River-Cayoosh rocks traversed so far on Highway 99, and far higher than rocks of these packages exposed to the north along Carpenter Lake.

64.8–66.6 km Nearly continuous roadcuts in Chism Creek Schist in the lower plate of the Twin Lakes fault zone. The section here is dominated by metavolcanic wacke, phyllite and quartz-rich sandstone. Phyllite near the east end of Duffey Lake (66.5 km) contains late- and post-kinematic porphyroblasts of garnet, staurolite ± andalusite.

67.2–70.2 km Roadcuts in phyllitic quartzite, graphitic phyllite and minor greenschist. Spectacular views to the west are into high peaks of the Joffre Group along the southern margin of the mid-Cretaceous Spetch Creek pluton.

70.5 km Roadcut in granodiorite

Stop 4-3 Intrusive contact of Mount Rohr Pluton with Chism Creek Schist

70.6 km Pullout on right in parking area by Duffey Lake. The bedrock here is in granodiorite along the northeast margin of the Mount Rohr Pluton. The Mount Rohr Pluton is contiguous with, and part of the Scuzzy plutonic suite of Late Cretaceous age (U-Pb dated at ~85 Ma), and is part of the third (from northeast) plutonic suite. Walk back (to east) for ~400 m across the contact, into phyllitic quartzite and locally disrupted thin-bedded quartzite and graphitic pelitic schist. The graphitic schist layers locally contain porphyroblasts of euhedral garnet and staurolite. Minor structures include two generations of tight asymmetric folds and late stage brittle gouge zones. These rocks are typical of the Chism Creek Schist in this area.

72.1–72.4 km Roadcuts are granitic rock of the Mount Rohr Pluton, considered to be part of the Late Cretaceous (±85 Ma) Scuzzy plutonic suite.

73.7 km Road crosses Van Horlick Creek.

73.9 km Fine-grained banded gneiss and metasedimentary rock correlated with the Chism Creek schist belong to an elongate pendant within the eastern part of the Mount Rohr pluton. Pendants along strike to the south contain a distinctive suite of phyllitic quartzite, more massive orthoquartzite, graphitic schist, minor amphibolite and thin lenses of marble. Still farther south these rocks can tracked into the Settler Schist, east of Harrison Lake, where they are in the upper plate of the Coast Belt thrust system. Similar rocks can be tracked for at least 60 km northwestward to near Bralorne.

74.9 km Road crosses Cayoosh Creek; at 75.8 km there is a view of Mount Joffre ahead.

76.6–81.1 km Discontinuous outcrops of medium to coarse grained hornblende quartz diorite of the Mount Rohr pluton; 81.1 km is roughly the summit on Highway 99 in the southeastern Coast Mountains

82.9 km Bridge across upper Cayoosh Creek. Cross at ~84 km the western contact of the Mount Rohr pluton with the mid-Cretaceous (U-Pb 102 ± 1 Ma) Spetch Creek pluton, which belongs the oldest and southwesternmost of the four plutonic suites in the southeastern Coast Belt. The Spetch Creek Pluton is 33 km long and up to 16 km wide. It consists mainly of homogenous, medium and coarse grained, biotite hornblende granodiorite. The nearly concordant date of 102 ± 1 Ma is from the south end of the pluton, and K-Ar dates range from 86 (hornblende) and 77 (biotite) and reflect a complex cooling history.

The high peaks to the northwest are underlain by rocks of Spetch Creek pluton, which in this area intrudes Cayoosh assemblage and Triassic and younger strata previously assigned to Cadwallader terrane by Journeay and Mahoney (1994) and Monger and Journeay (1994a). However, recent mapping to the north by Schiarizza et al. (1997) suggest that the latter (which they call Lillooet Lake-Nuit Lake belt) cannot be directly connected with "type Cadwallader" and may either be be a detached and duplicated structural sliver of Cadwallader or else a different belt of rocks. They may be more closely linked to rocks of the southwestern Coast Belt, because they contain Lower to Middle Jurassic volcanics correlated by Journeay and Mahoney (1994) with the Harrison Lake terrane, and deposited on top of rocks linked by them with the Cadwallader terrane. Regardless, the boundary between southwestern and southeastern Coast Belts is somewhere between here (~83 km) and the bottom of the Pemberton-Harrison Lake valley (~100 km).

Coast Belt Thrust System: the "Suture" Between Intermontane and Insular Terranes

West of the summit, Highway 99 descends to the Pemberton-Harrison Lake valley. The granitic rocks passed in the descent unfortunately mask a major structural boundary. The details of the structures in that boundary, called the *Coast Belt thrust system* by Journeay and Friedman (1993), are well exposed along strike ~80 km to the south southeast on the east side of Harrison Lake, where they have been studied by many workers over the past 40 years (e.g. Brown and Walker, 1993; Journeay and Friedman, 1993).

Integration of near-surface geology with Lithoprobe deep seismic reflection images provides new insights on the crustal structure and history of deformation of this part of the Coast Belt. As noted earlier, Line 13 of the southern Canadian Cordillera Lithoprobe reflection study was run along this part of Highway 99, and lines 14 and 17 along the Carpenter Lake road to the north. Interpretations of these data (Fig. 5; Varsek et al., 1993; Journeay, *in* Monger and Journeay, 1994a, b) show the crustal structure of the southeastern Coast Belt as a stack of imbricated fault slices or tectonic flakes. Structures within the stack record a history of dominant west-directed thrusting, more local east-directed thrusting, orogen-parallel displacements, and normal faulting (Journeay and Northcote, 1992; Schiarizza, et al., 1997).

The western margin of the southeastern Coast Belt features the Coast Belt thrust system. This delineates a crustal-scale ramp across and above which terranes of the southeastern Coast Belt were thrust southwestward over Middle Jurassic through Early Cretaceous (~170–95 Ma) rocks of the Gravina-Gambier arc (Fig. 8) emplaced in and on the eastern part of Wrangellia (Journeay and Friedman, 1993). High-pressure metamorphic rocks (7–9 kb) exposed in folded thrust sheets east of Harrison Lake indicate that imbricate faults of the Coast Belt thrust system rooted eastward at deep (25–30 km) crustal levels.

The southwestern Coast Mountains apparently acted as a relatively rigid block, as first suggested by Crickmay (1930), that was incorporated into the footwall of the southeastern Coast Belt accretionary complex in the late Early Cretaceous (~100 Ma). The western part the southwestern Coast Belt-Wrangellian block, located near the west coast of Vancouver Island, overlies a Late Cretaceous-Holocene subduction complex (Fig. 5; Pacific Rim, Siletzia, Olympic Core, Hoh terranes of Fig. 7). Incorporation of the Wrangellia-southwestern Coast Belt block into the upper plate of what is today called the Cascadia subduction zone appears to have had a buttressing effect on subsequent deformation in southeastern Coast Mountains, which led to renewed shortening, detachment and backthrusting in southeastern Coast Belt.

The southwestern Coast Belt-Wrangellian block ends not far south of the city of Vancouver. Rocks correlative with those in southwestern Coast Belt are recognized only in one place southeast of Vancouver, in a structural window (Mount Baker window) low in the thrust stack of the northwest North Cascades. Accretionary complex rocks probably correlative with those exposed locally on westernmost and southern Vancouver Island (Pacific Rim terrane) form the western foothills of the North Cascade Ranges in northwestern Washington State.

On a Cordilleran scale, a series of faults aligned with the Coast Belt thrust system of Journeay and Friedman (1993) can be traced for a distance of ~1600 km, from the northwest Cascades in Washington State at latitude ~47°30′N, to the west side of the Coast Belt near latitude 60°N (Fig. 8; Rubin et al., 1990; Wheeler and McFeely, 1991). North of this, the faults merge with the Tertiary Denali fault system. Not all of the aligned faults are thrusts, but most are of Late Cretaceous (≤95 Ma) age. The aligned structures trend acutely across the Coast Belt, from a central location in the south, to its western side near Prince Rupert (latitude 54°N). North of this, in southeastern Alaska, they form the western boundary of the Coast Belt. Rocks west of this system of aligned structures include the Paleozoic and early Mesozoic rocks of Wrangellia and Alexander terrane, which carry the magmatic and sedimentary rocks of the Jura-Cretaceous (~170–95 Ma) Gravina-Gambier arc. Rocks to the east include Bridge River and Methow terranes south of latitude 52°N, and Stikinia and a sliver of pericratonic Yukon-Tanana rocks farther the north.

86.0–97.9 km Granitic rocks of the Spetch Creek Pluton are exposed in series of roadcuts as Highway 99 descends into the great valley containing, from south to north, Harrison Lake, Lillooet River, Lillooet Lake and Lillooet River. This valley is the physiographic boundary between eastern and western parts of the southern Coast Mountains.

99.2–99.7 km Highway 99 runs northwestward along east shore of Lillooet Lake with roadcuts in rusty weathering mid-Cretaceous (102 Ma) Spetch Creek quartz diorite.

100.0 km Cross bridge over the Lillooet River, which typically is cloudy with rock flour as it drains a region carrying alpine glaciers. The peaks to the southwest across the valley, are underlain by the Early Cretaceous (113 Ma) Pemberton Diorite Complex.

109.3 km Village of Mount Currie. Turn left at junction and continue along Highway 99.

To left (southwest), prominent north-northeast-trending linear scarps highlighted by snow filled gullies are visible on the high alpine ridges of Mount Currie. These structures offset erosional surfaces along the crest of the ridges and are probably gravitational collapse structures.

To right (east), the bluffs are of probable Triassic volcanic rocks, that formerly were correlated with Triassic arc volcanics of Cadwallader terrane (Journeay and Mahoney, 1994), but currently are of uncertain terrane affinity (Schiarizza et al., 1997; their Fig. 44). These rocks are imbricated along a steep north-northwest trending, northeast-side-up reverse fault in a complex zone of brittle deformation nearly 2 km wide called the Owl Creek Fault, with Cretaceous volcanic and sedimentary rocks to the west, which are correlated with the Early Cretaceous Gambier Group. The latter is an integral part of the southwestern Coast Belt.

113.9 km Cross bridge over Lillooet River.

115.9 km Junction of Highway 99 with turnoff to the village of Pemberton. Turn left and continue southwards on Highway 99.

PEMBERTON TO VANCOUVER — SOUTHWESTERN COAST BELT

From here to the conclusion of the field trip in Vancouver, Highway 99 crosses the southwestern Coast Belt. Most of the route is in Late Jurassic to mid-Cretaceous (~170–95 Ma) granitic rocks that are at least in part intrusive equivalents of the volcanic rocks of the Gravina-Gambier arc. In contrast with the regionally penetrative deformation in the southeastern

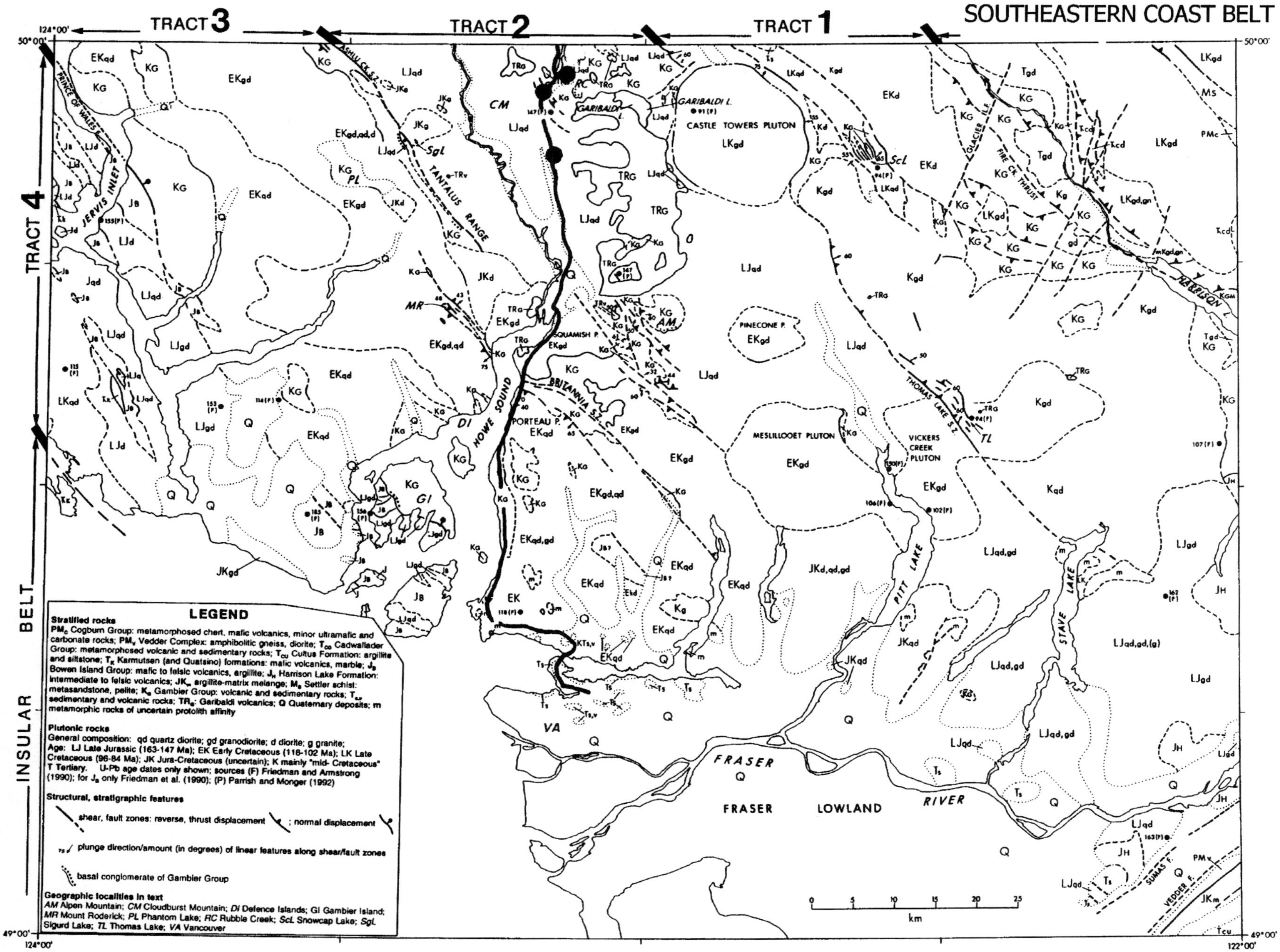
SOUTHEASTERN COAST BELT
TRACT 1
TRACT 2
TRACT 3
TRACT 4
INSULAR BELT
CASTLE TOWERS PLUTON
MESLILLOOET PLUTON
VICKERS CREEK PLUTON
PINECONE P.
SQUAMISH P.
PORTEAU P.
HOWE SOUND
JERVIS INLET
PRINCE OF WALES R.
TANTALUS RANGE
GARIBALDI L.
PITT LAKE
STAVE LAKE
HARRISON L.
FRASER RIVER
FRASER LOWLAND
THOMAS LAKE S.Z.
BRITANNIA S.Z.
ASHLU CK S.Z.
GLACIER L.F.
FIRE CK THRUST
SUMAS F.
VEDDER F.
km
LEGEND
Stratified rocks
PM_c Cogburn Group: metamorphosed chert, mafic volcanics, minor ultramafic and carbonate rocks; PM_v Vedder Complex: amphibolitic gneiss, diorite; T_{co} Cadwallader Group: metamorphosed volcanic and sedimentary rocks; T_{cu} Cultus Formation: argillite and siltstone; T_K Karmutsen (and Quatsino) formations: mafic volcanics, marble; J_B Bowen Island Group: mafic to felsic volcanics, argillite; J_H Harrison Lake Formation: intermediate to felsic volcanics; JK_m argillite-matrix melange; M_s Settler schist: metasandstone, pelite; K_G Gambier Group: volcanic and sedimentary rocks; $T_{s,v}$ sedimentary and volcanic rocks; TR_G: Garibaldi volcanics; Q Quaternary deposits; m metamorphic rocks of uncertain protolith affinity
Plutonic rocks
General composition: qd quartz diorite; gd granodiorite; d diorite; g granite;
Age: LJ Late Jurassic (163-147 Ma); EK Early Cretaceous (118-102 Ma); LK Late Cretaceous (96-84 Ma); JK Jura-Cretaceous (uncertain); K mainly "mid- Cretaceous" T Tertiary. U-Pb age dates only shown; sources (F) Friedman and Armstrong (1990); for J_B only Friedman et al. (1990); (P) Parrish and Monger (1992)
Structural, stratigraphic features
shear, fault zones: reverse, thrust displacement ; normal displacement
plunge direction/amount (in degrees) of linear features along shear/fault zones
basal conglomerate of Gambier Group
Geographic localities in text
AM Alpen Mountain; CM Cloudburst Mountain; DI Defence Islands; GI Gambier Island; MR Mount Roderick; PL Phantom Lake; RC Rubble Creek; ScL Snowcap Lake; SgL Sigurd Lake; TL Thomas Lake; VA Vancouver

Coast Mountains, the deformation in southwestern Coast Mountains is concentrated in discrete shear zones spaced ~10–20 km apart (Fig. 4-11). These separate areas in which the rocks have undergone little penetrative deformation, even though they are metamorphosed mainly in greenschist-lower amphibolite facies (Roddick and Woodsworth, 1975; Monger, 1991, 1993). The shear zones are mostly steeply dipping (>45°) towards the northeast, although some dip to the southwest, although some have the opposite sense. The shear zones appear to be mostly of early Late Cretaceous age (one is 94–91 Ma). The time of intrusion of the mid-Cretaceous granitic rocks (~106–95 Ma) widespread in southwestern Coast Belt, overlaps the time of compression (95–85 Ma) in the Coast Belt thrust system on its eastern margin. Some plutons have concordant fabrics that range from flow fabrics to mylonites to sheared granite with broken crystals, features that suggest emplacement and cooling during compression. Although the southwestern Coast Mountains block acted as the *foreland* to the Coast Belt thrust system, it was not completely rigid.

Where crossed on Highway 99, the southwestern Coast Mountains can be divided into four north northwest-trending "tracts," each of which is dominated by granitic rock of a particular age, and most of which are separated from flanking tracts by shear zones (Fig. 4-11).

Tract 1

Tract 1 is crossed on Highway 99, between the Pemberton valley and a location just south of the village of Whistler at ~158 km. It comprises mainly Early Cretaceous granitic rocks, with Middle Jurassic (~167–162 Ma) granitic rocks along strike towards the southwest, east of Harrison Lake. Near Harrison Lake, Jurassic granitic rocks intrude the Triassic-Jurassic rocks of Harrison Lake terrane (probably correlative with the Jurassic rocks of Wrangellia), that are stratigraphically overlain by Cretaceous volcanic and sedimentary rocks correlated with the Gambier Group, and all are intruded by Cretaceous (106–91 Ma) granitic rocks. These rocks are the footwall of the Coast Belt thrust system, but in places (northeast of Harrison Lake) are themselves involved in the thrusting (Lynch, 1992; Journeay and Friedman, 1993). The southwestern boundary of this tract is the Thomas Lake shear zone (Fig. 4-11).

Figure 4-11 (facing page). Map of the southwestern Coast Belt "foreland" showing its division into "tracts" 1–4 separated by shear zones; between shear zones the rocks are mostly little deformed. The route across the mainly granitic southwestern Coast Belt between Pemberton and Vancouver is shown by heavy line. Heavy dark dots indicate stop locations.

Tract 2

Tract 2 is crossed on Highway 99 between 158 km and ~200 km, just north of the town of Squamish. It consists mainly of Middle and Late Jurassic (162–147 Ma) granitic rocks, which locally are non-conformably overlain by Early Cretaceous clastic and volcanic strata of the Gambier Group, and locally intruded by mid-Cretaceous (106–91 Ma) granitic plutons. Within this tract, an "intra-Gravina-Gambier" episode of uplift and erosion is recorded by an erosional unconformity of latest Jurassic-earliest Cretaceous age. Late Tertiary-Holocene flows and volcanic centres of the Garibaldi Group, the northernmost part of the *Cascade magmatic arc,* occur mainly in this tract along Highway 99. The southwestern boundary of the tract is the Ashlu Creek shear zone crossed near the town of Squamish at the north end of Howe Sound.

Tract 3

Tract 3 is crossed on Highway 99 between ~200 km and ~245 km near the south end of Howe Sound. It consists mainly of mid-Cretaceous (103–95 Ma) plutonic rocks of the Howe Sound Batholith and subordinate septa of Gambier Group. The southwestern boundary of tract 3 is marked by probable Early Cretaceous northeast-side-down normal faults along Prince of Wales Reach of Jervis Inlet and, along strike, on Gambier Island in Howe Sound, although continuity between the two fault segments is obscured by intrusions as old as 114 Ma.

Tract 4

Tract 4 comprises mainly granitic rocks of Middle and Late Jurassic (164–153 Ma) and Early Cretaceous ages (118–102 Ma). The tract contains pendants of undated, locally pillowed metabasalt and marble, which probably are metamorphosed equivalents of the Triassic Karmutsen and Quatsino formations on Vancouver Island (*the* characteristic Wrangellian units!) as well as Early Jurassic metavolcanic and sedimentary rock of the Bowen Island Group. The latter unit is correlative with the latest Triassic-Middle Jurassic Bonanza arc volcanics of Vancouver Island (Roddick and Woodsworth, 1975; Friedman et al., 1990). Isoclinal folds in the Bowen Island Group are intruded by 155 Ma plutons, and record an episode of

Early to Middle Jurassic contractional deformation not recognized farther east (Fig. 4-1). The southwestern limit of this tract is the mainly submerged, western boundary of the Coast Belt, which where exposed on islands at the north end of the Strait of Georgia, is a Middle Jurassic (164 Ma) magmatic front that intrudes Wrangellia (Nelson, 1979; Monger and McNicoll, 1993).

116.7–117.1 km Roadcuts on west (right) side of Highway 99 are in rocks of the Pemberton diorite complex. Near 116.2 km the road crosses the trace of the north northwest-trending Miller Creek Fault, which along strike this is imbricated with a pendant of metavolcanic rocks correlated with the Gambier Group.

118.4–138.7 km Highway 99 continues up the valley of the northeast-flowing Soo River; at 119.9 km it crosses B.C. Rail track; at 121.4 it crosses Rutherford Creek, a tributary of the Soo River; at 121.9 km it is crossed by B.C. Rail overpass; and at 122.4 km it crosses Green River. Until ~129 km all roadcuts are in rocks of the Pemberton diorite complex, with a long roadcut on left in Pemberton diorite at 128.1 km.

Pemberton Diorite Complex

The complex ranges from variably foliated hornblende diorite to quartz diorite. Throughout the complex are abundant inclusions and schlieren of amphibolite and meta-sedimentary and meta-volcanic rock. Dykes and sills of hornblende granodiorite and granite occur throughout the complex. The complex is U-Pb dated at 113 ± 2 Ma (Friedman and Armstrong, 1990).

128.9–129.6 km Pass a large rock slide; the blocks are from a heterogenous "migmatitic" phase of the Pemberton complex. At 131.3 km Cross B.C. Rail track, and Highway 99 continues past numerous small rock cuts in an undated body of quartz diorite which flanks the western margin of the Pemberton diorite complex.

139.5 km Green Lake on left. Granitic rocks exposed in roadcuts are fractured and altered. The fracturing is believed by Journeay (*in* Monger and Journeay, 1994a) to be associated with the *Green Lake Fault.* This is one of a family of northeast-striking, at least partly right-lateral oblique-slip faults of Oligocene-Miocene age (~25–14 Ma) that occur widely within the southern Coast and northern Cascade Mountains. They are associated with deformation within early and possibly late stages of the Cascade magmatic arc (Coish and Journeay, 1992).

143.8 km Cross Nineteen-mile Creek, and at 145.2 km cross the bridge over over B.C. Rail track.

147.9 km Stop lights at Villagegate junction in the village of Whistler.

Capping the ridge west of the village of Whistler is the young Mount Cayley volcanic complex, which is one of the eruptive centres in the northernmost part of the Cascade magmatic arc. The high amplitude seismic reflections recorded on Lithoprobe line 15 at depths of 4.5–5 s from just west of Mount Cayley are interpreted to indicate either presence of a magma chamber at depth, a mafic sill of very different composition from the country rock, or else a concentration of fluids trapped at this depth (Lewis et al., 1992).

148.8–155.3 km A series of roadcuts in chlorite and sericite schist, rare talc schist, phyllite, and metavolcanic breccia, together with more massive volcanics and minor limestone, probably are derived from the Early Cretaceous Gambier Group.

154.2 km Stop lights in Whistler village; at 156.5 km cross B.C. Rail track.

156.5–158.1 km Rock cuts on right are in sheared and broken green granodiorite, here undated but probably part of the Late Jurassic Cloudburst quartz diorite.

~157.7 km Approximate location of the trace of the Thomas Lake shear zone on Highway 99.

Boundary Between Tracts 1 and 2

Highway 99 crosses the northwesterly extension of the Thomas Lake shear zone, which is the boundary between Tracts 1 and 2. The shear zone can be traced southeast of here for over 100 km to the north end of Stave Lake (Fig. 4-11). It crosses the ridge near Corrie Peak, ~12 km southeast of here, where undated granitic rocks are thrust southwestward over metamorphosed sandstone and pelite of the Helm Formation, which is the upper unit of the Early Cretaceous Gambier Group in this area. The shear zone dips 60° to the northeast. Fault zone fabrics (sigmoidal structures and down-dip stretching lineations) record northeast-side-up displacement. The time of movement on the shear zone can be dated at between 91 and 94 Ma. About 16 km southeast of here, the Castle Towers pluton (U-Pb dated at 91 ± 3 Ma) intrudes the fault, and about ~60 km southeast of here near Thomas Lake, granitic rocks dated at 94 Ma are in the hanging wall of the shear zone.

160.1 km Road cuts in Cheakamus valley basalt. The columnar jointed, brownish weathering, basalt exposed on left of Highway 99 is part of an olivine basalt flow that descends Callaghan Creek valley to the northwest for ~20 km, and continues down the main Cheakamus River valley for at least another 10 km (Green et al., 1988). It belongs to the younger part of the Cascade magmatic arc. Intercalated till and fluvioglacial sediments yield radiocarbon dates of 34 000 years B.P. The flow is exposed in roadcuts for the next 10 km, and in places lies on ice-scoured surfaces on chlorite schist (meta-Gambier?), Jurassic granitic rock, and elsewhere on fluvioglacial sediments.

161.3 km Cross Callaghan Creek.

162.0–163.9 km The columnar jointed lava flow near is near highway. The "shark fin"-like peak called The Black Tusk is visible on the ridge to the southeast (left), at a distance of 11 km and an elevation of 2500 m. It comprises the volcanic products of two eruptive stages. Older andesitic lavas are K-Ar dated at ~1.2 Ma. Younger hypersthene andesite lavas and a dome that probably rose through and above the Pleistocene ice sheet are dated at 0.09–0.21 ma (Green et al., 1988).

163.9 km Cross Brandywine River.

164.0 km Schistose rocks, probably derived from the Gambier Group.

165.8 km Brandywine Falls Park

166.5 km Cross B.C. Rail track. South of here, both along the railway tracks and for ~3 km along Highway 99, is a sequence of chlorite schist, schistose metasandstone with local conglomerate beds containing granitic clasts, local dark grey phyllite and thin discontinuous marble horizons.

Gambier Group Stratigraphy

The schistose rocks can be traced across Daisy Lake to the east and up on to the ridges forming the eastern side of Cheakamus valley, where they underlie the young volcanics of The Black Tusk. Between the valley floor and an elevation of ~2000 m they lose their penetrative foliation and pass into unfoliated but greenschist-grade rocks belonging to the Cheakamus, Empetrum and Helm formations of Mathews (1958), which make up the Early Cretaceous Gambier Group in this area (Fig. 4-12). The basal Cheakamus Formation consists of coarse, green, locally cross-bedded arkosic sandstone, lesser argillite and local conglomerate, and in places contains large, well preserved bivalves, *Inoceramus* c.f. *paraketzovi* Efimova of probable Hauterivian age (re-collected with W.H. Mathews in 1991, and identified by J.W. Haggart; G.S.C. Catalogue No. C-187629). The Cheakamus Formation appears to be a shallow-marine deposit. Similar strata are observed in three localities to overlie quartz diorite that at least locally is as young as 147 Ma: near Rubble Creek (Stop 4-4; Mathews, 1958); on Alpen Mountain (15 km east of Squamish; Lynch, 1991); and near Sigurd Lake in the Tantalus Range, 30 km northwest of Squamish (Fig. 4-11; Monger, 1993). The overlying Empetrum Formation is a distinctive olistostromal(?) coarse breccia and conglomerate, containing granitic, dioritic and volcanic clasts and blocks of carbonate up to 100 m in length. One granitic clast was U-Pb dated at 161 ± 7 Ma (R.M. Friedman, pers. comm. 1992). Uppermost, the Helm Formation comprises graded, dark slaty argillite and quartzitic sandstone.

The stratigraphic relationships demand that uplift, erosion and exposure of granite, and deposition of overlying shallow marine strata took place over a period of 15–20 million years (Hauterivian age range is 132–127 Ma). The lithologies of the overlying formations suggest that subsidence continued after initial deposition. The Empetrum Formation possibly formed by mass flow from submarine fault scarps, and the Helm Formation appears to be a turbiditic, deep water deposit. Normal faults of probable Early Cretaceous age occur to the southwest along the boundary of tracts 3 and 4. Speculatively, deposition of the Gambier Group was in a rapidly subsiding extensional basin, in contrast with the contractional and transtensional setting in which the mid- and Late Cretaceous plutons were emplaced.

168.6–170.7 km Columnar jointed Cheakamus valley basalt along Highway 99 overlies chlorite schist, granitic rock and unconsolidated syn-glacial sediments.

169.7 km View of Cloudburst Mountain to southwest; Daisy Lake just east (left) of highway is ponded above a dam on the Cheakamus River. Water from Daisy Lake runs for ~12 km through a penstock under Cloudburst Mountain to generate electricity at its outflow on the Squamish River; the hydraulic head is ~330 m.

171.8–172.4 km Roadcuts in columnar jointed lava; at 171.4 km, cross the south-flowing Cheakamus River.

172.6 km Bridge over the Cheakamus River below Daisy Lake dam.

173.2 km Side trip Take a sharp turnoff to left (northeast) by park sign; turn right (to southeast) after 0.4 km onto a side road leading to the trailhead

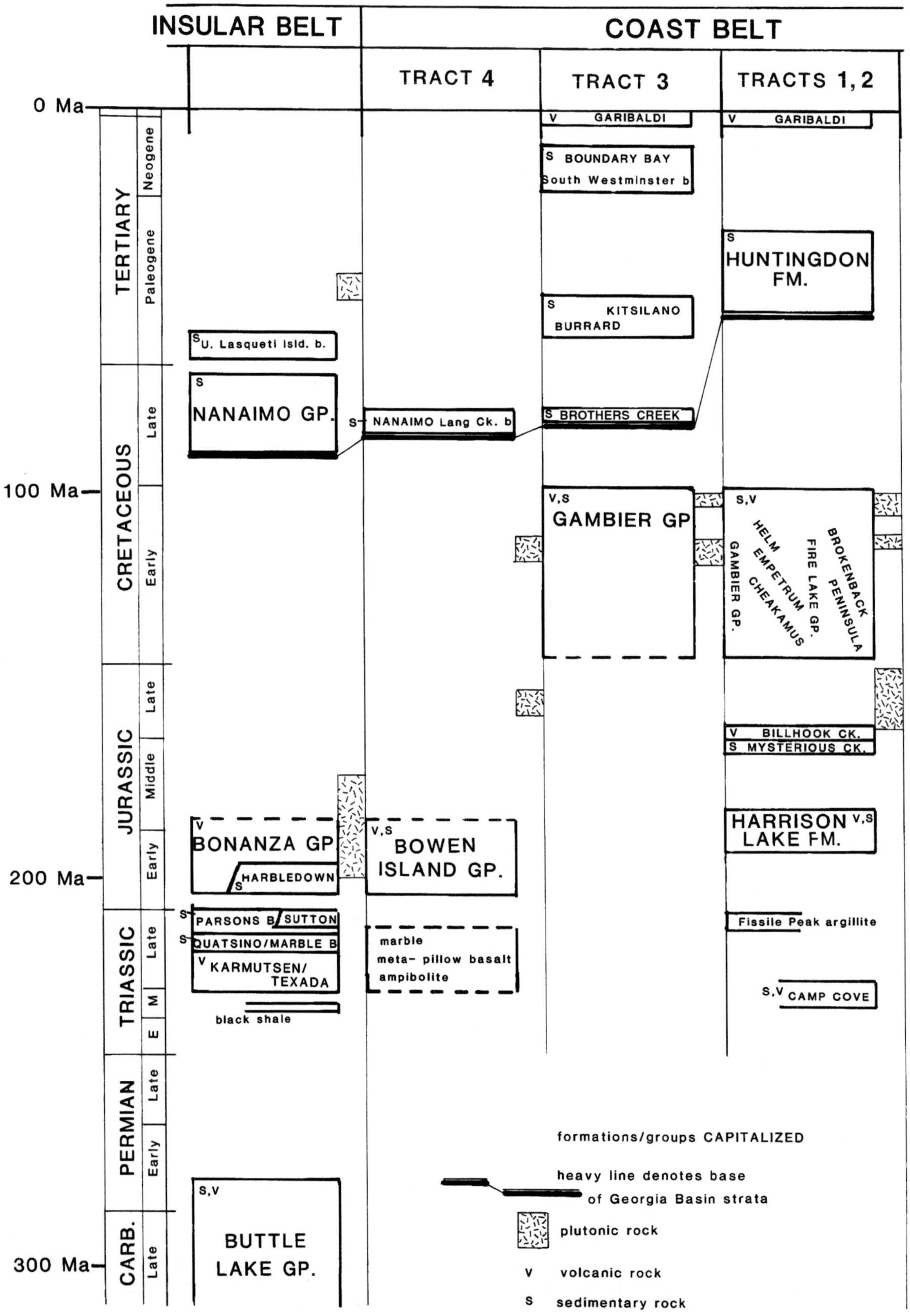
INSULAR BELT
COAST BELT
TRACT 4
TRACT 3
TRACTS 1,2
0 Ma
100 Ma
200 Ma
300 Ma
TERTIARY
Neogene
Paleogene
CRETACEOUS
Late
Early
JURASSIC
Late
Middle
Early
TRIASSIC
Late
M
E
PERMIAN
Late
Early
CARB.
Late
V GARIBALDI
V GARIBALDI
S BOUNDARY BAY
South Westminster b
S HUNTINGDON FM.
S KITSILANO
BURRARD
S U. Lasqueti Isld. b.
S NANAIMO GP.
S NANAIMO Lang Ck. b
S BROTHERS CREEK
V,S GAMBIER GP
S,V
HELM
EMPETRUM
CHEAKAMUS
GAMBIER GP.
BROKENBACK
PENINSULA
FIRE LAKE GP.
V BILLHOOK CK.
S MYSTERIOUS CK.
V BONANZA GP
S HARBLEDOWN
V,S BOWEN ISLAND GP.
HARRISON V,S LAKE FM.
S PARSONS B/SUTTON
S QUATSINO/MARBLE B
V KARMUTSEN/ TEXADA
black shale
marble
meta- pillow basalt
ampibolite
Fissile Peak argillite
S,V CAMP COVE
S,V BUTTLE LAKE GP.
formations/groups CAPITALIZED
heavy line denotes base of Georgia Basin strata
plutonic rock
V volcanic rock
S sedimentary rock

parking lot for the trail which goes to Black Tusk Meadows in Garibaldi Provincial Park. The distance to the parking lot is 2.8 km. The road is over boulders of the Rubble Creek slide, which is a debris flow that formed in the winter of 1855–56 by collapse of material from the volcanic complex exposed in the cliff above Rubble Creek (Mathews, 1958).

Stop 4-4 Contact between Cheakamus Formation and Cloudburst diorite

From the parking lot, walk up the north bank of Rubble Creek for ~100 m to examine outcrops of cobble conglomerate containing a variety of granitic and non-granitic clasts of the Cheakamus Formation. The actual basal contact of these rocks, nonconformably overlying the Cloudburst quartz diorite, is exposed on the knoll 1 km south of here, above the valley wall of Rubble Creek. There, bouldery conglomerate, metamorphosed in greenschist grade, overlies green, altered quartz diorite.

To the east, up Rubble Creek, the cliff called The Barrier is composed of young (≥6000 B.P.) lava flows that were ponded against ice occupying the Cheakamus River valley. Subsequent melting of the ice created an oversteepened unstable mass. One of the flows collapsed to fill the valley with the 30 million cubic metres of bouldery rubble crossed on the way to Stop 4-4 (Mathews, 1958). Comparable features involving collapse of young volcanic rocks following deglaciation occur in several places on the west side of the Cheakamus River valley south of here for about 30 km.

From Stop 4-4 return to Highway 99.

179.04 km Cross Rubble Creek. South of here, Highway 99 crosses the Rubble Creek debris flow; logs caught in the debris flow are visible in places to the right in the Cheakamus River bed.

181.2 km Roadcut on left side of road in mylonite of the Cloudburst quartz diorite.

South of here for 4.5 km are more-or-less continuous roadcuts in mylonite.

182.8 km Pass first pull-out on right.

Figure 4-12 (facing page). Correlation of rock units in the predominantly granitic southwestern Coast Belt. Wrangellian strata in the Insular Belt can be traced into the southwestern Coast Belt, which is the site of a Middle Jurassic to mid-Cretaceous arc. All of these rocks are overlain by the "molasse" of the Late Cretaceous Nanaimo Group, which is mostly derived from the tectonically uplifted southeastern Coast Belt, and preserved in the erosional remnant called Georgia Basin.

Stop 4-5 Mylonite in Late Jurassic Cloudburst quartz diorite

183.2 km Stop in the second pull-out on right. Mylonite within the Cloudburst quartz diorite first was recognized by Mathews (1958), and forms a north northwest trending zone that is up to 1.5 km wide and has been traced for ~8 km along strike. It may well extend farther, as the continuation south southeastward is concealed below the cover of young volcanic rocks. The mylonites exposed in the roadcuts along Highway 99 have a northeast-dipping foliation and only a weakly developed linear fabric, but generally record a sense of northeast-side-up movement. The age of mylonitization is unknown but at least has a movement sense consistent with that of the early Late Cretaceous faults, although mylonitic fabrics are known in some granitic pebbles in the basal conglomerate of the Cheakamus Formation. A U-Pb date of 145 ± 2 Ma was obtained by Friedman and Armstrong (1990) from mylonitic quartz diorite in a roadcut on Highway 99 about 1.5 km south of this locality.

185.5 km Cross the sharp contact between mylonite and fine grained hornblende quartz diorite. From this point south for the next 15 km, as far as the bridge over the Cheekye River, the numerous roadcuts are in massive and generally greenish coloured Late Jurassic Cloudburst quartz diorite, with chloritized mafic minerals. In many places the quartz diorite is intruded by dark green dykes, which are probably feeders to the Early Cretaceous Gambier volcanic rocks, and rare brown dykes that are probably feeders to the late Tertiary-Holocene volcanics.

186.1 km Cross bridge over Culliton Creek; note the artificial levees above the bridge. The upper parts of this creek contain volcanic rocks that once were ponded against ice, as in Rubble Creek, and the creek presents similar hazards.

Stop 4-6 View of stratigraphic and intrusive relationships across the Ashlu Creek Shear Zone

193.9 km Pull into parking area on right. The Ashlu Creek shear zone is the boundary between tracts 2 and 3 (Fig. 4-11).

To the west, across the Cheakamus River and Squamish River valleys, the high, rugged Tantalus Range is composed mainly of Late Jurassic granitic rock (U-Pb date 150 Ma; V. McNicoll, pers. comm. 1994). Southwest of this are mainly mid-Cretaceous granitic rocks. A narrow discontinuous band of locally highly strained Gambier Group rocks crosses from the south side of the southeast end of the Tantalus Range west of Squamish, to the northeast side of the range in the upper part of Ashlu Creek. Near Sigurd Lake, in the northwestern Tantalus Range (Fig. 4-11), Gambier strata unconformably overlie Late Jurassic quartz diorite, and both are intruded by a small mid-Cretaceous pluton (K-Ar date 94.7 Ma; P. Hunt, pers. comm. 1994). Although the Ashlu Creek zone is a large shear zone, the sense of movement along it is equivocal, although many strain indicators, mainly broken and rotated feldspars in both Jurassic plutons and Gambier strata indicate southwest-side-up movement

To the southeast, the valley to the east of the head of Howe Sound is the Stawamus River valley, which follows the trace of the southwestern extension of the Ashlu Creek shear zone. To the east (left) of Stawamus River, the mountain is composed of Late Jurassic granitic rocks cut by abundant dykes, that are feeders to the stratigraphically overlying Cretaceous Gambier volcanics. The basal contact is on Alpen Mountain (Fig. 4-11). In the Stawamus River valley, subhorizontal slickensides on fault surfaces indicate local dextral movement. To the right of Stawamus River valley, mid-Cretaceous granitic rocks intrude the gently west-dipping Gambier flows and volcaniclastics visible near the skyline.

197.7 km Brohm Lake on right.

201.1 km Cross bridge over Cheekye River; and pass the southernmost roadcuts along Highway 99 of Jurassic granitic rocks.

South of here, Highway 99 runs for >5 km down the gently south-sloping surface of Cheekye Fan. The fan is composed of debris resulting from the partial collapse the volcanic cone of Mount Garibaldi, a Quaternary (0.22–0.26 Ma) volcano constructed above the east side of the Cheakamus River valley and erupted through Wisconsin ice (Mathews, 1958; Green et al., 1988). The 2900 m high "half cone" of Mount Garibaldi is located 12 km northeast of the Cheekye River bridge, and was built on an irregular bedrock surface at elevations of 1500–2000 m. Melting of the ice filling the valley led to collapse of the unsupported western flank of Mount Garibaldi. Small debris flows down the Cheekye valley occurred as recently as 1958.

201.7 km Junction with road on left to Alice Lake; 206.0 km, stop lights at Garibaldi Way.

208.3 km Cross bridge over the Mamquam River. East of here, the bottom of the Mamquam River valley contains a young, post-glacial flow (post-6670 B.P., Green et al., 1988). From this point is a good view to the southeast of the "half dome" of Stawamus Chief Mountain, composed of massive mid-Cretaceous (U-Pb date 101 Ma; Friedman and Armstrong, 1990) granodiorite. As noted earlier (Stop 4-6), this body intrudes Early Cretaceous Gambier Group rocks which are visible near the skyline. About 7 km east of here and northeast of the Ashlu Creek shear zone, latest Jurassic Cloudburst quartz diorite (U-Pb date 147 ± 0.5 Ma; Parrish and Monger, 1992) is intruded by numerous green, Gambier(?) dykes.

211.9 km Stop lights at junction of Highway 99, with the turnoff to right to the town of Squamish, located at the north end of Howe Sound.

212.3 km Cross Mamquam Blind Channel; at 213.6 km cross over Stawamus River.

214.7 km View on left of west face of Stawamus Chief, the local rock climbers' Mecca. The clean granodiorite with rare dykes exposed here contrasts markedly with the generally altered Jurassic granite rock intruded by abundant dykes seen previously.

Howe Sound Batholith

Older maps (e.g. Roddick and Woodsworth, 1979) show the rock forming Stawamus Chief as a discrete intrusion called the Squamish Pluton (Fig. 4-11). Monger and McNicoll (1993) argued on the basis of newer mapping and U-Pb dating that it probably was an apophysis of a much larger, mid-Cretaceous (103–96 Ma) body they called Howe Sound Batholith, which also contains the partly fault-bounded Porteau pluton crossed farther south (231 km) along Howe Sound. Their batholith is at least 120 km in length along the regional, north northwest grain, and is ~40 km wide. The available dates suggest that it was emplaced immediately before the time that the Coast Belt thrust belt was active.

215.9 km Pass Shannon Creek with its spectacular waterfall on left. South of this for the next 6 km, rock cuts are in mid-Cretaceous granodiorite, locally with rusty weathering magnetite layers.

217.5 km Cross Gonzales Creek; at 220.3 km pass Petgill Lake on right.

222.1 km Highway 99 descends hill and crosses the intrusive contact between the granitic rock and a pendant of hornfelsed Early Cretaceous Gambier volcanic breccia which at this place retains its primary fabric.

223.0 km Bridge over Britannia Creek at Britannia Beach.

Britannia Mine and Shear Zone

The settlement of Britannia Beach is the townsite and former millsite of the Britannia Mine, which was discovered in 1888, worked from 1905 to 1974, and now is the home of the B.C. Museum of Mining. The mine yielded primarily copper, with lesser amounts of lead and zinc, and minor gold and silver, that were derived from pyrite, chalcopyrite and sphalerite formed as volcanogenic massive sulphides (Payne et al., 1980). The deposits are located in chlorite and sericite schist derived from the Gambier Group, and formed within a complex northwest-striking shear zone that mostly dips to the southwest.

On the west side of Howe Sound, along strike from the rocks at Britannia Beach, there is a 1 km wide exposure of deformed Gambier strata. Its northern contact is intrusive, whereas the southern contact is a reverse fault which dips southwest at ~75°, across which the Gambier Group is juxtaposed with mid-Cretaceous granitic rock of the Howe Sound batholith. Near the contact, the granitic rock features not only planar fabrics produced by aligned mafic minerals and elongate xenoliths flattened in the plane of foliation that presumably were formed during intrusion, but also cataclastic fabrics with broken feldspars, and mylonite with the same orientation (Monger, 1993). Similar southwest-dipping fabrics also occur 8 km along strike to the northwest, near the summit of Mount Roderick, entirely within the plutonic rock, there U-Pb dated at 95.8 ± 1 Ma (Monger and McNicoll, 1993). In this same area, northeast-dipping bands of ultramylonite also are present within the pluton. The fabrics probably all formed during emplacement and cooling of the Howe Sound Batholith (106–95 Ma). The age is just before the time of contraction on the Coast Belt thrust system to the east, and the batholith presumably was cooling during a period of strong regional compression. The early intrusive flow fabrics seemingly were replaced by cataclastic and mylonitic fabrics, and ultimately by ultramylonites as cooling of the batholith progressed.

224.8 km Highway 99 crosses the approximate position of the southern contact of the pendant of Gambier volcanics within the mid-Cretaceous granitic rock. The contact is exposed on the shore ~300 m southwest of here. It is an intrusive contact that dips southwest at 70°, and along which reverse faulting has taken place, and is similar to that noted above across Howe Sound.

South of here along Highway 99 for ~7 km, roadcuts are in granitic rock of the Howe Sound Batholith. The granitic rock in places contains many dark inclusions, and locally has an east-dipping (~40°), locally cataclastic, foliation.

228.7 km Bridge over Furry Creek.

230.1 km Large roadcuts in granitic rock north of Porteau Cove. Note the prominent jointing which dips toward Howe Sound. This is a major cause for concern for slope stability, and necessitates extensive rock bolting and scaling.

231.3 km Porteau Cove. From this part of Howe Sound batholith (here called Porteau pluton), Symons (1973) and Irving et al. (1985) obtained paleomagnetic data that were interpreted (together with data from other mid-Cretaceous plutons in the southern Coast Belt) to suggest that the rocks had moved northwards for ~2400 km and had rotated clockwise since mid-Cretaceous. As noted by Irving et al. (1985) and argued strongly by Butler et al. (1989), the absence of paleohorizontal markers in the intrusions makes it possible that the anomalous paleomagnetic directions resulted from systematic southwest tilting, rather than from northward displacement.

232.0 km Granitic rock on left side of road. This is close to the contact with another pendant of Gambier strata to the south. The contact is in places intrusive but elsewhere is a fault.

232.1–243.9 km Hornfelsed and mostly well bedded silicified argillites and tuffaceous siltstones of the Gambier Group. Between 236.1–236.7 km the bedding outlines an anticline with gently dipping limbs. The southern ~5 km of the roadcuts along this section of the highway are in more massive volcaniclastic rock ranging from breccia to lithic tuff. These rocks form a thin screen along the eastern wall of the fiord that is Howe Sound, and the hosting granitic rocks are exposed higher up and ~2 km to the east (Fig. 4-11).

On the shoreline near Brunswick Point, ~4 km south of the northern contact, Gambier Group strata contain the ammonite *Cleoniceras (Grycia?) perezianium* of probable middle Albian age (~105 Ma), which is ≤10 million years older than the intruding granitic rocks.

238.4 km Cross Deeks Creek; at 238.7 km cross M Creek; at 242.2 km cross Magnesia Creek.

243.9 km Highway 99 runs through the village of Lions Bay. Note the extensive (and expensive!) engineering undertaken to control debris flows in the several creeks crossed in this part of the highway, which traverses the steep eastern side of Howe Sound. Many buildings in Lions Bay are built upon unconsolidated debris flow and overbank deposits; hence the need for careful control of the creek channels.

Southwestern Boundary of Tract 3

The large island in the western side of Howe Sound is Gambier Island. The northern part the island is composed of rocks of the Gambier Group, and the southern part consists of the Lower Jurassic Bowen Island Group intruded by the granitic Thornborough Intrusions (U-Pb date 156 ± 2 Ma; Friedman and Armstrong, 1990). The contact between the Gambier and Bowen Island groups was long thought to be stratigraphic, but at least locally it is a northeast-side-down normal fault, which is probably a continuation or splay of the normal fault to the northwest along the Prince of Wales Reach of Jervis Inlet (Fig. 4-11). The fault marks the southwest boundary of tract 3. Exactly where it crosses Highway 99 is not clear, but it is probably between ~247.2 km (the southernmost exposure of Gambier volcanics) and ~250.2 km (possible Bowen Island rocks).

244.2 km Highway 99 crosses Harvey Creek; at 246.0 km crosses Lonetree Creek.

247.2 km Cross Newman Creek and southernmost exposure of probable Gambier Group on highway. On BC Rail track below highway about 200 m south of this locality are outcrops of altered green granodiorite of unknown age.

248.2 km Bridge across Charles Creek; at 250.2 km cross Montizambert Creek. There are scattered outcrops of altered granitic rock along this part of highway. South of Montizambert Creek are exposures of fine grained amphibolite, gneiss and small granitic, dioritic and pegmatitic bodies, and local breccia with clasts of these lithologies in a plagioclase-mafic matrix. The rocks are described by Roddick (1965) and included in his Twin Islands Group. According to Roddick (1965) the protolith apparently was andesitic tuff. The age and stratigraphic affinity of these rocks are unknown, although they are of higher metamorphic grade than the rocks of the Gambier Group to the north. Speculatively they could belong to the Early Jurassic Bowen Island Group.

Bowen Island Group and Wrangellian Linkage

Bowen Island to the west in the entrance to Howe Sound, is underlain by penetratively deformed, fine-grained altered andesite(?), thinly laminated to massive, white, grey, green and rarely pink andesitic tuff(?), and light to dark grey laminated chert and argillite. Felsic tuff from the Bowen Island Group on Mount Elphinstone on the west side of southernmost Howe Sound, is U-Pb dated at 185 Ma (Friedman et al., 1990). Ammonites of Early Jurassic age (probably Sinemurian, ~195 Ma) occur in septa with a similar lithological range, northwest of Jervis Inlet, ~70 km along strike to the north northwest.

The Bowen Island Group shows that Wrangellian stratigraphy extends at least this far east within the Coast Belt (Fig. 4-12). On Sechelt Peninsula, ~50 km north northwest of Bowen Island, the Bowen Island Group is associated with locally pillowed basalt and marble, which are lithologically correlated with the classic Wrangellian stratigraphy of Upper Triassic Karmutsen volcanics and Quatsino limestone of Vancouver Island and on nearby Texada and South Thormanby islands in the eastern Strait of Georgia. Overlying the Karmutsen and Quatsino on Vancouver Island are uppermost Triassic to Middle Jurassic arc-related volcanic rocks of the Bonanza Group, and their associated intrusions (age range 203–178 Ma; G. Nixon, pers. comm. 1998). The Bowen Island Group is correlated with the Bonanza Group (Fig. 4-12). Furthermore, Mahoney and DeBari (1995) suggest that Bowen Island and Bonanza volcanics also can be correlated with the Harrison Lake volcanics, exposed ~90 km east of here, if so, the basement of the entire southwestern Coast Belt basement was part of Wrangellia by Early Jurassic time.

250.9 km Outcrop of granodiorite. This is probably a western, marginal phase of a granodiorite body that about 4 km east southeast of here (on the Cyprus Provincial Park road at an elevation of 250 m) is U-Pb dated at 118 ± 2 Ma by Friedman and Armstrong (1990). Plutons of about this age occur widely in the westernmost part of the Coast Belt northwest of here (in tract 4; Fig. 4-11).

254.8 km Junction with Highway 1 at the turnoff to Horseshoe Bay and B.C. Ferries terminals. Continue southeastwards through West Vancouver on Highway 1 (Trans-Canada Highway) through roadcuts in probable Early Cretaceous granitic rocks. Ahead (to east) the southern slopes of the Coast Mountains dip southward at ~10°, a surface roughly conformable with that of the Late Cretaceous-early Tertiary erosion surface and of the bedding in overlying Late Cretaceous-early Tertiary molassic strata. The surface was upwarped in Late Miocene to Recent time (Roddick, 1965; Parrish, 1983).

In the far distance, ~100 km to the southeast, are the North Cascade Ranges capped by the 3000 m volcanic cone of Mount Baker (latest eruption in 1872), which is the northernmost volcano of the Cascade Magmatic Arc in the United States.

Strait of Georgia and Nanaimo "Molasse"

Visible to the southwest (right rear) across the Strait of Georgia are the mountains of Vancouver Island and the lower Gulf Islands.

The Strait of Georgia occupies a late Tertiary-Holocene structural depression, that is one of a series of depressions which extend from Alaska to Oregon and lie ~150 km east of the margin of the North American Plate. Evidence for active faulting is provided by marine seismic profiles showing that Holocene sediments are offset locally on faults in the floor of the Strait.

Vancouver Island features mainly Wrangellian stratigraphy. It is overlapped by the Late Cretaceous marine and non-marine, coal-bearing, clastic strata of the Nanaimo Group which also form most of the Gulf Islands. Nanaimo correlatives also overlie rocks of the southwestern Coast Mountains in a few places, such as North and West Vancouver, (Fig. 4-13). Nanaimo strata physically resemble classic *molasse*, and are similar in age and general depositional style to the Belly River sandstone of the Rocky Mountain Foothills east of Calgary (seen on Day 1), although the sandstones in each area are of very different composition.

The apparent basinal configuration of "Georgia Basin" in Figure 4-13 appears to result from Tertiary uplift and erosion. With the exception of some basal deposits, the provenance of the Nanaimo clastics is the Coast Belt. The strata appear to be *foreland basin* deposits related to tectonic uplift and subsequent erosion associated with the Coast Belt thrust system (Brandon et al., 1988; Mustard and Rouse, 1994). They also can be considered as *forearc basin* deposits related to the mid- to Late Cretaceous arc magmatism in the southeastern Coast and Cascade mountains (England and Calon, 1991). The oldest known Nanaimo strata are Turonian (~90 Ma; Haggart, 1991), and P-T-t considerations suggest that southeastern Coast Belt rocks were uplifted by ~15 km between 95 and 85 Ma (Journeay, *in* Monger and Journeay, 1994a). The age similarity between times of uplift and initial deposition suggests that thrust activity is the major control. The Nanaimo Group

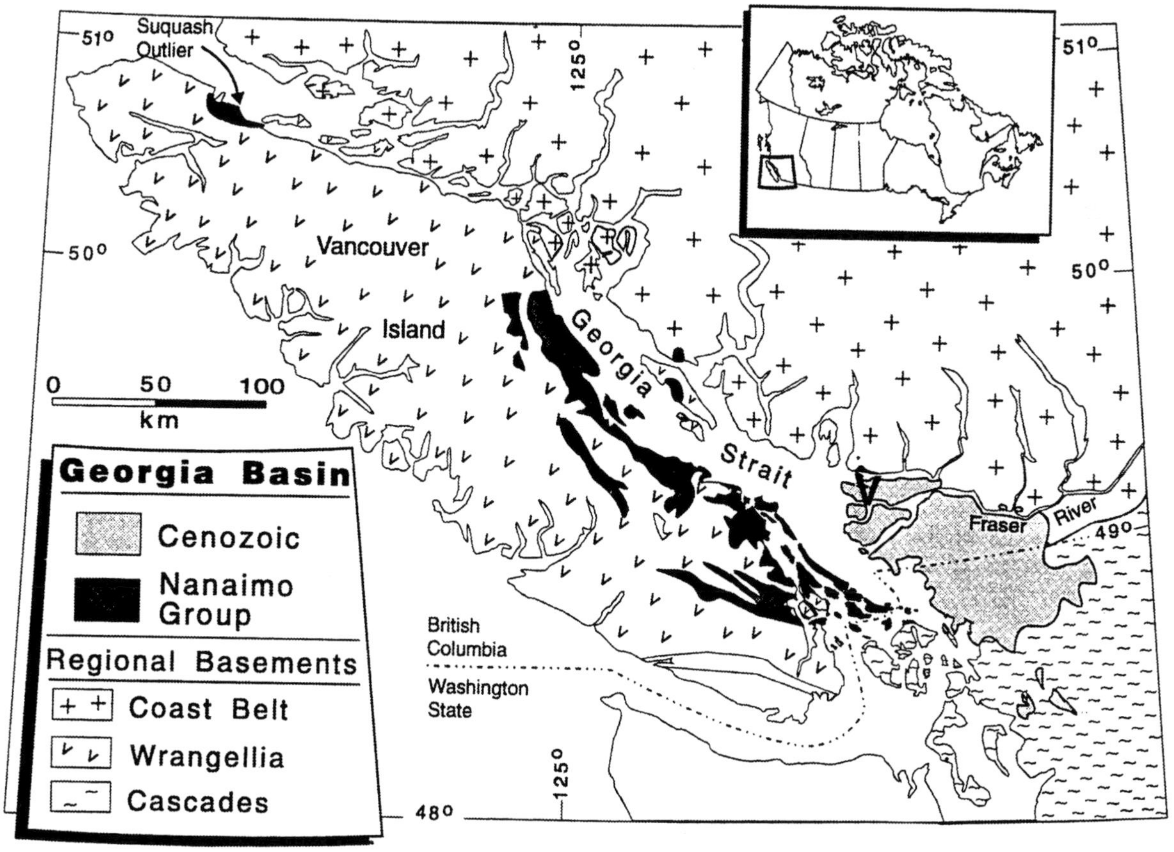

Figure 4-13. Distribution of rocks in Georgia Basin; "V" denotes city of Vancouver (from Mustard, 1994).

was deformed later in the Eocene Cowichan fold and thrust system, which is exposed on Vancouver Island (Fig. 4-5; England and Calon, 1991).

257.3 km Bridge across Cypress Creek.

266.9 km The knoll south (right) of highway is called Sentinel Hill and is composed of Oligocene mafic volcanic rocks.

267.2 km Turn right on Taylor Way, descend hill; at 268.8 km turn left on Marine Drive, cross bridge over the Capilano River, and at 269.3 km turn right on to the ramp leading to Lions Gate Bridge, which crosses the First Narrows of Burrard Inlet.

271.0 km The sea cliff at the south abutment of Lions Gate Bridge consists of Late Cretaceous continental sandstone and shale of the Brothers Creek Formation, which is correlative with the Nanaimo Group. The basal contact of the clastics with underlying granitic rocks is exposed ~2.5 km north northeast of here, in the Capilano River canyon. The sedimentary rocks are intruded by early Oligocene mafic sills and dykes (K-Ar whole rock dates 31–34 Ma; R.L. Armstrong, pers. comm. 1980). The mafic rocks are exposed in several places in the Vancouver area and are considered to be westernmost parts of the Late Eocene to Holocene Cascade Magmatic Arc (Hamilton and Dostal, 1994).

271.0–273.2 km The route is through Stanley Park, whose northern part is underlain by palynologically dated Late Cretaceous strata and whose southern part by lithologically similar continental sandstone of Eocene age (Burrard, Kitsilano formations; Mustard and Rouse, 1994). On the left, near the exit from Stanley Park is Coal Harbour, named from the presence of Eocene coal seams.

273.2 km The city of Vancouver is founded on Eocene sandstone and shale, which outcrops in several places along the south shore of the harbour (Burrard Inlet) and may be temporarily exposed in building excavations downtown. The Late Cretaceous and Tertiary clastic strata thicken southward beneath the Quaternary cover of Fraser Lowland to a thickness of 4400 m in exploratory hydrocarbon boreholes near the International Boundary. They are correlative with the early Tertiary Chuckanut Formation, exposed near Bellingham in northwestern Washington State.

274.9 km Intersection of Georgia and Burrard streets in Vancouver; end of field trip.

REFERENCES

Aitken, J.D., 1966a, Sub-Fairholme Devonian rocks of the eastern Front Ranges, southern Rocky Mountains, Alberta: Geological Survey of Canada Paper 64-33, 88 p.

Aitken, J.D., 1966b, Middle Cambrian to Middle Ordovician cyclic sedimentation, southern Rocky Mountains of Alberta: Bulletin of Canadian Petroleum Geology, v. 14, p. 405–441.

Aitken, J.D., 1971, Control of lower Paleozoic sedimentary facies by the Kicking Horse Rim, southern Rocky Mountains, Canada: Bulletin of Canadian Petroleum Geology, v. 19, p. 557–569.

Aitken, J.D., 1978, Revised models for depositional Grand Cycles, Cambrian of Western Canada: Bulletin of Canadian Petroleum Geology, v. 26, p. 515–542.

Aitken, J.D., 1997, Stratigraphy of the Middle Cambrian platformal succession, southern Rocky Mountains, Geological Survey of Canada Bulletin 398, 322 p.

Aitken, J.D., and Norford, B.S., 1967, Lower Ordovician Survey Peak and Outram Formations, southern Rocky Mountains of Alberta: Bulletin of Canadian Petroleum Geology, v. 15, p. 150–207.

Anderson, R.L., 1993, A Mesozoic stratigraphic and plutonic framework for northwestern Stikinia (Iskut river area), northwestern British Columbia, Canada, *in* Dunne, G., and McDougall, K., eds., Mesozoic paleogeography of the western United States-II: Pacific Section, Society of Economic Paleontologists and Mineralogists, v. 71, p. 447–494.

Archibald, D.A., Glover, J.K., Price, R.A., Farrar, E., and Carmichael, D.M., 1983, Geochronology and tectonic implications magmatism and metamorphism, southern Kootenay arc and neighbouring regions, southeastern British Columbia. Part I: Jurassic to mid-Cretaceous: Canadian Journal of Earth Sciences, v. 20, p. 1891–1913.

Argand, E., 1924, La Tectonique de l'Asie: Comptes Rendus del la XIII[e] Session Congres Geologique International, en Belgique, 1922, p. 171–372.

Armstrong, R.L., 1988, Mesozoic and early Cenozoic magmatic evolution of the Canadian Cordillera, *in* Clark Jr., S.P., Burchfiel, B.C., and Suppe, J., eds., Processes in continental lithospheric deformation: Geological Society of America Special Paper 218, p. 55–91.

Arnott, R.W. and Hein, F.J., 1986, Submarine canyon fills of the Hector Formation, Lake Louise, Alberta: Late Proterozoic synsedimentary rift deposits of the proto-Pacific miogeocline: Bulletin of Canadian Petroleum Geology, v. 34, p. 395–406.

Balkwill, H.R., 1972, Structural geology, lower Kicking Horse River region, Rocky Mountains, British Columbia: Bulletin of Canadian Petroleum Geology, v. 20, p. 608–633.

Balkwill, H.R., Price, R.A., and Mountjoy, E.W., 1980a, Geology, Golden, east half, British Columbia: Geological Survey of Canada Map 1596A (1:50,000, 82 N/7 E).

Balkwill, H.R., Price, R.A., and Mountjoy, E.W., 1980b, Geology, Golden, west half, British Columbia: Geological Survey of Canada Map 1497A (1:50,000, 82 N/7 W).

Bally, A.W., Gordy, P.L., and Stewart, G.A., 1966, Structure, seismic data and orogenic evolution of the southern Canadian Rockies: Canadian Society of Petroleum Geologists Bulletin, v. 14, p. 337–381.

Beaumont, C., 1981, Foreland basins: Royal Astronomical Society Geophysical Journal, v. 65f, p. 291–329.

Beaumont, C., Quinlan, G.M., and Stockmal, G.S., 1993, The evolution of the Western Canada Interior Basin: causes, consequences and unsolved problems, *in* Caldwell, W.G.E., and Kauffman, E.G., eds., Evolution of the western Interior Basin: Geological Association of Canada Special Paper 39, p. 97–117.

Benvenuto, G.L., and Price, R.A., 1979, Structural evolution of the Hosmer thrust sheet, southeastern British Columbia: Bulletin of Canadian Petroleum Geology, v. 27, p. 360–394.

Beyers, J.M., and Orchard, M.J., 1991, Upper Permian and Triassic conodont faunas from the type area of the Cache Creek Complex, south-central British Columbia, Canada: Geological Survey of Canada Bulletin 417, p. 269–297.

Bond, G., and Kominz, M.A., 1984, Construction of tectonic subsidence curves for the early Paleozoic miogeocline, southern Canadian Rocky Mountains: Implications for subsidence mechanisms, age of breakup, and crustal thinning: Geological Society of America Bulletin, v. 95, p. 155–173.

Brandon, M.T., Cowan, D.S., and Vance, J.A., 1988, Stratigraphic and structural framework of the Late Cretaceous San Juan Islands and Cascades thrust system: Geological Society of America Special Paper 221, 79 p.

Brown, R.L., Journeay, J.M., Lane, L.S., Murphy, D.C., and Rees, C.J., 1986, Obduction, backfolding and piggyback thrusting in the metamorphic hinterland of the southeastern Canadian Cordillera: Journal of Structural Geology, v. 8, p. 255–268.

Brown, R.L., Carr, S., Coleman, V., Johnson, B., Cook, F., and Varsek, J., 1992, The Monashee décollement of the southern Canadian Cordillera: a crustal-scale shear zone linking the Rocky Mountain foreland belt to lower crust beneath accreted terranes, *in* McClay, K., ed., Thrust tectonics: London, Chapman and Hall, p. 357–364.

Brown, R.L., Beaumont, C., and Willet, S., 1993, Comparison of the Selkirk fan structure with mechanical models: Implications for interpretation of the southern Canadian Cordillera: Geology, v. 21, p. 1015–1018.

Brown, R.L., Crowley, J.L., and Johnson, B.J., 1993, Part 2: A transect of the southern Omineca belt, *in* Brown,

R.L., McDonough, M.R., Crowley, J.L., Johnson, B.J., Mountjoy, E.W., and Simony, P.S., eds., Cordilleran transect, Rocky Mountain foreland and Omenica hinterland, southeastern British Columbia: Geological Association of Canada/Mineralogical Association of Canada Joint Annual Meeting 1993, Field Trip B-9, Guidebook: Edmonton, Alberta, Edmonton Geological Society, Department of Geology, University of Alberta.

Bruce, C.J., Frey, F.R., Graff, G.W., and Tippet, C.R., 1995, Geological Guidebook, 1995 C.S.P.G. Student-Industry Field Trip, in Calgary, Alberta: Canadian Society of Petroleum Geologists, 261 p.

Butler, R.F., Gehrels, G.E., McClelland, W.C., May, S.R., and Klepacki, D., 1989, Discordant paleomagnetic poles from the Canadian Coast Plutonic Complex: regional tilt rather than large-scale displacement: Geology, v. 17, p. 691–694.

Butler, R.F., Gehrels, G.E., and Bazard, D.R., 1997, Paleomagnetism of Paleozoic strata of the Alexander terrane, southeastern Alaska: Geological Society of America Bulletin, v. 109, p. 1372–1388.

Cant, D.J., and Stockmal, G.S., 1999, A field guide to portions of the Jurassic Fernie-Kootenay Trans-Canada Highway section, Banff, Alberta: new sedimentological and structural observations and interpretations: Bulletin of Canadian Petroleum Geology, v. 47, p. 1–18.

Carr, J.M., and Read, A.J., 1976, Afton: a supergene copper deposit, *in* Sutherland Brown, A., ed., Porphyry deposits of the Canadian Cordillera: Canadian Institute of Mining and Metallurgy Special Volume 15, p. 376–387.

Carr, S., 1992, Tectonic setting and U-Pb geochronology of the early Tertiary Ladybird leucogranite suite, Thor-Odin-Pinnacles area, southern Omineca belt, British Columbia: Tectonics, v. 11, p. 258–278.

Clowes, R.M., Zelt, C.A., Amor, J.R., and Ellis, R.M., 1995, Lithospheric structure in the southern Canadian Cordillera from a network of seismic refraction lines: Canadian Journal of Earth Sciences, v. 32, p. 1485–1513.

Coish, R.A., and Journeay, J.M., 1992, The Crevasse Crag volcanic complex, southwestern British Columbia: structural control on the geochemistry of arc magmas: Current Research, Part A, Geological Survey of Canada, Paper 91-1A, p. 95–105.

Coleman, M.E., and Parrish, R.R., 1991, Eocene dextral strike-slip and extensional faulting in the Bridge River Terrane, southwest British Columbia: Tectonics, v. 10, p. 1222–1238.

Colpron, M., 1996, Stratigraphy, structure and thermotectonic evolution of the western flank of the Selkirk fan structure, northern Selkirk Mountains, British Columbia [Ph.D. thesis]: Queen's University, Kingston, Ontario.

Colpron, M., and Price, R.A., 1995, Tectonic significance of the Kootenay terrane, southeastern British Columbia: an alternative model: Geology, v. 23, p. 25–28.

Colpron, M., Price, R.A., Archibald, D.A., and Carmichael, D.M., 1996, Middle Jurassic exhumation along the western flank of the Selkirk fan structure: Thermobarometric and thermochronometric constraints from the Illecillewaet syclinorium, southern British Columbia: Geological Society of America Bulletin, v. 108, p. 1372–1392.

Colpron, M., Warren, M.J., and Price, R.A., 1998, Selkirk fan structure, southeastern Canadian Cordillera: Tectonic wedging against an inherited basement ramp: Geological Society of America Bulletin, v. 110, p. 1060–1074.

Colpron, M., Price, R.A., and Archibald, D.A., 1999, $^{40}Ar/^{39}Ar$ thermochronometric constraints on the tectonic evolution of the Clachnacudainn complex, southeastern British Columbia: Canadian Journal of Earth Sciences, v. 36, p. 1989–2006.

Coney, P.J., 1972, Cordilleran tectonics and North American plate motions: American Journal of Science, v. 272, p. 462–465.

Coney, P.J., Jones, D.L., and Monger, J.W.H., 1980, Cordilleran suspect terranes: Nature, v. 288, p. 329–333.

Cook, D.G., 1970, A Cambrian facies change and its effect on structure, Mount Stephen-Mount Denis area, British Columbia, *in* Wheeler, J.O., ed., Structure of the southern Canadian Cordillera: Geological Association of Canada Special Paper 6, p. 27–39.

Cook, D.G., 1975, Structural style influenced by lithofacies, Rocky Mountain Main Ranges, Alberta-British Columbia: Geological Survey of Canada Bulletin 233, 73 p.

Cook, F., Green, A., Simony, P., Price, R.A., Parrish, R.R., Milkereit, B., et al., 1988, LITHOPROBE seismic reflection structure of the southeastern Canadian Cordillera: Initial results: Tectonics, v. 7, p. 157–180.

Cook, F.A., 1995, The reflection Moho beneath the southern Canadian Cordillera: Canadian Journal of Earth Sciences, v. 32, p. 1520–1530.

Cook, F.A., Varsek, J.L., Clowes, R.M., Kanasewich, E.R., Spencer, C.S., Parrish, R.R., et al., 1992, Lithoprobe crustal reflection cross-section of the southern Canadian Cordillera, 1, foreland thrust and fold belt to Fraser River: Tectonics, v. 11, p. 12–35.

Cook, F.A., and van der Velden, A.J., 1995, Three-dimensional crustal structure of the Purcell anticlinorium in the Cordillera of southwestern Canada: Geological Society of America Bulletin, v. 107, p. 642–664.

Cordey, F., and Schiarizza, P.F., 1993, Long-lived Panthalassic remnant: the Bridge River accretionary complex, Canadian Cordillera: Geology, v. 21, p. 263–266.

Cordey, F., Mortimer, N., De Wever, P., and Monger, J.W.H., 1987, Significance of Jurassic radiolarians from the Cache Creek terrane, British Columbia: Geology, v. 15, p. 1151–1154.

Craw, D., 1978, Metamorphism, structure and stratigraphy in the Southern Rark Ranges, British Columbia: Canadian Journal of Earth Sciences, v. 15, p. 86–98.

Crickmay, C.H., 1930, The structural connection between the Coast Range of British Columbia and the Cascade Range of Washington: Geological Magazine, v. 67, p. 482–491.

Crowley, J.L., 1997, U-Pb geochronologic constraints on

the cover sequence of the Monashee Complex, Canadian Cordillera: Peleoproterozoic deposition on basement, Canadian Journal of Earth Sciences, v. 34, p. 1008–1022.

Crowley, J.L., 1999, U-Pb geochronologic constraints on Paleoproterozoic tectonism in the Monashee Complex, Canadian Cordillera: elucidating an overprinted geologic history: Geological Society of America Bulletin, v.111, p. 560–577.

Crowley, J.L., and Brown, R.L., 1994, Tectonic links between the Clachnacudainn terrane and Selkirk allochthon, southern Omineca belt, Canadian Cordillera: Tectonics, v. 13, p. 1035–1051.

Cui, Y., and Russell, J.K., 1995, Magmatic origins of calc-alkaline intrusions from the Coast Plutonic Complex, southwestern British Columbia: Canadian Journal of Earth Sciences, v. 32, p. 1643–1667.

Currie, K.L., 1975, The geology and petrology of the Ice River alkaline complex, British Columbia: Geological Survey of Canada Bulletin 245, 67 p.

Daly, R.A., 1926, Our mobile Earth: Charles Scribner's Sons, New York and London, 342 p.

Davis, G.F., and Richards, M.A., 1992, Mantle convection: Journal of Geology, v. 100, p. 131–206.

Dawson, G.M., 1879, Preliminary report on the physical and geological features of the southern part of the interior of British Columbia: Geological Survey of Canada Report of Progress for 1877–78, 174 p.

Dawson, G.M., 1881, Sketch of the geology of British Columbia: Geological Magazine New Series, v. 8, p. 156–162, and 214–227.

Devlin, W.J., and Bond, G.C., 1988, The initiation of the early Paleozoic Cordilleran miogeocline: Evidence from the uppermost Proterozoic-Lower Cambrian Hamill Group of southeastern British Columbia: Canadian Journal of Earth Sciences, v. 25, p. 1–19.

Devlin, W.J., Brueckner, H.K., and Bond, G.C., 1988, New isotopic data and a preliminary age for volcanics near the base of the Windermere Supergroup in northeastern Washington, U.S.A.: Canadian Journal of Earth Sciences, v. 25, p. 1906–1911.

Douglas, R.J.W., Gabrielse, H., Stott, D.F., and Belyea, H.R., 1972, Geology of Western Canada, *in* Douglas, R.J.W., ed., Geology and economic minerals of Canada: Geological Survey of Canada Economic Geology Report No. 1, p. 365–546.

Duffell, S., and McTaggart, K.C., 1952, Ashcroft map area, British Columbia: Geological Survey of Canada, Paper 67-10.

Eisbacher, G.H., Carrigy, M.A., and Campbell, R.B., 1974, Paleo-drainage pattern and late-orogenic basins of the Canadian Cordillera, *in* Dickinson, W.R., ed., Tectonics and sedimentation: Society of Economic Paleontologists and Mineralogists Special Publication 22, p. 143–166.

Elder, W.P., and Saul, L.R., 1993, Paleogeographic implications of molluscan assemblages in the Upper Cretaceous (Campanian) Pigeon Point formation, California, *in* Dunn, G. and Mcdougall, K., eds., Mesozoic paleogeography of the western United States-II: Pacific Section, Society of Economic Paleontologists and Mineralogists, Book 71, p. 171–186.

Engebretson, D.C., Cox, A., and Gordon, R.G., 1985, Relative motions between oceanic and continental plates in the Pacific Basin: Geological Society of America Special Paper 206, 59 p.

Engebretson, D.C., Kelley, K.P., Cashman, H.P., and Richards, M.A., 1992, 180 million years of subduction: GSA Today, v. 2, p. 93–100.

England, T.J.D., and Calon, T.J., 1991, The Cowichan fold and thrust system, Vancouver Island, southwestern British Columbia: Geological Society of America Bulletin, v. 103, p. 336–362.

Evenchick, C.A., 1997, Contractional structures in the western Skeena fold belt; an indication of Cretaceous sinistral convergence within the Coast Belt; Geological Society of America, Abstracts with programs 29, p. A-82.

Ewing, T.E., 1980, Paleogene tectonic evolution of the Pacific Northwest: Journal of Geology, v. 88, p. 619–638.

Fitch, T.J., 1972, Plate convergence, transcurrent faults, and internal deformation adjacent to Southeast Asia and the western Pacific: Journal of Geophysical Research, v. 77, p. 4432–4460.

Frebold, H., 1962, The Devonian-Jurassic contact and the subdivision of the Fernie Group in the Banff Area., Alberta: Geological Survey of Canada Paper 62-3, 19 p.

Friedman, R.M., and Armstrong, R.L., 1990, U-Pb dating, southern Coast Belt (49°-51°), southwestern British Columbia: Proceedings of LITHOPROBE: Southern Canadian Cordillera Transect Workshop, University of Calgary, Alberta, p. 146–155.

Friedman, R.M., and Armstrong, R.L., 1995, Jurassic and Cretaceous geochronology of the southern Coast Belt, B.C., 49-51°N, *in* Miller, D., and Busby, C., eds., Jurassic magmatism and tectonics of the North American Cordillera: Geological Society of America Special Paper 299, p. 95-139.

Friedman, R.M., and van der Heyden, P., 1992, Late Permian dates for the Farwell and northern Mount Lytton bodies: Current Research, Part A, Geological Survey of Canada Paper 92-1A, p. 137–144.

Friedman, R.M., Monger, J.W.H., and Tipper, H.W., 1990, Age of the Bowen Island Group, southwestern Coast Mountains, British Columbia: Canadian Journal of Earth Sciences, v. 27, p. 1456–1461.

Friedman, R.M., Mahoney, J.B., and Cui, Y., 1995, Magmatic evolution of the southern Coast Belt: constraints from Nd-Sr isotopic systematics and geochronology of the southern Coast Plutonic Complex: Canadian Journal of Earth Sciences, v. 32, p. 1681–1698.

Fritz, W.H., Cecile, M.P., Norford, B.S., Morrow, D., Geldsetzer, H.H.J., and Morrow, D.W., 1991, Chapter 7: Cambrian to Middle Devonian assemblages, *in* Gabrielse, H., and Yorath, C.J., eds., Geology of the Cordilleran Orogen in Canada: Geological Survey of Canada, v. 4, p. 151–218.

Fyson, W.K., 1970, Structural relations in metamorphic rocks, Shuswap Lake area, British Columbia, *in* Wheeler, J.O., ed., Structure of the Southern Canadian Cordillera: The Geological Association of Canada Special Paper 6, p. 107–122.

Gabrielse, H., and Campbell, R.B., 1991, Chapter 6: Upper Proterozoic Assemblages, *in* Gabrielse, H., and Yorath, C.J., ed., Geology of the Cordilleran Orogen in Canada: Geological Survey of Canada, v. Geology of Canada, No. 4 (also Geological Society of America, Geology of North America, v. G-2), p. 125–150.

Gabrielse, H., Monger, J.W.H., Wheeler, J.O., and Yorath, C.J., 1991, Part A. Morphogeological belts, tectonic assemblages, and terranes, *in* Gabrielse, H., and Yorath, C.J., eds., Chapter 2 of Geology of the Cordilleran Orogen in Canada: Geological Survey of Canada, Geology of Canada, No. 4, p. 15–28 (also Geological Society of America, The Geology of North America, v. G-2).

Gardner, D., 1977, Structural geology and metamorphism of calcareous Lower Paleozoic slates, Blaeberry River-Redburn Creek area, near Golden, British Columbia [Ph.D. thesis]: Ontario, Kingston, Queen's University, 224 p.

Gehrels, G.E., and Ross, G.M., 1998, Detrital zircon geochronology of Neoproterozoic to Permian miogeoclinal strata in British Columbia and Alberta: Canadian Journal of Earth Sciences, v. 35, p. 1380–1402.

Ghent, E.D., Simony, P.S., and Raeside, R.P., 1981, Metamorphism and its relation to structure within the core zone west of the southern Rocky Mountains, *in* Thompson, R.I., and Cook, D.G., eds., Field Guides to Geology and Mineral Deposits—Calgary '81: Geological Association of Canada, p. 373–391.

Ghosh, D.K., 1995, Nd-Sr isotopic constraints on the interactions of the Intermontane Superterrane with the western edge of North America in the southern Canadian Cordillera: Canadian Journal of Earth Sciences, v. 32, p. 1740–1758.

Gilman, R.A., 1972, Geology of the Clachnacudainn salient near Albert Canyon, British Columbia: Canadian Journal of Earth Sciences, v. 9, p. 1447–1454.

Gordey, S.P., Geldsetzer, H.H.J., Morrow, D.W., Bamber, E.W., Henderson, C.M., Richards, B.C., et al., 1991, Chapter 8: Upper Devonian to Middle Jurassic Assemblages, *in* Gabrielse, H., and Yorath, C.J., eds., Geology of the Cordilleran Orogen in Canada: Geological Survey of Canada, The Geology of Canada, v. 4, p. 219–327.

Gordon, R.G., 1998, The plate tectonic approximation: plate nonrigidity, diffuse plate boundaries, and global plate reconstructions: Annual Reviews of Earth and Planetary Sciences, v. 26, p. 615–642.

Gough, D.I., and Majorowicz, J.A., 1992, Magnetotelluric soundings, structure, and fluids in the southern Canadian Cordillera: Canadian Journal of Earth Sciences, v. 29, p. 609–620.

Grand, S.P., van der Hilst, R.D., and Widiyantoro, S., 1997, Global seismic tomography: a snapshot of convection in the Earth: GSA Today, v. 7, no. 4, p. 1–7.

Green, N.L., Armstrong, R.L., Harakal, J.E., Souther, J.G., and Read, P.B., 1988, Eruptive history and K-Ar geochronology of the late Cenozoic Garibaldi volcanic belt, southwest British Columbia: Geological Society of America Bulletin, v. 100, p. 563–579.

Greig, C.J., 1991, Jurassic and Cretaceous plutonic and structural styles of the Eagle Plutonic Complex, southwestern British Columbia, and their regional significance: Canadian Journal of Earth Sciences, v. 29, p. 793–811.

Haggart, J.W., 1991, A new assessment of the age of the basal Nanaimo Group, Gulf Islands, British Columbia: Current Research, Part E, Geological Survey of Canada, Paper 91-1E, p. 77–82.

Hall, R.L., Poulton, T.P., and Monger, J.W.H.,1998, Chapter 2, Field Trip Al: Calgary-Vancouver, *in* Smith, P.L., ed., Field Guide for the Fifth International Symposium on the Jurassic System: International Union of Geological Sciences/Subcommission on Jurassic Stratigraphy, Vancouver, Canada, August 12-25, 1998, p. 29–103.

Hamilton, T.S., and Dostal, J., 1994, Middle Tertiary eruptive rocks in the Vancouver area, British Columbia, *in* Monger, J.W.H., ed., Geology and geological hazards of the Vancouver region, southwestern British Columbia: Geological Survey of Canada Bulletin 481.

Harms, T.A., 1986, Structural and tectonic analysis of the Sylvester Allochthon, northern British Columbia: Implications for paleogeography and accretion [Ph.D thesis]: Arizona, Tuscon, University of Arizona, 88 p.

Helwig, J., 1974, Eugeosynclinal basement and a collage concept of orogenic belts, *in* Dott, R.H., and Shaver, R.H., eds., Modern and ancient geosynclinal sedimentation: Society of Economic Paleontologists and Mineralogists Special Publication, v. 19, p. 359–380.

Hoffman, P.F., 1989, Precambrian geology and tectonic history of North America (Chapter 16), *in* Bally, A.W., and Palmer, A.R., The Geology of North America – An overview: The Geological Society of America, v. A, p. 447–512.

Holland, S.S., 1964, Landforms of British Columbia, a physiographic outline: British Columbia Ministry of Energy, Mines and Petroleum Resources, Bulletin 48, 138 p.

Hurlow, H., 1993, Mid-Cretaceous strike-slip and contractional fault zones in the western Intermontane terrane, Washington, and their relation to the north Cascades-southeastern Coast Belt orogen: Tectonics, v. 12, p. 1240–1257.

Hyndman, R.D., 1972, Plate motions relative to the deep mantle and the development of subduction zones: Nature, v. 238, p. 263–265.

Hyndman, R.D., 1995, Giant earthquakes in the Pacific Northwest: Scientific American, December 1995, p. 68–75.

Hyndman, R.D., and Lewis, T.J., 1995, Review: The thermal regime along the southern Canadian Cordillera Lithoprobe Corridor: Canadian Journal of Earth Sciences, v. 32, p. 1611–1617.

Irving, E., Woodsworth, G.J., Wynne, P.J., and Morrison, A., 1985, Paleomagnetic evidence for the displacement from the south of the Coast Plutonic Complex, British Columbia: Canadian Journal of Earth Sciences, v. 22, p. 584–598.

Irving, E., Thorkelson, D.J., Wheadon, P.M., and Enkin, R., 1995, Paleomagnetism of the Spences Bridge Group and northward displacement of the Intermontane Belt, British Columbia: a second look: Journal of Geophysical Research B; v. 100, p. 6057–6071.

Jarrard, R.D., 1986, Relations among subduction parameters: Reviews of Geophysics, v. 24, p. 217–284.

Johnson, B.J., and Brown, R.L., 1996, Crustal structure and early Tertiary extensional tectonics of the Omineca belt at 51° latitude, southern Canadian Cordillera: Canadian Journal of Earth Sciences, v. 33, p. 1596–1611.

Jones, A.G., 1957, Vernon map-area, British Columbia: Geological Survey of Canada Memoir 296, 190 p.

Journeay, J.M., 1993, Tectonic assemblages of the eastern Coast Belt, southwestern British Columbia: implications for the history and mechanisms of terrane accretion: Current Research, Part A, Geological Survey of Canada Paper 93-1A, p. 221–233.

Journeay, J.M., and Northcote, B.R., 1992, Tectonic assemblages of the eastern Coast Belt, southwest British Columbia: Current Research, Part A, Geological Survey of Canada Paper 92-1A, p. 215–224.

Journeay, J.M., and Friedman, R.M., 1993, The Coast Belt thrust system: evidence of Late Cretaceous shortening in southwest British Columbia: Tectonics, v. 12, p. 756–773.

Journeay, J.M., and Mahoney, J.B., 1994, Cayoosh Assemblage: regional correlations and implications for terrane linkages in the southern Coast Belt, British Columbia: Current Research 1994-A, Geological Survey of Canada, p. 165–175.

Journeay, J.M., Sanders, C., Van-Konijnenburg, J.H., and Jaasma, M., 1992, Fault systems of the Eastern Coast Belt, southwest British Columbia: Current Research, Part A, Geological Survey of Canada Paper 92-1A, p. 225–235.

Kleinspehn, K.L., 1985, Cretaceous sedimentation and tectonics, Tyaughton-Methow Basin, southwestern British Columbia: Canadian Journal of Earth Sciences, v. 22, p. 154–174.

Klepacki, D.W., 1985, Stratigraphy and structural geology of the Goat Range area, southeastern British Columbia [Ph.D. thesis]: Massachusetts, Boston, Massachusetts Institute of Technology.

Klepacki, D.W., and Wheeler, J.O., 1985, Stratigraphy and structural relations of the Milford, Kaslo and Slocan groups, Goat Range, Lardeau and Nelson map areas, British Columbia: Current Research, Part A, Geological Survey of Canada, Paper 85-1A, p. 277–300.

Kubli, T.E., and Simony, P.S., 1992, The Dogtooth High, northern Purcell Mountains, British Columbia: Bulletin of Canadian Petroleum Geology, v. 40, p. 36–51.

Kubli, T.E., and Simony, P.S., 1994, The Dogtooth Duplex, a model for the structural development of the northern Purcell Mountains: Canadian Journal of Earth Sciences, v. 31, p. 1672–1686.

Lane, L.S., 1984, Brittle deformation in the Columbia River fault zone near Revelstoke, southeastern British Columbia: Canadian Journal of Earth Sciences, v. 21, p. 584–598.

Leech, G.B., 1960, Geology, Fernie (west half), Kootenay District, British Columbia: Geological Survey of Canada Map 11-1960.

Leech, G.B., 1966, The Rocky Mountain Trench, *in* Irvine, T.N., ed., The World rift system: Geological Survey of Canada Paper 66-14, p. 307–329.

Leitch, C.H.B., 1989, Geology, wall rock alteration and characteristics of the ore-fluid at the Bralorne mesothermal gold deposit, southwestern British Columbia [Ph.D. thesis]: University of British Columbia.

Leitch, C.H.B., van der Heyden, P., Godwin, C.I., Armstrong, R.L., and Harakal, J.E., 1991, Geochronometry of the Bridge River Camp, southwestern British Columbia, Canadian Journal of Earth Sciences, v. 28, p. 195–208.

Lewis, T.J., Bentkowski, W., and Hyndman, R.D., 1992, Crustal temperatures near the southern Canadian Cordilleran transect: Canadian Journal of Earth Sciences, v. 29, p. 1197–1214.

Lickorish, W.H., 1993, Structural evolution of the Porcupine Creek anticlinorium, Western Main Ranges, Rocky Mountains, British Columbia: Journal of Structural Geology, v. 15, p. 477–489.

Lickorish, W.H., and Simony, P.S., 1995, Evidence for late rifting of the Cordilleran margin outlined by stratigraphic division of the Lower Cambrian Gog Group, Rocky Mountain Main Ranges, British Columbia and Alberta: Canadian Journal of Earth Sciences, v. 32, p. 860–874.

Lynch, J.V.G., 1991, Georgia Basin Project: stratigraphy and structure of Gambier Group rocks in the Howe Sound-Mamquam River area, southwest Coast Belt, British Columbia: Current Research, Part A, Geological Survey of Canada Paper 91-1A, p. 49–57.

Lynch, J.V.G., 1992, Deformation of Early Cretaceous volcanic-arc assemblages, southern Coast Belt, British Columbia: Canadian Journal of Earth Sciences, v. 29, p. 2706–2721.

Mahoney, J.B., and DeBari, S.M., 1995, Geochemical and isotopic characteristics of Early to Middle Jurassic volcanism in southern British Columbia: relationship of the Bonanza and Harrison arc systems: Geological Association of Canada Abstracts with Program, v. 20, p. A-65.

Mahoney, J.B., Monger, J.W.H., and Hickson, C.J., 1996, Albian-Cenomanian conglomerates along the Intermontane/Insular terrane boundary: a critical test for large-scale terrane translation: Current Research 1996-A, Geological Survey of Canada, p. 101–109.

Mathews, W.H., 1958, Geology of the Mount Garibaldi map-area, southwestern British Columbia: Geological

Society of America Bulletin, v. 69, p. 161–177.

Mathews, W.H., 1986, Physiography of the Canadian Cordillera: Geological Survey of Canada Map 1701A, scale 1:5,000,000.

Mathews, W.H., and Rouse, G.E., 1984, The Gang Ranch-Big Bar area, south-central British Columbia: stratigraphy, geochronology, and palynology of the Tertiary beds and their relationships to the Fraser Fault: Canadian Journal of Earth Sciences, v. 21, p. 1132–1144.

McConnell, R.G., 1887, Report of the geological structure of a portion of the Rocky Mountains, accompanied by a section measured near the 51st parallel: Annual Report, 1886, Geological Survey of Canada, p. 1D–41D.

McIlreath, I.A., 1977, Stratigraphic and sedimentary relationships at the western edge of the Middle Cambrian carbonate facies belt, Field, British Columbia [Ph.D. thesis]: Alberta, Calgary, University of Calgary.

McMechan, M.E., and Thompson, R.I., 1989, Chapter 4: Structural style and history of the Rocky Mountain fold and thrust belt, *in* Ricketts, B.D., ed., Western Canada Sedimentary Basin: A case history: Calgary, Canadian Society of Petroleum Geologists, p. 47–71.

Mihalynuk, M.G., Nelson, J., and Diakow, L.J., 1994, Cache Creek terrane entrapment: Oroclinal paradox within the Canadian Cordillera: Tectonics, v. 13, p. 575–595.

Misch, P., 1966, Tectonic evolution of the northern Cascades of Washington State—a west Cordilleran case history, *in* Gunning, H.C., ed., Tectonic history and mineral deposits of the western Cordillera: Canadian Institute of Mining and Metallurgy Special Volume 8, p. 101–148.

Misch, P., 1977, Dextral displacements at some major strike-slip faults in the north Cascades: Geological Association of Canada Annual Meetings 1977, Vancouver, British Columbia, Program with Abstracts, v. 2, p. 37.

Monger, J.W.H., 1977, Upper Paleozoic rocks of the western Canadian Cordillera and their bearing on Cordilleran evolution: Canadian Journal of Earth Sciences, v. 14, p. 1832–1859.

Monger, J.W.H., 1985, Structural evolution of the southwestern Intermontane Belt, Ashcroft and Hope map-areas, British Columbia: Current Research, Part A, Geological Survey of Canada Paper 85-1A, p. 349–358.

Monger, J.W.H., 1989, Geology of Hope and Ashcroft map-areas: Geological Survey of Canada Maps 41-1989, 42-1989, scale 1:250,000.

Monger, J.W.H., 1991, Georgia Basin Project: structural evolution of parts of southern Insular and southwestern Coast Belts, British Columbia: Current Research, part A, Geological Survey of Canada Paper 91-1A, p. 219–228.

Monger, J.W.H., 1993, Georgia Basin Project—Geology of Vancouver map area, British Columbia: Current Research Part A, Geological Survey of Canada, Paper 93-1A, p. 149–157.

Monger, J.W.H., and Ross, C.A., 1971, Distribution of fusulinaceans in the Canadian Cordillera: Canadian Journal of Earth Sciences, v. 8, p. 770–791.

Monger, J.W.H., and McNicoll, V.J., 1993, New U-Pb dates from southwestern Coast Belt, British Columbia: Radiogenic Age and Isotopic Studies: Report 7, Geological Survey of Canada Paper 93-2, p. 119–126.

Monger, J.W.H., and Journeay, J.M., 1994a, Guide to the geology and tectonic evolution of the southern Coast Mountains: Geological Survey of Canada, Open File 2490, 77 p.

Monger, J.W.H., and Journeay, J.M., 1994b, Basement geology and tectonic evolution of the Vancouver region, *in* Monger, J.W.H., ed., Geology and geological hazards of the Vancouver region, southwestern British Columbia: Geological Survey of Canada Bulletin 481, p. 3–25.

Monger, J.W.H., and Nokleberg, W.J., 1996, Evolution of the northern North American Cordillera: generation, fragmentation, displacement and accretion of successive North American plate-margin arcs, *in* Coyner, A.R., and Fahey, P.L., ed., Geology and ore deposits of the American Cordillera: Geological Society of Nevada Symposium Proceedings, Reno/Sparks, Nevada, April 1995, p. 1133–1152.

Monger, J.W.H., and Price, R.A., 1996, Comment on "Paleomagnetism of the upper Cretaceous strata of Mount Tatlow: Evidence for 3000 km of northward displacement of the eastern Coast Belt, British Columbia" by P.J. Wynne et al., and on "Paleomagnetism of the Spences Bridge Group and northward displacement of the Intermontane Belt, British Columbia: as second look" by E. Irving et al.: Journal of Geophysical Research, v. 101, p. 13,793–13,799.

Monger, J.W.H., Price, R.A., and Tempelman-Kluit, D.J., 1982, Tectonic accretion and the origin of two major metamorphic and plutonic welts in the Canadian Cordillera: Geology, v. 10, p. 70–75.

Monger, J.W.H., Clowes, R.M., Price, R.A., Simony, P.S., Riddihough, R.P., and Woodsworth, G.J., 1985, Centennial Continent/Ocean Transect #7, B-2 Juan de Fuca Plate to Alberta Plains, *in* Speed, R., ed., Decade of North American Geology Continent/Ocean Transects: Geological Society of America, 2 sheets and 21 p.

Monger, J.W.H., van der Heyden, P., Journeay, J.M., Evenchick, C.A., and Mahoney, J.B., 1994, Jurassic-Cretaceous basins along the Canadian Coast Belt: their bearing on pre-mid-Cretaceous sinistral displacements: Geology, v. 22, p. 175–178.

Morrison, G.W., 1980, Stratigraphic control of Cu-Fe skarn ore distribution and genesis at Craigmont, British Columbia: Canadian Institute of Mining and Metallurgy Bulletin 73, p. 109–123.

Mortensen, J.K., 1992, Pre-mid-Mesozoic tectonic evolution of the Yukon-Tanana terrane, Yukon and Alaska: Tectonics, v. 11, p. 836–853.

Mortimer, N., 1987, The Nicola Group: Late Triassic and Early Jurassic subduction-related volcanism in British Columbia: Canadian Journal of Earth Sciences, v. 24, p. 2521–2536.

Mortimer, N., van der Heyden, P., Armstrong, R.L., and Harakal, J., 1990, U-Pb and K-Ar dates related to the timing of magmatism and deformation in the Cache Creek terrane and Quesnellia, southern British Columbia: Canadian Journal of Earth Sciences, v. 27, p. 117–123.

Murphy, D.C., Parrish, R.R., Klepacki, D.W., McMillan, W., Struik, L.C., and Gabites, J., 1995, New geochronological constraints on Jurassic deformation of the western edge of North America, southeastern Canadian Cordillera, *in* Miller, D.M., and Busby, C., eds., Jurassic magmatism and tectonics of the North American Cordillera: Geological Society of America Special Paper 299, p. 159–171.

Mustard, P.S., 1994, The Upper Cretaceous Nanaimo Group, Georgia Basin, southwestern British Columbia, *in* Monger, J.W.H., ed., Geology and geological hazards of the Vancouver region, southwestern British Columbia: Geological Survey of Canada Bulletin 481, p. 27–95.

Mustard, P.S., and Rouse, G.E., 1994, Stratigraphy and evolution of Tertiary Georgia Basin and subjacent Upper Cretaceous sedimentary rocks, southwestern British Columbia and northwestern Washington State, *in* Monger, J.W.H., ed., Geology and geological hazards of the Vancouver region, southwestern British Columbia: Geological Survey of Canada Bulletin 481, p. 97–169.

Nassichuk, W.W., 1977, Upper Permian ammonoids from the Cache Creek Group in western Canada: Journal of Paleontology, v. 51, p. 557–590.

Nelson, J.L., 1979, The western margin of the Coast Plutonic Complex on Hardwicke and West Thurlow Islands: Canadian Journal of Earth Sciences, v. 16, p. 1166–1175.

Ney, C.S., 1954, Monarch and Kickinghorse Mines, Field, British Columbia, *in* Scott, J.C., and Fox, F.G., eds., Guidebook, Fourth Annual Field Conference, Banff-Golden-Radium, August 26, 27, 28, 1954: Alberta Society of Petroleum Geologists, p. 119–136.

North, F.K., and Henderson, G.G.L., 1954, Summary of the Geology of the southern Rocky Mountains of Canada, *in* Scott, J.C., and Fox, F.G., eds., Guidebook, Fourth Annual Field Conference, Banff-Golden-Radium, August 26, 27, 28, 1954: Calgary, Alberta Society of Petroleum Geologists, p. 15–81.

O'Brien, J.A., Gehrels, G.E., and Monger, J.W.H., 1992, U-Pb geochronology of plutonic clasts from conglomerates in the Ladner and Jackass Mountain groups and the Peninsula formation, southwestern British Columbia: Current Research, Part A, Geological Survey of Canada Paper 92-1A, p. 209–214.

Okulitch, A.V., 1989, Revised stratigraphy and structure in the Thompson-Shuswap-Okanagan map area, southern British Columbia: Current Research, Part E, Geological Survey of Canada Paper 89-1E, p. 51–60.

Okulitch, A.V., Wanless, R.K., and Loveridge, W.D., 1975, Devonian plutonism in south-central British Columbia: Canadian Journal of Earth Sciences, v. 12, p. 1760–1769.

Oldow, J.S., Bally, A.W., and Avé Lallement, H.G., 1990, Transpression, orogenic float, and lithospheric balance: Geology, v. 18, p. 991–994.

Ollerenshaw, N.C., 1978, Calgary, Alberta-British Columbia: Geological Survey of Canada, Map 1457A, 1:250 000 scale map and structure sections.

Orchard, M.J., and Danner, W.R., 1991a, The Paleontology of Quesnellia, *in* Smith, P.L., ed., A Field Guide to the Paleontology of Southwestern Canada: Canadian Paleontology Conference 1, Vancouver, British Columbia, p. 139–168 [with contributions by Crasquin-Soleau, S., Nassichuk, W.W., Nelson, S.J., Norris, A.W., Sakagami, S., and Tozer, E.T.].

Orchard, M.J., and Danner, W.R., 1991b, Paleontology of the Cache Creek Terrane, *in* Smith, P.L., ed., A Field Guide to the Paleontology of Southwestern Canada: Canadian Paleontology Conference 1, Vancouver, British Columbia, p. 169–188 [with contributions by Beyers, J.M., and Cordey, F.].

Pàlfy, J., Smith, P.L., and Mortensen, J.K., 2000, A revised numeric time scale for the Jurassic, in Hall, R.L., and Smith, P.L., eds., Advances *in* Jurassic Research 2000: GeoResearch Forum v. 6, p. 181–821.

Parrish, R.R., 1983, Cenozoic thermal evolution and tectonics of the Coast Range of British Columbia: I. Fission track dating, apparent uplift rates, and patterns of uplift: Tectonics, v. 2, p. 601–631.

Parrish, R.R., 1992, Miscellaneous U-Pb zircon dates from southeast British Columbia, Radiogenic and Isotopic Studies: Report 5: Geological Survey of Canada Paper 91-2, p. 143–153.

Parrish, R.R., and Monger, J.W.H., 1992, New U-Pb dates from southwestern British Columbia, Radiogenic Age and Isotope Studies: Report 5: Geological Survey of Canada Paper 91-2, p. 87–108.

Parrish, R.R., Heinrich, S., and Archibald, D., 1987, Age of the Ice River complex, southeastern British Columbia, Radiogenic age and isotope studies: Report 1: Geological Survey of Canada Paper 87-2, p. 33–37.

Parrish, R.R., Carr, S.D., and Parkinson, D.L., 1988, Eocene extensional tectonics and geochronology of the southern Omineca Belt, British Columbia and Washington: Tectonics, v. 7, p. 181–212.

Patchett, P.J., and Gehrels, G.E., 1998, Continental influence on Canadian Cordilleran terranes from Nd isotopic study, and significance for crustal growth processes: The Journal of Geology, v. 106, p. 269–280.

Payne, J.G., Bratt, J.A., and Stone, B.G., 1980, Deformed Mesozoic Cu-Zn sulphide deposits in the Britannia district, British Columbia: Economic Geology, v. 75, p. 700–721.

Pfiffner, O.A., Levato, L., and Valasek, P., 1991, Crustal reflections from the Alpine orogen: results from deep seismic profiling, *in* Meissner, R., Brown, L., Curbaum, D-J., Franke, W., Fuchs, K., and Seifert, F., eds., Continental lithosphere: deep seismic reflections: American Geophysical Union Geodynamics Series, v. 22, p. 185–193.

Pitcher, M.G., 1960, Fusulinids of the Cache Creek Group, Stikine River area, Cassiar District, British Columbia, Canada: Brigham Young University Research Studies, Geology Series, v. 7, no. 7, 64 p.

Poulton, T.P., and Simony, P.S., 1980, Stratigraphy, sedimentology, and regional correlation of the Horsethief Creek Group (Hadrynian, Late Precambrian) in the northern Purcell and Selkirk Mountains, British Columbia: Canadian Journal of Earth Sciences, v. 17, p. 1708–1714.

Preto, V.A., 1979, the geology of the Nicola Group between Merritt and Princeton: British Columbia Ministry of Energy, Mines and Petroleum Resources, Bulletin 69.

Preto, V.A., Ostatenko, M.J., McMillan, W.J., and Armstrong, R.L., 1979, Isotopic dates and strontium isotopic ratios for plutonic and volcanic rocks in the Quesnel Trough and Nicola Belt, south-central British Columbia: Canadian Journal of Earth Sciences, v. 16, p. 1658–1672.

Price, R.A., 1973, Large-scale gravitational flow of supracrustal rocks, southern Canadian Rockies, *in* De Jong, K.A., and Scholten, R., eds., Gravity and tectonics: New York, Wiley and Sons, p. 491–502.

Price, R.A., 1979, Intracontinental ductile crustal spreading linking the Fraser River and northern Rocky Mountain Trench transform fault zones: Geological Society of America Programs and Abstracts, v. 11, p. 499.

Price, R.A., 1981, The Cordilleran foreland thrust and fold belt in the southern Canadian Rocky Mountains, *in* McClay, K.R., and Price, N.J., eds., Thrust and nappe tectonics: The Geological Society of London, Special Publication Number 9, p. 427–448.

Price, R.A., 1986, The southeastern Canadian Cordillera: thrust faulting, tectonic wedging, and delamination of the lithosphere: Journal of Sedimentary Geology, v. 8, p. 239–254.

Price, R.A., 1994, Cordilleran tectonics and the evolution of the Western Canada sedimentary basin, *in* Mossop, G.D., and Shetsen, I., eds., Geological Atlas of Western Canada: Calgary, Canadian Society of Petroleum Geologists/Alberta Research Council, p. 13–24.

Price, R.A., and Fermor, P.R., 1985, Structure section of the Cordilleran foreland thrust and fold belt west of Calgary, Alberta, Geological Survey of Canada, 1 sheet.

Price, R.A., and Mountjoy, E.W., 1970, Geologic structure of the Canadian Rocky Mountains between Bow an Athabasca Rivers: A progress report, *in* Wheeler, J.O., ed., Structure of the southern Canadian Cordillera: Geological Association of Canada Special Paper 6, p. 7–25.

Price, R.A., and Mountjoy, E.W., 1970a, Geology, Canmore east half, Alberta: Geological Survey of Canada Map 1265A, (1:50,000, 82 O/3 E).

Price, R.A., and Mountjoy, E.W., 1970b, Geology, Canmore west half, Alberta: Geological Survey of Canada Map 1266A (1:50,000, 82 O/3 W).

Price, R.A., and Mountjoy, E.W., 1971, Geology, Lake Minnewanka west half, Alberta: Geological Survey of Canada Map 1272A (1:50,000, 82 O/6 W).

Price, R.A., and Mountjoy, E.W., 1972a, Geology, Banff east half, Alberta: Geological Survey of Canada Map 1294A (1:50,000, 82 O/4 E).

Price, R.A., and Mountjoy, E.W., 1972b, Geology, Banff west half, Alberta: Geological Survey of Canada Map 1295A (1:50,000, 82 O/4 W).

Price, R.A., and Mountjoy, E.W., 1972c, Geology, Mount Eisenhower east half, Alberta: Geological Survey of Canada Map 1296A (1:50,000, 82 O/5 E).

Price, R.A., and Mountjoy, E.W., 1972d, Geology, Mount Eisenhower west half, Alberta: Geological Survey of Canada Map 1297A (1:50,000, 82 O/5 W).

Price, R.A., and Mountjoy, E.W., 1978a, Geology, Hector Lake, east half, Alberta: Geological Survey of Canada Map 1463A (1:50,000, 82 N/9 E).

Price, R.A., and Mountjoy, E.W., 1978b, Geology, Hector Lake, west half, Alberta: Geological Survey of Canada Map 1464A (1:50,000, 82 N/9 W).

Price, R.A., and Sears, J.W., 2000, A preliminary palinspastic map of the Mesoproterozoic Belt-Purcell Supergroup, Canada and USA: Implications for the tectonic setting and structural evolution of the Purcell anticlinorium and the Sullivan deposit, *in* Lydon, J.W., Höy, T. Slack, J.F., and Knapp, M.E., The Geological Environment of the Sullivan deposit, British Columbia: Geological Association of Canada, Mineral Deposits Divsion, Special Publication No. 1, p. 61-81.

Price, R.A., Balkwill, H.R., Charlesworth, H.A.K., Cook, D.G., and Simony, P.S., 1972, The Canadian Rockies and tectonic evolution of the southeastern Canadian Cordillera: 24th International Geological Congress, Guidebook for Excursion A15-C15, Geological Survey of Canada, 129 p.

Price, R.A., Mountjoy, E.W., and Cook, D.G., 1979a, Geology, Mount Goodsir, east half, British Columbia and Alberta: Geological Survey of Canada Map 1476A (1:50,000, 82 N/1 E).

Price, R.A., Mountjoy, E.W., and Cook, D.G., 1979b, Geology, Mount Goodsir, east half, British Columbia: Geological Survey of Canada Map 1477A (1:50,000, 82 N/1 W).

Price, R.A., Balkwill, H.R., and Mountjoy, E.W., 1980, Geology, McMurdo, east half, British Columbia: Geological Survey of Canada Map 1510A (1:50,000, 82 N/2 E).

Price, R.A., Cook, D.G., Aitken, J.D., and Mountjoy, E.W., 1980a, Geology, Lake Louise, east half, British Columbia: Geological Survey of Canada Map 1482A (1:50,000, 82 N/8 E).

Price, R.A., Cook, D.G., Aitken, J.D., and Mountjoy, E.W., 1980b, Geology, Lake Louise, west half, British Columbia: Geological Survey of Canada, Map 1483A (1:50,000, 82 N/8 W).

Price, R.A., Monger, J.W.H., and Muller, J.E., 1981, Cordilleran cross-section: Calgary to Victoria, *in* Thompson, R.I., and Cook, D.G., eds., Field Guides to Geology and Mineral Deposits—Calgary '81: Geological Association of Canada, p. 261–334.

Price, R.A., Monger, J.W.H., and Roddick, J.A., 1985, Cordilleran cross-section. Calgary to Vancouver; *in* Tempelman-Kluit, D.J., ed., Field Guides to Geology and Mineral Deposits of the Southern Canadian Cordillera:

Geological Society of America, Cordilleran Section Meeting, Vancouver, British Columbia, May '85, Trip 3, p. 3:1–3:85.

Read, P.B., 1990, Cretaceous and Tertiary stratigraphy and industrial minerals, Hat Creek, British Columbia (92I/12, 13, 14): British Columbia Ministry of Energy, Mines and Petroleum Resources Open File 1990-23.

Read, P.B., and Okulitch, A.V., 1977, The Triassic unconformity of south-central British Columbia: Canadian Journal of Earth Sciences, v. 14, p. 606–638.

Read, P.B., and Brown, R.L., 1981, Columbia River fault zone: southeastern margin of the Shuswap and Monashee complexes, southern British Columbia: Canadian Journal of Earth Sciences, v. 18, p. 1127–1145.

Read, P.B., Woodsworth, G.J., Greenwood, H.J., Ghent, E.D., and Evenchick, C.A., 1991, Metamorphic Map of the Canadian Cordillera: Geological Survey of Canada Map 1714A, scale 1:2,000,000.

Richards, B.C., Bamber, E.W., Henderson, C.M., Higgins, A.C., Johnston, D.I., Mamet, B., and Meijer Drees, N.C., 1994, Uppermost Devonian (Famennian) and Lower Carboniferous (Tournaisian) at Jura Creek and Mount Rundle, southwestern Alberta: Geological Survey of Canada Open File 2866, 80 p. (also Field Trip #7, Carboniferous to Jurassic, PANGEA, Calgary, Alberta, August 15–19, 1993).

Richards, D.R., Butler, R.F., and Harms, T.A., 1993, Paleomagnetism of the late Paleozoic Slide Mountain terrane, northern and central British Columbia: Canadian Journal of Earth Sciences, v. 30, p. 1898–1913.

Riddihough, R.P., and Hyndman, R.D., 1991, Modern plate tectonic regime of the continental margin of western Canada, Chapter 13, *in* Gabrielse, H., and Yorath, C.J., eds., Geology of the Cordilleran Orogen in Canada, The Geology of Canada, Geological Survey of Canada, v. 4, p. 435–455 (also Geological Society of America, The Geology of North America, v. G-2).

Riddell, J., Schiarizza, P., Gaba, R., McLaren, G., and Rouse, J., 1993, Geology of the Mount Tatlow map area (92O/5, 6, 12): British Columbia Ministry of Energy, Mines and Petroleum Resources Open File 1993-8.

Roback, R.C., Sevigny, J.H., and Walker, N.W., 1994, Tectonic setting of the Slide Mountain terrane, southern British Columbia: Tectonics, v. 13, p. 1242–1258.

Robinson, R.A., 1960, Lower and Middle Cambrian stratigraphy of the eastern Great Basin, *in* Guidebook to the Geology of East Central Nevada: Intermountain Association of Petroleum Geologists, p. 43–52.

Roddick, J.A., 1965, Vancouver North, Coquitlam and Pitt Lake map areas, with special emphasis on the evolution of the plutonic rocks: Geological Survey of Canada Memoir 335, 276 p.

Roddick, J.A., 1967, Tintina trench: Journal of Geology, v. 75, p. 23–33.

Roddick, J.A., 1983, Geophysical review and composition of the Coast Plutonic Complex, south of latitude 55°N, *in* Roddick, J.A., ed., Circum-Pacific plutonic terranes: Geological Society of America, Memoir 159, p. 195–211.

Roddick, J.A., and Woodsworth, G.J., 1975, Coast Mountains Project: Pemberton (92J W/2) map area, British Columbia: Report of Activities, Part A, Geological Survey of Canada Paper 75-1, p. 37–40.

Roddick, J.A., and Woodsworth, G.J., 1979, Vancouver (west half) and mainland parts of Alberni: Geological Survey of Canada Open File 611, scale 1:250,000.

Roots, C.F., and Parrish, R.R., 1988, Age of Mount Harper volcanic complex, southern Ogilvie Mountains, Yukon: Radiogenic age and isotopic studies: Report 2, Geological Survey of Canada Paper 88-2, p. 29–35.

Ross, C.A., 1969, Upper Paleozoic Fusilinacea: *Eowaeringella* and *Wedekindellina* from Yukon Territory and giant *Parafusulina* from British Columbia: Contributions to Canadian Paleontology, Geological Survey of Canada Bulletin 182, p. 129–134.

Ross, C.A., and Nassichuk, W.W., 1970, *Yabeina* and *Waagenoceras* from the Atlin Horst area, northwestern British Columbia: Journal of Paleontology, v. 44, p.779–781.

Ross, C.A., and Ross, J.R.P., 1983, Late Paleozoic accreted terranes of western North America, *in* Stevens, C.H., ed., Pre-Jurassic rocks of western North American suspect terranes: Pacific Section, Society of Economic Paleontologists and Mineralogists, p. 7–22.

Ross, G.M., 1991, Precambrian basement in the Canadian Cordillera: an introduction: Canadian Journal of Earth Sciences, v. 28, p. 1133–1139.

Ross, G.M., and Murphy, D.C., 1988, Transgressive stratigraphy, anoxia, and regional correlations within the late Precambrian Windermere grit of the southern Canadian Cordillera: Geology, v. 16, p. 139–143.

Ross, G.M., and Bowring, S.A., 1990, Detrital zircon geochronology of the Windermere Supergroup and the tectonic assembly of the southern Canadian Cordillera: Journal of Geology, v. 98, p. 879–893.

Ross, G.M., Bloch, J.D., and Krouse, H.R., 1995, Neoproterozoic strata of the southern Canadian Cordillera and the isotopic evolution of seawater sulfate: Precambrian Research, v. 73, p. 71–99.

Ross, G.M., Gehrels, G.E., and Patchett, P.J., 1997, Provenance of Triassic strata in the Cordilleran miogeocline, western Canada: Bulletin of Canadian Petroleum Geology, v. 45, p. 461–473.

Ross, J.V., 1968, Structural relations at the eastern margin of the Shuswap complex, near Revelstoke, southeastern British Columbia: Canadian Journal of Earth Sciences, v. 5, p. 831–849.

Rubin, C.M., Saleeby, J.B., Cowan, D.S., Brandon, M.T., and McGroder, M.F., 1990, Regionally extensive mid-Cretaceous west-vergent thrust system in the northwestern Cordillera: implications for continent-margin tectonism: Geology, v. 18, p. 276–280.

Rui, L., and Nassichuk, W.W., 1996, Upper Permian (Wordian) fusulinaceans from the Cache Creek terrane, northern British Columbia: Canadian Journal of Earth Sciences, v. 33, p. 1022–1036.

Rusmore, M.E., 1985, Geology and tectonic significance of the Upper Triassic Cadwallader Group and its bounding faults, southwestern British Columbia [Ph.D. thesis]: Washington, Seattle, University of Washington.

Russo, R.M., and Silver, P.G., 1996, Cordillera formation, mantle dynamics, and the Wilson cycle, Geology, v. 24, p. 511–514.

Sager, W.W., and Koppers, A.P., 2000, Late Cretaceous polar wander of the Pacific Plate: evidence of a rapid true polar wander event, Science, v. 287, p. 455–459.

Samson, S.D., and Patchett, P.J., 1991,. The Canadian Cordillera as a modern analogue of Proterozoic crustal growth: Australian Journal of Earth Sciences, v. 38, p. 595–611.

Schiarizza, P., and Preto, V.A., 1984, Geology of the Adams Plateau-Clearwater area (82L/13; 82M/3,4,5,6,12; 92P/1,8,9): British Columbia Ministry of Energy, Mines and Petroleum Resources Preliminary Map 56, scale 1:100,000.

Schiarizza, P., and Preto, V.A., 1987, Geology of the Adams Plateau-Clearwater-Vavenby area: Province of British Columbia Ministry of Energy, Mines and Petroleum Resources, Geological Survey Branch Paper 1987-2, 88 p.

Schiarizza, P., Gaba, R.G., Glover, J.K., Garver, J.I., and Umhoefer, P.J., 1997, Geology and mineral occurrences of the Taseko-Bridge River area: British Columbia Ministry of Employment and Investment, Geological Survey Branch Bulletin 100, 292 p.

Sears, J.W., 1979, Tectonic contrasts between infrastructure and suprastructure of the Columbian orogen, Albert Peak area, western Selkirk Mountain, British Columbia [Ph.D. thesis]: Ontario, Kingston, Queen's University.

Sears, J.W., and Price, R.A., 2000, A new look at the Siberian connection: No SWEAT, Geology, v. 28, No. 5, p. 423–426.

Selwyn, A.R.C., 1872, Journal and report of preliminary explorations in British Columbia: Geological Survey of Canada, Report of Progress for 1871–72, p. 16–72.

Simony, P.S., 1998, Transect of the Rocky Mountain fold and thrust belt, Canmore to Revelstoke: *in* Simony, P.S., ed., Canadian Tectonics Group NUNA field trip in honour of Paul F. Williams: University of Calgary, 12 p.

Smith, R.B., 1979, Geology of the Harper Ranch Group (Carboniferous-Permian) and Nicola Group (Upper Triassic) northeast of Kamloops, British Columbia [M.Sc. thesis]: British Columbia, Vancouver, University of British Columbia, 211 p.

Smith, P.L., (editor), 1991, A Field Guide to the paleontology of Southwestern Canada: First Canadian Paleontology Conference, University of British Columbia, Vancouver, 1991.

Smith, P.L., (editor), 1998, Field guide for the Fifth International Symposium on the Jurassic System: International Union of Geological Sciences/Subcommission on Jurassic Stratigraphy, Vancouver, Canada.

Smith, P.L., and Tipper, H.W., 1986, Plate tectonics and paleobiogeography: Early Jurassic (Pliensbachian) endemism and diversity: Palaios, v. 1, p. 399–412.

Smith, P.L., Monger, J.W.H., Arthur, A., Poulton, T.P., and Tipper, H.W., 1998, Field Trips B2 and B3: southwestern British Columbia, *in* Smith, P.L., ed., Field Guide for the Fifth International Symposium on the Jurassic System: Vancouver, Canada, p. 230–307.

Stewart, D., 1997, Cambrian Section, *in* Styan, B., Stewart, D., and Tippett, C., ed., Rocky Mountain transect: Stratigraphic interfaces and hydrocarbon charge, Field Trip, CSPG–SEPM Joint Convention, p. 55–106.

Stewart, J.H., 1972, Initial deposits of the Cordilleran geosyncline: Evidence of late Precambrian (<850 m.y.) continental separation: Geological Society of America Bulletin, v. 83, p. 1345–1360.

Stewart, W.D., 1991, Stratigraphy and sedimentology of the Chancellor succession (Middle and Upper Cambrian) southeastern Canadian Rocky Mountains [Ph.D. thesis]: Ontario, Ottawa, University of Ottawa, 534 p., 2 vols.

Stewart, W.D., Dixon, O.A., and Rust, B.R., 1993, Middle Cambrian carbonate-platform collapse, southeastern Canadian Rocky Mountains: Geology, v. 21, p. 687–689.

Struik, L.C., 1987, The ancient western North American margin: an alpine rift model for the east-central Canadian Cordillera, Geological Survey of Canada, Paper 87-15, 19 p.

Struik, L.C., 1988, Crustal evolution of the eastern Canadian Cordillera: Tectonics, v. 7, p. 727–748.

Symons, D.T.A., 1973, Concordant Cretaceous paleolatitudes from felsic plutons in the Canadian Cordillera: Nature, v. 241, p. 59–61.

Thompson, R.I., 1978, Geology of the Akolkolex River area, British Columbia: British Columbia Ministry of Energy, Mines and Petroleum Resources Bulletin 60, 77 p.

Thompson, R.I., and Daughtry, K.L., 1997, Anatomy of the Neoproterozoic-Palaeozoic continental margin, Vernon map area, British Columbia: Current Research 1997-A, Geological Survey of Canada, p. 145–150.

Thompson, R.I., and Daughtry, K.L., 1998, Stratigraphic linkages across Vernon map area, British Columbia: Current Research 1998-A, Geological Survey of Canada, p. 181–187.

Thompson, M.L., and Wheeler, H.E., 1942, Permian fusulinids from British Columbia, Washington and Oregon: Journal of Paleontology, v. 16, p. 700–7811.

Tozer, E.T., 1982, Marine Triassic faunas of North America; their significance for assessing plate and terrane movements: Geologische Rundschau, v. 71, p. 1077–1104.

Travers, W.B., 1982, Possible large-scale overthrusting near Ashcroft, British Columbia: Implications for petroleum prospecting: Canadian Petroleum Geology Bulletin, v. 30, p. 1–8.

Travers, W.B., 1987, Cache Creek-Ashcroft area, western Intermontane Belt, British Columbia, *in* Hill, M.L., ed., Centennial Field Guide Volume 1: Cordilleran Section of the Geological Society of America, p. 407–412.

Uyeda, S., and Kanamori, H., 1979, Back-arc opening and the mode of subduction: Journal of Geophysical Research, v. 84, p. 1049–1061.

van der Velden, A.J., and Cook, F.A., 1996, Structure and tectonic development of the southern Rocky Mountain trench: Tectonics, v. 15, p. 517–544.

Varsek, J.L., Cook, F.A., Clowes, R.M., Journeay, J.M., Monger, J.W.H., Parrish, R.R., Kanasewich, E.R., and Spencer, C.S., 1993, Lithoprobe crustal reflection structure of the southern Canadian Cordillera II: Coast Mountains transect: Tectonics, v. 12, p. 334–360.

Villeneuve, M.E., Ross, G.M., Parrish, R.R., Theriault, R.J., Miles, W., and Broome, J., 1993, Geophysical subdivision, U-Pb geochronology and Sm-Nd isotope geochemistry of the crystalline basement of the Western Canada Sedimentary Basin, Alberta and northeastern British Columbia: Geological Survey of Canada Bulletin 447, 86 p.

Warren, M.J., 1997, Crustal extension and subsequent crustal thickening along the Cordilleran rifted margin of ancestral North America, western Purcell Mountains, southeastern British Columbia [Ph.D. thesis]: Ontario, Kingston, Queen's University, 361 p.

Wheeler, J.O., 1965, Big Bend map-area, British Columbia: Geological Survey of Canada Paper 64-32, Map 12-1064 and Report, 37 p.

Wheeler, J.O., and McFeely, P., 1991, Tectonic assemblage map of the Canadian Cordillera and adjacent parts of the United States of America: Geological Survey of Canada Map 1712A, scale 1:2,000,000.

Wynne, P.J., Irving, E., Maxson, J.A., and Kleinspehn, K.L., 1995, Paleomagnetism of the Upper Cretaceous strata of Mount Tatlow: Evidence for 3000 km of northward displacement of the eastern Coast Belt, British Columbia: Journal of Geophysical Research, v. 100, p. 6073–6091.